Cancer Genetics and Therapeutics

Madhumita Roy • Amitava Datta

Cancer Genetics and Therapeutics

Focus on Phytochemicals

Springer

Madhumita Roy
Environmental Carcinogenesis
and Toxicology
Chittaranjan National Cancer Institute
Kolkata, India

Amitava Datta
Department of Computer Science
and Software Engineering
The University of Western Australia
Perth, WA, Australia

ISBN 978-981-13-9473-7 ISBN 978-981-13-9471-3 (eBook)
https://doi.org/10.1007/978-981-13-9471-3

This Springer imprint is published by the registered company Springer Nature Singapore Pte Ltd.
The registered company address is: 152 Beach Road, #21-01/04 Gateway East, Singapore 189721, Singapore

"Madhumita Roy dedicates this book to the four pillars who are behind her success in life: Debiprosad Pal (uncle), late Lakshmi Rani Pal (aunt), late Rama Prasad Pal (father), and Smritikana Pal (mother). She also dedicates the book to her sons, Kaustuv and Kausheyo."

"Amitava Datta dedicates the book to his parents, Pashupati Dutta and late Ila Dutta."

Foreword

I am delighted to write foreword to this well-achieved book on cancer genetics and the role of phytochemicals in prevention and management of cancer. While the authors have kept focus on a plethora of information available on the experimental evidence of a large number of phytochemicals on critical pathways of cellular, genetic and epigenetic mechanisms of malignant transformation, the importance of this book lies in smartly merging it with an overview of our current understanding in cancer genetics research.

Both Dr. Roy and Dr. Datta are erudite scientists in their respective fields of research. While Dr. Roy has outstanding contributions in the field of phytochemicals and their role in cancer prevention and control, Dr. Datta has an enviable standing in computational molecular biology and high-performance computing. Dr. Madhumita Roy has been my colleague and collaborator for more than two decades.

In order to appreciate the action of drugs and phytochemicals in cancer, it is necessary to understand the genetics and molecular biology that initiate and sustain malignant transformation. To begin with, the reader is introduced to the mechanism of malignant transformation of normal cell emphasizing the interplay of extracellular and intracellular signalling pathways and growth suppressors. Next, the book deals with genes where mutations lead to cascading consequences for **cancer initiation**, **tumorigenesis** and **metastasis**. The concept of driver and passenger genes and role of mutation in TP53 gene have been lucidly presented followed by the genomic landscapes for specific cancers, e.g. **breast, colorectal, liver, lung, melanoma, pancreatic and cervical cancers.**

The molecular mechanisms necessary to understand the action of phytochemicals for their role in **chemoprevention** and **chemosensitization** for cancer chemotherapy are elaborately dealt with. These include epigenetic control mechanisms, histone acetylation/deacetylation, DNA methylation and miRNA silencing, together with those involved in **oxidative stress** and **inflammation**. Then, the authors present an in-depth review on the role of dietary phytochemicals, e.g. curcumin, quercetin, resveratrol, EGCG (epigallocatechin gallate) and genistein in oxidative stress, inflammation and intervention in the Nrf2 pathway for chemoprevention. The chapter concludes with an inclusive review on molecular mechanism of chemosensitization and the role of phytochemicals.

The isolation, structure-function and classification of phytochemicals and the scientific evidence for their role in prevention of cancer and other diseases are well-presented. The anti-oxidant, anticarcinogenic, antimicrobial, antidiabetic and anti-inflammatory role of different phytochemicals and mechanism of action has also been well-analysed. In order to introduce the reader to the possible use of phytochemicals in cancer treatment and/or **anti-cancer drug development**, an elaborate classification of commonly used chemotherapeutic drugs and their mode of action are followed by a detailed assessment of phytochemicals used in the treatment of several diseases with particular reference to their possible use in cancer treatment. This leads to a comprehensive discussion on **pharmacogenomics** and the scope of **'personalized medicine'** in chemotherapeutic treatment of cancer and the challenges it faces due to **genetic variations** in individuals. This is complemented by a review on the possible role of different phytochemicals in pharmacogenomics.

The authors have discussed the challenges in prognosis of cancer, the mechanism of **drug resistance** during chemotherapeutic treatment and **metastasis**. This is followed by a highly relevant description on different phytochemicals showing **mitigation of drug resistance** and improving prognosis by way of preventing metastasis.

Since the positive effects of phytochemicals are generally marred by poor bioavailability of phytochemicals in disease condition, the book provides a welcome discussion on the concepts of **bioavailability** and **bioenhancers** of drugs. The role of phytochemicals as bioenhancers in improving bioavailability of cancer drugs with particular reference to piperine, quercetin, genistein, curcumin, gingerol and resveratrol is an important feature of the last chapter.

For the benefit of readers, the authors provide highly imaginative diagrams to explain various concepts and pathways. Thus, it is a distinctive book of high scientific value on the current status of research evidence on the possible role of phytochemicals in prevention and treatment of cancer in view of current concepts in cancer genetics.

Kolkata, India Maqsood Siddiqi

Preface

Plant-derived molecules or phytochemicals have been used as therapeutic agents from ancient times. Usually, whole plant parts were used as traditional medicines for treating many diseases. Phytochemicals are now routinely extracted from these plant parts and used for synthesizing drugs by the modern pharmaceutical industry, including drugs and bioenhancers for chemotherapeutic treatment of cancer.

The aim of this book is to present cancer genetics and therapeutics with a focus on the use of phytochemicals in cancer therapy. The literature on cancer genetics, cancer therapy, and phytochemicals is vast, and it is impossible to cover all these areas comprehensively in one book. We have tried to capture the key aspects of these three areas and the interconnections among them in this book. We have concentrated on the latest understanding of cancer genetics and the conversion of that understanding into therapeutic drugs, with a focus on phytochemicals.

The book has eight chapters, almost equally divided between cancer genetics and therapeutics. Chapter 1 gives an overview, the hallmarks, and the treatments of cancer. The discussion on the hallmarks of cancer examines the main genetic factors influencing the growth of tumors and the apoptosis pathway in details. Chapter 2 analyzes cancer genetics and the important genetic pathways implicated in cancer in more details. We discuss the concepts of driver mutations and the key driver genes involved in some of the most important forms of cancer. These genes are usually organized in key pathways that we discuss next. Genes and pathways involved in cancer are some of the most active areas of research in cancer biology, and we have tried to capture the important parts of the current understanding in these areas.

Chapter 3 gives an overview of phytochemicals and a classification of phytochemicals. Though there is no universally accepted classification scheme, we have adopted a classification that is based on the chemical structures of phytochemicals. Chapter 4 covers the drugs used in chemotherapy, as the focus of this book is on chemotherapeutic treatment of cancer. Many phytochemicals have shown promise as anticancer agents in in vitro studies and animal models but have not progressed to clinical trials. We have reviewed these phytochemicals in this chapter. Chapter 5 covers the molecular mechanisms of phytochemicals in cancer, in particular as agents that mitigate the effects of inflammation and oxidative stress, two of the fundamental causes of cancer. We have also discussed in details the epigenetic mechanisms involved in cancer and the beneficial actions of phytochemicals on these mechanisms. Chemosensitization is an important issue in cancer therapy as quite often cancer cells develop drug resistance. We have discussed how phytochemicals act as chemosensitizing agents.

We have discussed drug resistance in more details and the action of phytochemicals in reversing drug resistance in Chap. 6. Metastasis is the invasive and most dangerous phase of cancer, and we have discussed how phytochemicals can prevent metastasis. In Chap. 7, we discuss the important topic of pharmacogenomics, the effect of genetic variations in individuals that may hinder the treatment of cancer, and the diverse roles phytochemicals play in pharmacogenomics in many diseases including cancer. Finally, in Chap. 8, we discuss the important topic of bioavailability. The mechanisms affecting bioavailability and the role of phytochemicals as bioenhaning agents for other cancer drugs are discussed.

We expect the book to be useful for advanced undergraduate and postgraduate students in understanding the genetics of cancer, the important genetic pathways involved in cancer, and the role of phytochemicals in cancer therapy.

We also expect it to be useful for cancer researchers who want to understand the roles of phytochemicals in cancer treatment and for researchers primarily in the area of phytochemicals who want to expand their understanding on how phytochemicals are finding its use in cancer therapy.

Kolkata, India — Madhumita Roy
Perth, WA, Australia — Amitava Datta

Contents

About the Authors

Madhumita Roy completed her MSc in Physics at Calcutta University. She pursued her PhD studies at the Department of Crystallography and Molecular Biology, Saha Institute of Nuclear Physics, Kolkata, and was awarded her PhD by the University of Calcutta in 1994. She is currently the Assistant Director and Head of the Department of Environmental Carcinogenesis and Toxicology, Chittaranjan National Cancer Institute, Kolkata. Her research focuses on cancer biology, particularly radiation biology, cytogenetics, cell biology, tumor cell apoptosis, cancer chemoprevention, and environmental toxicology, including the arsenic disaster in West Bengal, India, identification of important tumor biomarkers, and chemo-enhancing efficacy of tumor cells by plant-derived molecules towards apoptosis by conventional chemotherapeutic drugs. She has published over 74 research papers in respected international journals and has attended more than 100 conferences and workshops in India and abroad. She is a Life Member of several societies, such as the Environmental Mutagen Society of India (EMSI), Indian Association for Cancer Research (IACR), Electron Microscope Society of India, Indian Science Congress Association, All India Congress of Cytology & Genetics, and Society of Nutraceuticals and Chronic Diseases.

Amitava Datta completed his MTech and PhD in Computer Science at the Indian Institute of Technology, Madras. He pursued postdoctoral research at the Max Planck Institute for Computer Science and the University of Freiburg, both in Germany. He is currently a Professor at the Department of Computer Science and Software Engineering at the University of Western Australia. He was a Visiting Professor at the University of Freiburg and is currently a Visiting Professor at the University of Life Sciences, Warsaw, Poland, and Southwest University, China. His scientific interests include computational molecular biology, data mining, and parallel and distributed computing, and his computational molecular biology research focuses on RNA structural biology and miRNA gene and target prediction, particularly in relation to cancer biology. He has jointly developed the DotKnot software tool for predicting RNA tertiary structures, including pseudoknots, which is used by researchers worldwide. He has published over 150 research papers in respected international journals and in international conference proceedings.

1 Cancer: Types and Hallmarks

Abstract

We review some of the main characteristics of cancer in this chapter. Our aim is to give the reader a general overview and a molecular biological or genetic understanding of the disease. However, our review is by no means exhaustive, and the interested reader should explore these topics from the references provided at the end of the chapter. We start with explaining cancer as a neoplastic disease and some reasons for its occurrence. Next, we discuss the different categories of cancers depending on the cell types where they originate. This is a convenient categorization as there are many different types of cancers with many common attributes that can be captured through this categorization. Then we discuss some of the major hallmarks of cancer and the current genetic and biochemical understanding of these hallmarks. We explain some of the key genetic pathways that support these hallmarks and also some interconnections between these pathways. Next, we discuss some of the enabling characteristics like genetic mutation and tumor-promoting inflammation that are responsible for the occurrence of cancer. Finally, we conclude the chapter with an overview of symptoms, diagnosis, and different types of treatments of cancer.

1.1 What Is Cancer?

Cancer is a group of neoplastic diseases with some significant characteristic features. Before proceeding further, it is good to define the term neoplasm. An abnormal growth of tissue forming a mass is called a tumor. Cells in a neoplastic tumor may contain more than one type of cell, i.e., they may be heterogeneous.

Tumor heterogeneity has been observed in various cancers. Due to tumor heterogeneity, cells may have different cellular morphology and characteristics. Heterogeneity may be between tumors, i.e., inter-tumor heterogeneity, or within a tumor, i.e., intra-tumor heterogeneity. During cell division, mutations may arise,

M. Roy, A. Datta, *Cancer Genetics and Therapeutics*,
https://doi.org/10.1007/978-981-13-9471-3_1

resulting in a heterogeneous or diverse population of cancer cells, culminating in poor treatment outcome and hence prognosis. Models may explain this heterogeneity: one is the cancer stem cell (CSC) model and the other is the clonal evolution model. CSC model is based on the notion that only a group of cells in a tumor are tumorigenic. These cancer stem cells, which may self-renew or may differentiate into defective or mutated cells, are present in the tumor microenvironment. Clonal evolution model suggests that genetic and epigenetic changes occur in individual cancer cells, thus accumulating mutations imparting selective advantage to these clones, which competitively exclude other tumor cells.

Not all tumors are cancers; they may be benign, potentially malignant, or malignant. Benign tumors are those that are confined to a particular location and are not transformed to cancer. Like the benign tumors, potentially malignant tumors are also localized, i.e., they do not spread to other parts of the body, but they are cancerous. This type of neoplasm is coined carcinoma in situ. The neoplasms that invade adjoining tissues and metastasize are known as malignant neoplasm.

Carcinogenesis is a multistep process, involving four stages, namely, initiation, promotion, progression, and malignant conversion. Interaction of any cancer-causing agent or a carcinogen with DNA leads to damages of the genetic material, which, if remains unrepairable, results in mutation. Therefore, an efficient DNA repair machinery may alter the initiation step of carcinogenesis. The initiated cells sometimes have a selective growth advantage. Initiation step is therefore induction of a mutation in a vital gene responsible for cell division. Initiation is an irreversible process. The mutation may not be sufficient to initiate cancer in a cell. In order to get an effective mutation, it has to be permanent. A class of substances that influence the efficiency of carcinogenesis are called promoters. This step of carcinogenesis involves selective clonal expansion of initiated cells. Tumor promoters are normally non-mutagenic and non-carcinogenic and elicit their biological effect without metabolic activation. There may be a long gap between addition of an initiating agent/carcinogen and a promoting agent. Generally, this is an epigenetic event, as manifestation is seen in gene expression, not in the genetic material like DNA. The time interval between promotion and formation of a malignant tumor is known as progression. Duration of the progression step is variable, and the cells divide a number of times, and an increasing population of defective cells is obtained. The multiplying cells confer a permanent genetic growth advantage. The progression step, which is an irreversible one, leads to malignant conversion. All these steps of carcinogenesis are depicted in Fig. 1.1.

Cancer Cell vs Normal Cell

When normal cells are transformed into cancer cells, the entire event consists of cellular, genetic, and epigenetic changes, which impart a growth advantage to the cancer cells. Involvements of a number of mutations support the process [1–3]. The genetic changes that are acquired may take time, and during this time behavior of the cells gradually changes. There are several distinctive features that mark the cancer cells. Size and shape of the nucleus as well as the cell are distinctive, specialized features get lost, and tissue organization gets hampered. Normal cells are controlled

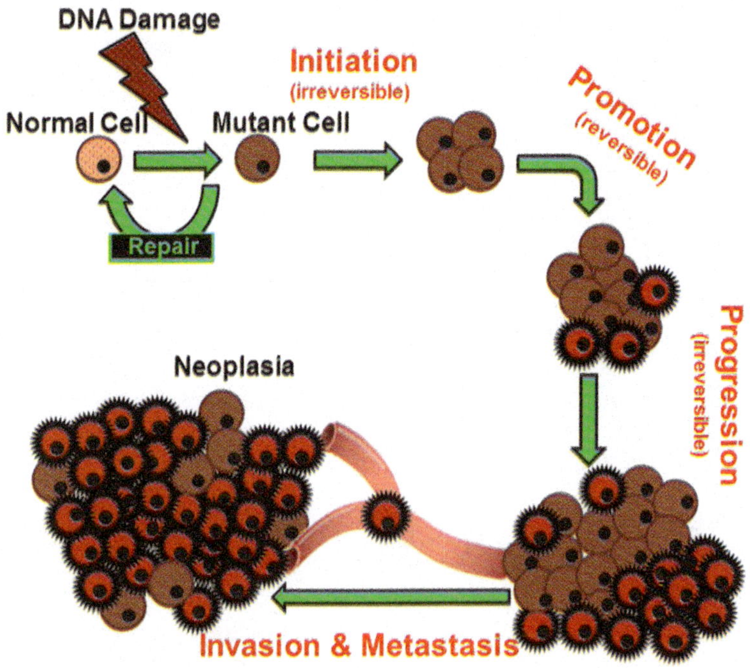

Fig. 1.1 The steps involved in carcinogenesis

by growth suppressors, which are mainly of three types; they instruct cell (**i**) to slow down and stop division, (**ii**) to fix changes in damaged cells, and (**iii**) to apoptosize. If mutation occurs in any of these genes, then the suppressive action gets lost, and cancer cells start dividing. Normal cells are guided by the signals of the neighboring cells, and due to the property of contact inhibition, they stop growing when they come in contact with another cell in near vicinity. Cancer cells, on the other hand, ignore "contact inhibition" and invade other tissues, which a normal cell does not. Normal cells obtain their energy through Krebs cycle and also through glycolysis. In contrast to the normal cells, which produce energy in the presence of oxygen, cancer cells produce energy in hypoxic conditions. Normal cells have a definite lifespan, guided by the telomere, which gets shorter with each division, and finally a cell is destined to die when the length is too short. On the contrary, cancer cells, for their own sustenance, devised a way to overcome this situation. Telomere in a cancer cell gets renewed, and an enzyme telomerase plays an active role in this regard. This enzyme maintains the length of the telomere, rendering immortality to the cancer cells. The number of chromosomes in a normal cell is unique, while a cancer cell may have abnormal number of chromosomes, and hence the genetic material in a cancer cell is also abnormal. This may be due to a number of mutations.

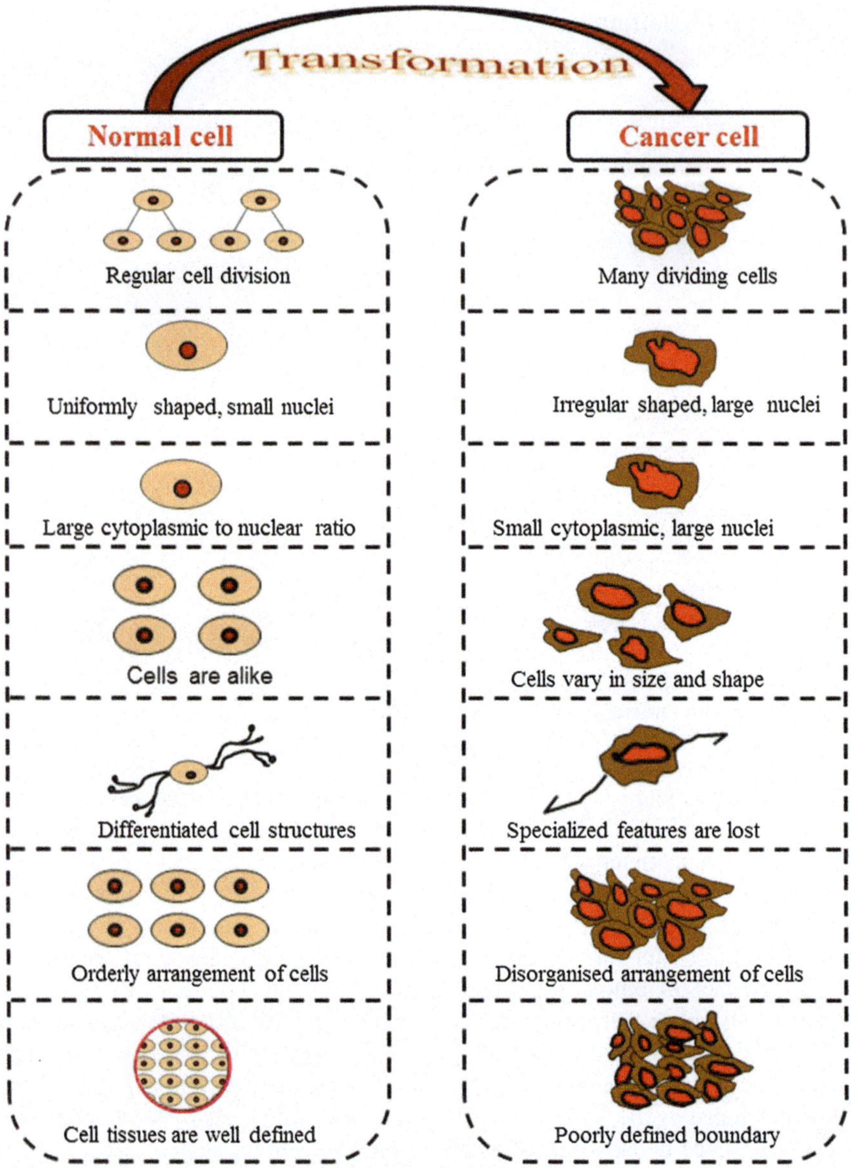

Fig. 1.2 Differences between normal and cancer cells

Some of these mutations may be "driver mutations," which drive a normal cell to be transformed to a cancerous one. Some of the mutations may be passenger mutations, having no direct function.

The difference between a normal and a cancer cell is shown in Fig. 1.2.

It is important to understand the hallmarks of cancer for a deeper understanding of the disease. In simple terms cancer is abnormal and uncontrolled growth and proliferation of cells that start at a particular organ and may spread all over the body. The initiation and progression of cancer can be divided into different stages. Cancer cells are confined to their organ of origin in the early stages. They form a tumor in the next stage through angiogenesis or growth of blood vessels inside the tumor. Cancer cells may migrate from their original site and spread to the lymph nodes and finally to other organs in the final and most critical stage of a cancer known as metastasis. Cancer cells originate through the alteration in their DNAs. These alterations may creep in due to incorrect copying of DNA during cell division. Carcinogen, or a cancer-causing agent, may initiate the process of carcinogenesis by causing errors in DNA. While there are robust cellular machineries for correcting such errors, some of them may escape the scrutiny of such machineries and result in alteration in the sequences of protein coding genes. This in turn results in the proteins with altered amino acid sequences, and such altered proteins may not fold correctly in their native forms, resulting in loss of their functionalities, as the correctly folded structures of proteins are responsible for their correct functioning.

Malignant transformation may be triggered due to changes or mutations in genome, which is called initiation. Mutations also play a leading role in progression, which results in the formation of malignant tumors. Genes, or the proteins transcribed from them, do not act in isolation. Genes are usually organized in complex pathways where different genes in a pathway are interdependent in terms of their expressions or transcriptions of proteins. A gene upstream in a pathway may be responsible for the expression of a gene that is downstream. In general, there are complex feedback mechanisms in genetic pathways controlling the appropriate levels of expressions of genes that allow the pathways to participate normally in cellular processes. Almost all the hallmarks that distinguish cancers from other neoplastic diseases are related to malfunctioning of key genetic pathways.

1.1.1 Types of Cancer

There are over 100 different cancers affecting different organs or tissues in the human body. A list of all known cancer types can be found at the US National Cancer Institute web site [1]. However, it is often easier to categorize cancers into types based on the specific cells where they originate. Next, we discuss the different categories of cancers depending on the site where they originate and also on the type of cells where the growth is initiated. The following classification is based on materials from the US National Cancer Institute web site [1]. Broadly cancer can be categorized as carcinoma, sarcoma, leukemia, lymphoma, multiple myeloma, melanoma, and brain and spinal cord tumors. We discuss these different categories below.

1.1.1.1 Carcinoma

Carcinomas are one of the most common types of cancers that originate in epithelial cells that form the lining inside hollow organs and also on the outer surface of the body. Carcinomas are further classified into different types depending on particular types of epithelial cells where they originate.

Adenocarcinoma: This is a type of cancer that develops in mucous secreting glandular tissues. Most cancers of the breast, colon, and prostate are adenocarcinomas.

Basal cell carcinoma: Epidermis is the outer layer or the skin of the human body, and basal cell carcinoma occurs in the deepest layer of the epidermis called basal cells. This is a type of skin cancer that hardly ever metastasizes but can invade tissues in the vicinity.

Squamous cell carcinoma: Squamous cells are epithelial cells that lie just beneath the outer surface of the skin. Squamous cells also line many other organs apart from the skin including the stomach, kidney, bladder, intestine, and lungs. Hence this type of cancer can occur in all these organs in addition to skin.

Transitional cell carcinoma: Transitional cell carcinoma occurs in the transitional epithelium, which consists of several layers of epithelium cells. This tissue lines the bladder, ureter, and parts of the kidney.

1.1.1.2 Sarcoma

Sarcomas are types of cancer that occur in bones and soft tissues like muscle, fat, nerves, blood vessels, cartilage, and lymph vessels. Sarcomas may be of two types, bone sarcoma and soft tissue sarcoma. These tumors mostly form in the chest, abdomen, arms, and legs. Rhabdomyosarcoma is the most common soft tissue sarcoma in children and occurs in muscles attached to bones. Gastrointestinal stromal sarcomas occur in soft tissues of the gastrointestinal tract, in the stomach or small intestine. Other types of sarcomas include Ewing sarcoma (common places are soft tissues attached to the pelvis, femur, clavicle, and humerus), Kaposi's sarcoma (lymph or blood vessels), and uterine sarcoma.

1.1.1.3 Leukemia

Leukemia does not form tumors; it is a cancer of the blood forming tissues of the bone marrow. A large number of abnormal and immature blood cells both in the bone marrow and in circulating blood result in leukemia. As a result, the count of normal blood cells drops, and it becomes increasingly difficult for the body to supply oxygen and nutrients to the cells. Leukemia may be categorized into acute and chronic forms, which are further classified according to the kind of blood cells affected.

1.1.1.4 Lymphoma

The lymph system is one of the most important parts of the immune system, and lymph tissues are found all over the body, with higher concentrations in the neck, armpits, and the groin area. Lymphoma is a cancer of the cells of the lymph system and can occur anywhere in the body. There are two main types of lymphoma,

Hodgkin lymphoma and non-Hodgkin lymphoma (NHL). The signature of normal Hodgkin lymphoma is an abundance of large and abnormal lymphocytes or white blood cells that compromise the ability of the immune system to fight infections. There is also a variant of Hodgkin lymphoma called the nodular lymphocyte-predominant Hodgkin lymphoma that is relatively rare and grows at a much slower rate compared to the normal version of Hodgkin lymphoma. There are many different types of NHL depending on the type of white blood cells they affect, for example, B-cell, T-cell, and NK-cell NHL. Large B-cell lymphoma is the most common form of NHL.

1.1.1.5 Myeloma

Myeloma, also called multiple myeloma, is a type of cancer that affects plasma cells, a type of immune cell. Plasma cells develop from B cells, and the responsibility of plasma cells is to produce antibodies for fighting infection. Plasma cells are transported by the blood plasma and the lymphatic system. Plasma cell neoplasm is a disease when a high concentration of abnormal plasma cells, called myeloma cells, deposits in the bone marrow. These abnormal plasma cells make abnormal antibodies that can damage the kidney due to thickening of blood.

1.1.1.6 Melanoma

Melanocytes are a type of cells in the epidermis responsible for producing the pigments that give the skin, eyes, and hairs their colors. Melanoma is a skin cancer that affects melanocytes, and it is more aggressive compared to squamous cell and basal cell carcinomas. Melanoma can spread to other parts of the body through metastasis unlike the other two types of skin cancer. One of the main causes of melanoma is exposure to ultraviolet rays of sunlight.

1.1.1.7 Brain and Spinal Cord Tumors

The brain and spinal cord together constitute the central nervous system (CNS); hence, tumors in these areas are also called CNS tumors. There are many different kinds of tumors of this class, and they are named depending on where in the nervous system they form or on which type of cell they form. For example, ependymal tumors begin in cells lining the spinal cord central canal, and astrocyte tumor begins in star-shaped brain cells called astrocytes.

1.2 Hallmarks of Cancer

Our review of the hallmarks of cancer is based on the seminal review article by Hanahan and Weinberg [4] and also on subsequent developments in understanding these hallmarks in recent literature. Cancer is a complex disease, and the complexity may be streamlined to a handful of underlying traits known as "hallmarks of cancer." These common features steer the transformation of a normal cell to a cancerous cell. As discussed above, all types of cancers have different stages in their development, and these hallmarks are associated with these different stages. The hallmarks

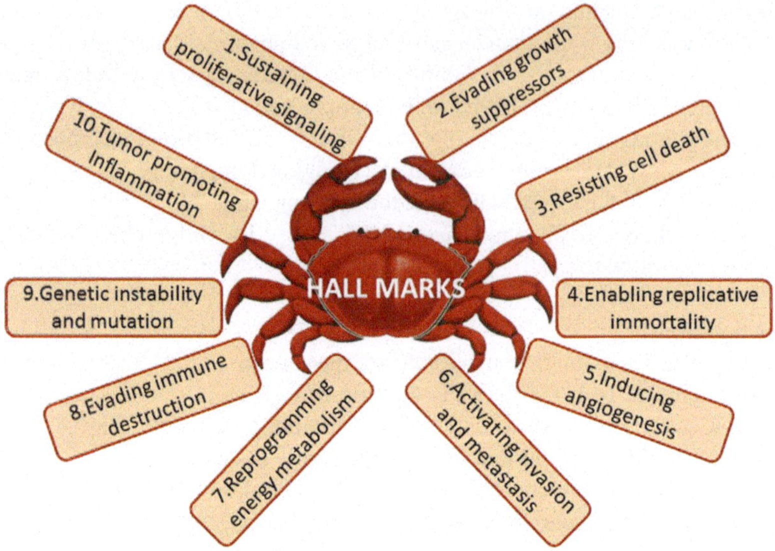

Fig. 1.3 The hallmarks of cancer

are [4] (1) sustaining proliferative signaling, (2) evading growth suppressors, (3) resisting cell death, (4) enabling replicative immortality, (5) inducing angiogenesis, (6) activating invasion and metastasis, (7) reprogramming energy metabolism, and (8) evading immune destruction. These eight characteristic features are typical of a cancer cell, but there are two other distinctive traits that contribute to the development of cancer. These are (9) genetic instability and mutation and (10) tumor-promoting inflammation. The first six of these hallmarks had been published in the journal *Cell* in 2000. Later in 2010 Hanahan and Weinberg proposed four more hallmarks [4]; therefore today there are altogether ten hallmarks of cancer. All ten hallmarks of cancer are pictorially represented in Fig. 1.3.

1.2.1 Sustaining Proliferative Signaling

Cells in eukaryotic tissues communicate with each other by transmitting and receiving hundreds of different signaling molecules, including amino acids, proteins, small peptides, and nucleotides, collectively called the growth factor signals. There are two different mechanisms in this signaling, extracellular and intracellular. The signaling molecules diffuse through the tissues before getting attached to the cell surface and are received by the intracellular receptor proteins. The intracellular pathways then use this information to maintain the concentration levels of a plethora of proteins participating in many pathways, so that there is no unwarranted level of cell division and abnormal growth.

While the propagation of the growth factor signals through the extracellular medium is dependent on temporal and spatial gradients, and is poorly understood, the intracellular part of growth factor signaling is relatively well understood, particularly in cancer cells. Cancer cells can participate in proliferative signaling in two different ways, either by secreting a higher concentration of growth factor ligands or by producing an elevated level of cognate receptor proteins or both. The higher concentration of receptor proteins makes cancer cells hypersensitive to growth factor signaling that subsequently results in abnormal levels of cell division and growth of tumors. This abnormal level of growth factor signaling is complemented by transcription of higher levels of proteins in the key pathways that constitute the intracellular part of growth factor signaling. However, it has also been observed that excessive proliferating signaling may result in the activation of senescence, or aging of cells, and eventually apoptosis or cell death. It is possible that cancer cells either may strike a balance so that proliferating signaling is kept at a level not to trigger senescence and apoptosis pathways or cancer cells may even interfere with these pathways to make proliferative signaling more effective for unconstrained cell division and growth in tissues.

There are also pathways that act as negative feedback loops. Such negative feedback loops try to mitigate the effects of proliferative signaling and are part of the innate cellular mechanisms for preventing the effects of excessive proliferative signaling. However, there is now growing evidence that cancer cells can interfere with such pathways and even deactivate them.

In summary, proliferative signaling is one of the fundamental hallmarks of cancer, and cancer cells develop various mechanisms for interfering with the innate abilities of cells to avoid the effect of proliferative signaling. In other words, cancer cells become independent of the normal homeostasis in tissues through these mechanisms and can divide and grow in an uncontrolled fashion.

1.2.2 Evading Growth Suppressors

Cancer cells also develop the capability of avoiding growth-suppressing pathways that prevent cell proliferation. Tumor-suppressing genes play central roles in pathways that repair DNA copying mistakes, slow down cell division, and initiate apoptosis or cell death. Dozens of tumor suppressor genes have been identified until now, and their effects have been validated in mouse models. Tumor suppressor genes may cause cancer due to their under expression or lack of expression, as opposed to oncogenes that cause cancer due to their overexpression. Two of the prototypical tumor suppressor genes are the RB1 gene located on chromosome 13 and the TP53 gene located on chromosome 17.

The pRB or RB protein, transcribed from the RB1 gene, plays a key role in deciding whether a cell should go through its growth and division cycle or how fast should such cycles occur. There is growing evidence that the RB protein accumulates a host of growth inhibitory signals from both extracellular and intracellular sources for this decision-making process. The main responsibility of the RB protein is to prevent the progression of the cell division cycle from the first gap phase (G1) to the synthesis (S) phase, by preventing the copying of DNA. This has a profound effect in preventing the uncontrolled division of cells and progression of tumors. Cancer cells often have defects in the RB pathways.

The TP53 gene transcribes the p53 protein that has diverse anticancer roles. TP53 mainly accumulates intracellular signals due to DNA damage and stress signals like reduced levels of nucleotide pools for creating new DNA. The p53 protein also has the ability to prevent the progression of the cell division cycle from the first gap phase to the synthesis phase by preventing copying of DNA. Moreover, p53 can activate an increased transcription of DNA repair proteins in case of excessive DNA damage. p53 can trigger apoptosis in cells as well in response to an alarming level of DNA damage. All of these functions of p53 are anticancer, and it has been observed that many types of cancer cells have defects in their p53-induced pathways.

1.2.2.1 Disruption of Contact Inhibition

Another kind of growth inhibition in eukaryotic cells, and in particular human cells is growth inhibition due to close contacts between cells. It has been observed in in vitro cultures that cell-to-cell contacts in densely packed normal cells prevent further cell proliferations, whereas this growth inhibition is absent in cultures of cancer cells. There is also evidence of similar growth inhibition in vivo. Proteins transcribed from the NF2 (neurofibromatosis type 2) gene in chromosome 22 and the LKB1 (liver kinase B1) gene in chromosome 19 are the main players in this kind of growth inhibition.

Merlin, the protein transcribed from the NF2 gene, has dual roles in preventing extracellular signals reaching the cell surface receptors in densely packed cells. Merlin has a major role in strengthening cell-to-cell adhesion by coupling cell surface adhesion molecules (cadherin family of proteins, in particular E-cadherin) to cell surface receptor kinases. This prevents extracellular growth signals from attaching to these receptor kinases. In addition, Merlin also has important roles in capturing and sequestering cell surface receptors. Once sequestered by Merlin, these molecules are unable to produce mitogenic signals. Loss-of-function mutations and/or deletions in the NF2 gene have been observed in neurofibromatosis type 2 (NF2), a multiple tumor-forming disease of the nervous system [5], as mutations in the NF2 gene disrupt these functionalities of Merlin.

LKB1 is a critical tumor suppressor gene that has been found to be mutated in many different cancers. LKB1 is responsible for regulating a broad spectrum of cellular functions including metabolism, growth, autophagy, and polarity [6]. For example, LKB1 has the ability to overrule the mitogenic signals of the Myc oncogene, resulting in growth regulation. On the other hand, the suppression of LKB1 expression destabilizes epithelial integrity and exposes epithelial cells to the mitogenic signals of the Myc oncogene facilitating unrestricted growth [7].

1.2.3 Resisting Cell Death

It is now well established that cancer cells overcome the mechanisms of programmed cell death or apoptosis in order to destroy the homeostatic balance of cell population in tissues. Apoptosis is as important as mitosis in maintaining a healthy balance in cell population and cell functions. The disruption of the apoptotic pathways is one of the crucial hallmarks of cancer cells.

We first discuss the different apoptotic pathways and their key mechanisms and subsequently discuss how cancer cells can disrupt these pathways. The execution or final phase of apoptosis can be triggered by both *extrinsic* and *intrinsic* pathways, which we discuss first.

Extrinsic Pathway

The extrinsic pathway is triggered due to extracellular ligands through transmembrane interactions with intracellular receptors. These receptors are proteins transcribed from the tumor necrosis factor (TNF) superfamily of genes. Members of this family of proteins have a domain of about 80 amino acids called the "death domain" which is responsible for sending death signal from the cell surface to the intracellular signaling pathways. Two of the most well understood extracellular death ligands and intracellular death receptor pairs are FasL/FasR (also known as CD95L/CD95) and TNF-α/TNFR1. In case of both of these pairs, the intracellular receptors cluster on the cell surface and bind with the homologous extracellular ligands. Cytoplasmic adapter proteins with matching death domains are then recruited after the receptor-ligand binding. FADD is the adapter protein in case of FasL/FasR pair, and TRADD, FADD, and RIP are the adapter proteins in case of the TNF-α/TNFR1 pair. Next, a death-inducing signaling complex (DISC) is formed by binding with procaspase-8, and the autocatalytic activation of procaspase-8 and other procaspases begins at this stage, and the execution pathway is triggered. The extrinsic pathway is illustrated in Fig. 1.4.

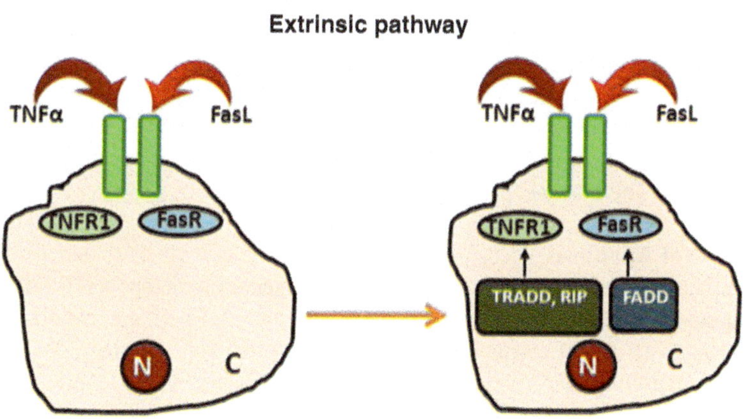

Fig. 1.4 An illustration of the extrinsic apoptosis pathway

Intrinsic Pathway

The intrinsic signaling pathways are mainly mitochondrial-initiated events, controlled by the Bcl-2 family of proteins. This family consists of about 20 members with either pro-apoptotic or anti-apoptotic functions. The members of Bcl-2 family can be divided into three groups [8] depending on the conserved Bcl-2 homology (BH) regions in the proteins. The pro-apoptotic members are again divided into two groups. Pro-apoptotic proteins Bax, Bak, and Bok share BH1, BH2, and BH3 regions, whereas BH3-only proteins share only the BH3 region. Members of this second group include Bad, Bim, Bid, Noxa, Puma, Bik/Blk, and Bmf. The anti-apoptotic members of the Bcl-2 family include Bcl-2, Bcl-x, Bcl-xL, Bcl-W, and BAG. The signals may be positive or negative. Positive signals are mainly activated by environmental factors like radiation, toxins, free radicals, viral infections, hypothermia, etc. Negative signals are usually lack of growth factors, hormones, and cytokines that are usually suppressors of apoptosis-related pathways in cells. Both positive and negative signals trigger the pro-apoptotic members of the Bcl-2 family of genes. Both positive and negative signals cause the opening of the mitochondrial permeability transition (MPT) pores and release of two groups of pro-apoptotic proteins in the cytosol. The first group initiates the death pathways, and the second group accelerates cell death when the apoptosis pathway has been activated already.

The first group of proteins from mitochondria consists of cytochrome c, Smac/DIABLO, and the serine protease HtrA2/Omi. These proteins activate the caspase-dependent execution pathway. In the first step, cytochrome c binds with procaspase-9. This activates caspase-9, and both autocatalytic and cascading activations of the caspase family of proteins start. On the other hand, Smac/DIABLO and HtrA2/Omi aid the apoptosis process by inhibiting a class of proteins called "inhibitors of apoptosis" (IAP) proteins. The second group of pro-apoptotic proteins is also released from the mitochondria; however, release of these proteins occurs when the apoptotic pathway has been activated already and the cell has committed to die. This group of proteins includes AIF, endonuclease G, and CAD. AIF and endonuclease G act independent of the caspase pathway; they translocate to the nucleus. The job of AIF is to cleave the DNA in smaller parts, into about a few hundred kilobase pieces. It also participates in condensation of the nuclear chromatin. Endonuclease G cleaves this condensed chromatin. CAD translocates to the nucleus after cleavage by caspase-3 and participates in more advanced chromatin condensation. The intrinsic pathway is illustrated in Fig. 1.5.

Execution of Apoptosis

Caspase family of proteins are central to the execution of apoptosis, they play an important role in destruction of cells, and hence they are called the key executioners. Different caspase proteins play different roles in degrading nuclear materials including DNA and protein cleavage. Caspases are expressed in an inactive zymogen or proenzyme form in all living cells. The activation of these procaspases is the final action in both the extrinsic and intrinsic apoptosis pathways.

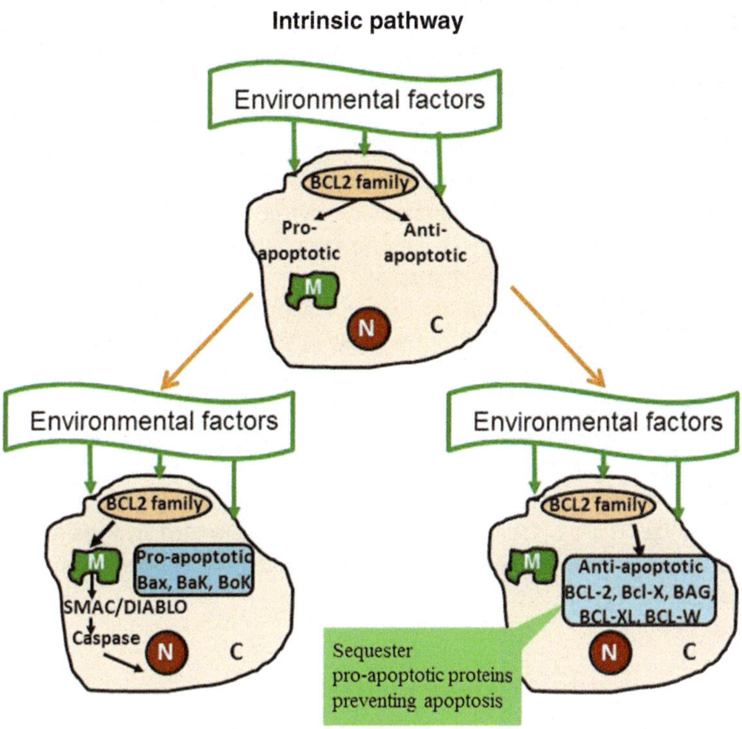

Fig. 1.5 An illustration of the intrinsic apoptosis pathway

Once activated, a caspase can activate other procaspases in a cascading manner. This cascading activation of caspases is a key feature of apoptosis, as once initiated, apoptosis signaling leads to rapid cell death. The initiator caspases for the apoptosis cascade are caspase-2, caspase-8, caspase-9, and caspase-10, while the executioner caspases that are responsible for cleaving DNA and proteins are caspase-3, caspase-6, and caspase-7. One of the key features of the execution pathway is the recognition of apoptotic cells by neighboring phagocytes for cleaning up waste products of the dying cells.

How Cancer Cells Avoid Death

Cancer cells overcome apoptosis by disabling both the intrinsic and the extrinsic apoptotic pathways. We first discuss the disruption of the intrinsic pathway. The TP53 gene and the corresponding protein p53 have a critical role to play in regulating the Bcl-2 family of proteins. Several studies in the mouse model have indicated that the BH3-only proteins, a key subfamily of the Bcl-2 family, inhibit the anti-apoptotic proteins and activate the pro-apoptotic proteins [9]. It has been observed that one of the most commonly mutated genes in human cancer is the TP53 gene, with mutation frequency exceeding 50% [10]. The TP53 gene is sometime called the "guardian of the genome" as it plays a central role in sensing DNA damage through multiple DNA damage response pathways. The mutation of the TP53 gene changes the structure of the p53 protein and inhibits its ability to sense DNA damage. One of the key actions of p53 is to trigger apoptosis when DNA damage is excessive. p53 induces apoptosis by upregulating the expressions of the BH3-only proteins Noxa and Puma. Again, this ability of p53 is hindered due to mutations in the TP53 gene. Alternatively, cancer cells can upregulate the anti-apoptotic genes Bcl-2 and Bcl-xL or downregulate the pro-apoptotic genes Bax, Bim, and Puma to achieve the same effect.

There is some evidence that the extrinsic apoptotic pathway is also regulated by p53; however, the details are not well understood. For example, the Fas death receptor loci and the gene encoding for the Fas ligand TNFSF6 are targets of the p53 protein [11]. Also, p53 may facilitate a cross talk between the intrinsic and the extrinsic apoptotic pathways by trans-activating the BH3-only protein Bid. However, it is also known CD95 (FasR) may behave like an oncogene and help in tumor progression once cancer cells have become immune to the apoptosis-inducing role of CD95L (FasL). Cancer cells may even use elevated expression of CD95 for killing lymphocytes in order to proliferate, through a process known as "tumor counterattack" [12].

1.2.4 Enabling Replicative Immortality

To grow macroscopic tumors, cancer cells acquire the ability for unlimited replication. Normal, non-cancerous cells go through a few cell growth and division cycles before entering senescence and eventual death. One of the main reasons for losing the ability to divide is the fusion of chromosomes that prevents cell division and endangers the viability of a normal cell. The ends of chromosomes are protected by several kilobase-long hexanucleotide repeat sequences like GGGTTA and TAACCC called *telomeres* [13]. These telomeres deplete every time a cell divides, with a loss of tens to hundreds of such repeat sequences with every cell division. As a result, the ends of the chromosomes get exposed after a small number of cell divisions and the chromosomes get fused end to end.

Telomeres can be repaired by the enzyme telomerase, a specialized DNA polymerase that is usually not expressed in non-cancerous cells but expressed at functionally significant levels in over 90% of cancerous cells. Telomerase extends the telomeric hexanucleotide repeat sequence so that the cancerous cells can continue to divide indefinitely and become resistant to senescence or apoptosis. This relationship between expression of telomerase and cell growth and division cycle has been observed in cell cultures where either telomerase has been engineered to be expressed in cultured cells or in cultures of spontaneously immortalized cells. This has also been observed in cancer tissues.

1.2.5 Inducing Angiogenesis

All cells require the supply of nutrients and oxygen for their metabolic activities and the removal of metabolic waste products and CO_2 for healthy functioning and cell division. These functions are performed by blood vessels in healthy tissues as well as in tumors. Angiogenesis is a major activity during embryogenesis through the birth of new endothelial cells and their assembly into veins and arteries. Angiogenesis is a controlled activity in adults and required to support growth as well as healing processes and turned on only transiently. However, tumors require uncontrolled angiogenesis to support the uncontrolled growth of cancerous cells into tumors. The angiogenesis switch is almost always activated during the neoplastic growth of tumors.

Angiogenesis is a balance between two kinds of factors that either induce or oppose it. The switch is tipped in favor of uncontrolled angiogenesis when the inducing factor dominates. It has been observed that the growth of blood vessels is not uniform in tumor tissues unlike in normal tissues, indicating the fact that factors opposing angiogenesis are still active in tumor tissues. An important angiogenesis-inducing factor is vascular endothelial growth factor-A (VEGF-A), and an important opposing factor is thrombospondin-1 (TSP-1) that binds to transmembrane receptors. Expression of TSP-1 is controlled by oncogenes like Ras and Myc. We refer the reader for more details of tumor angiogenesis to the paper by Hanahan and Weinberg [4].

The VEGF-A protein is transcribed from the VEGF-A gene located in human chromosome 6. This protein is responsible for generation of new blood vessels both during embryogenesis and also in adult bodies for maintaining homeostatic growth as well as wound repair. The VEGF-A signaling pathway is complex and regulated at multiple levels, an indication that transient angiogenesis in normal tissue growth is carefully controlled. However, the VEGF-A gene expression can be upregulated in various ways for uncontrolled angiogenesis. This can happen due to hypoxia, a condition when cells do not get sufficient oxygen, and also due to oncogene signaling [4]. In addition, the upregulation of the fibroblast growth factor (FGF) family of genes has been implicated in the upregulation of the VEGF-A gene.

TSP-1, also known as THBS1, is a protein transcribed from the THBS1 gene located in human chromosome 15. This protein also binds to endothelial cells and is a key suppressor of pro-angiogenic signals. The interplay between the angiogenesis-inducing and angiogenesis-suppressing factors creates aberrant vasculature in many tumors. This kind of aberrant vasculature results in excessive branching of arteries and also micro hemorrhaging [14]. Aberrant regulation of angiogenesis is one of the hallmarks of cancer. Vascular network is disrupted when there is lack of balance between pro- and anti-angiogenic signaling. Dysregulated vasculature leads to nonuniformity in tumor blood flow and oxygenation among others. These perturb the microenvironment and cater progression of tumor and dampen the potency of conventional therapeutics.

1.2.6 Activating Invasion and Metastasis

Metastasis is the process of migration of cancer cells from their primary site to other parts of the body. The extracellular matrix (ECM), a collection of extracellular molecules, is responsible for holding the cells in a tissue. ECM, apart from providing support to the cell, aids in regulation of intracellular communication and hence supports dynamic nature of cells. Formation of ECM is necessary for cell growth, healing of wound and fibrosis. Destruction of ECM by several enzymes proteases/proteinases is implicated in metastasis; hence, knowledge of ECM enables one to understand the dynamics of tumor invasion and metastasis. Carcinomas initially arise in epithelial tissues and invade first the local tissues and then nearby blood and lymphatic vessels. They eventually escape from blood and lymphatic vessels to tissues in other parts of the body. They initiate first micrometastatic lesions in distant tissues, which eventually grow into macroscopic tumors. Hence the process of metastasis is very complex and occurs in several stages.

Cell migration is an essential process in development as well as several other physiological processes, and cells need to detach from the ECM for migration to occur. However, cell migration is a carefully controlled mechanism in normal physiological processes. Cell migration requires alterations in cell shapes and their attachments to the ECM. E-cadherin (epithelial cadherin, also known as cadherin-1, the name cadherin stands for calcium-dependent adhesion) is one of the most important transmembrane proteins responsible for this alteration mechanism. E-cadherin is transcribed from the CDH1 gene located in chromosome 16. E-cadherin is a key protein for cell-to-cell adhesion and maintaining epithelial cell sheets. Downregulation of CDH1 gene and loss of E-cadherin is the key mechanism through which cancer cells detach from the epithelial sheets. The downregulation or mutation of the E-cadherin protein has been observed in most metastatic carcinomas. However, cadherin family of proteins controls the balance between suppression and promotion of invasion in cancer. N-cadherin is a protein of this family that is transcribed from the CDH2 gene located in chromosome 18. N-cadherin plays diverse roles in promoting cell motility during gastrulation and neural crest development. N-cadherin is upregulated in most aggressive carcinomas [3].

However, detachment from the ECM is just one of the requirements for cell motility during metastasis. It is now understood that cancer cells take the assistance of the *epithelial-mesenchymal transition* (EMT) program for metastasis [4, 15]. This is a crucial program during embryonic development as the differentiation and movement of mesenchymal cells are at the heart of growth, and development of all mammalian species and this program is highly conserved across species. It is known that a set of pleiotropically (a single gene affecting multiple phenotypic traits) acting transcription factors is responsible for the EMT, including Snail, Slug, Twist, and Zeb 1/2 [4]. These transcription factors are also expressed in various combinations in different malignant cancer types, indicating complex relationships between different genetic pathways during metastasis.

Growth signals trigger division of cells in healthy tissues. In ECM, these are detected by growth factor receptors, namely, vascular endothelial factor receptor (VEGFR) or epidermal growth factor receptor (EGFR), both of these transmembrane proteins belong to receptor tyrosine kinases (RTKs). EGFR when binds to a ligand, signal transduction pathways like Akt/PKB (protein kinase B) get activated. One of the functions of the ECM is to activate the protein Akt or protein kinase B (PKB) that is responsible for signaling cells for survival and division. The detachment of cells from the ECM ceases this signal, and anoikis, a form of programmed cell death in anchorage-dependent cells, is initiated.

Akt is a suppressor of Bad, a pro-apoptotic protein, and Bad initiates the intrinsic apoptosis pathway as we have noted before, leading to anoikis. Cancer cells must overcome anoikis before proliferating. It has been observed that downregulation of E-cadherin and upregulation of mesenchymal N-cadherin dampen the adhesion ability of cancer cells to epithelial cells, subsequently progressing to invasion; this leads to anoikis resistance, indicating that many of the processes related to metastasis are interrelated in ways that are yet to be understood [15].

The emerging picture for migration is that cancer cells need to activate genes that are necessary for accomplishing many functions including downregulation of transmembrane proteins that help in cell-to-cell attachment, upregulation of transmembrane proteins that help in cell mobility, degradation of cell-to-cell junctions, activation of anti-apoptotic mechanisms, and activation of proteases on the cell surface that help the cells to cut through the ECM. Another crucial condition for metastasis is the transition of cell characteristics from epithelial to mesenchymal cell. This is necessary for a cell to establish itself in a remote tissue after travelling through blood vessels, as a mesenchymal cell can get integrated into a different type of tissue by activating the developmental programs.

1.2.7 Reprogramming Energy Metabolism

Cells need energy for maintaining metabolic functions, growth and division, and glycolysis is one of the most important pathways for extracting energy from glucose. The glycolysis pathway and the proteins or enzymes involved in this pathway are highly conserved in almost all living organisms, indicating the ancient origin of this pathway. Cancer cells use the glycolysis pathway in a way that is quite mysterious, and this potential hallmark is based on this characteristic of cancer cells.

The glycolysis pathway is a collection of reactions that converts glucose into pyruvate. Free energy is released during this conversion, and this energy is used to form the intracellular energy-transporting molecules ATP and NADH [4, 16]. Glycolysis occurs in the cytosol and does not use oxygen in any of its reactions, and hence it is called anaerobic. However more energy is extracted from pyruvate using oxygen through oxidative phosphorylation, and an NADH molecule and a carbon dioxide molecule are produced in the process. This is an aerobic reaction and occurs in the mitochondria. Though glycolysis is the preferred reaction for extracting energy under anaerobic conditions, pyruvate is sent to mitochondria when oxygen is available as the energy extracted from a combination of glycolysis and processing pyruvate is far higher compared to glycolysis alone.

The anomaly observed in cancer cells is that these cells produce energy only through glycolysis even in the presence of oxygen and do not send pyruvate to the mitochondria. This is a mysterious metabolic switch used by cancer cells since the energy production in glycolysis alone is about 18-fold lower than a combination of glycolysis and pyruvate metabolism [4]. Moreover, cancer cells require far more energy for rapid proliferation and cell division. It has been observed that cancer cells compensate for this inefficiency in energy production by upregulating glucose transporter proteins so that more glucose is brought to cytosol. However, still this deliberate inefficiency in energy production in aerobic conditions is enigmatic, and a rationale has been elusive to find [4, 16]. Moreover, this glycolysis-dependent energy production has been shown to be associated with upregulation of oncogenes like MYC and RAS and mutation of TP53, the same factors that are responsible for facilitating the other hallmarks of cancer that we have discussed before [17].

1.2.8 Evading Immune Destruction

Another emerging hallmark is related to the understanding of the role of the immune system in eradicating incipient neoplastic lesions and micrometastases [4]. The normal functioning of the immune system posits that the cells of the immune system are constantly monitoring cells and tissues and this surveillance of the immune system should eliminate a vast majority of the incipient neoplastic lesions and early metastatic growths. Perhaps the neoplastic growths that survive this surveillance somehow avoid the actions of the immune system. However, there is evidence both for and against this theory.

It is known there is a significant increase of certain cancer types in immunocompromised individuals; however, almost all such cancers are of viral origin [18]. Hence perhaps the role of the immune system in such cases is to eliminate virus-infected cells, and it is hard to generalize this role for a vast majority of cancers. Experiments in the mouse model have shown that mice deficient in different components of the immune system can get cancers at a higher rate compared to mice that are immunocompetent. These results show that the immune system has some roles to play in eradicating cancers in certain experimental models at least. There is also evidence of antitumor immune system responses in some forms of human cancer. For example, patients with higher levels of CTL (cytotoxic T lymphocytes) and NK (natural killer) cells have better prognosis for colon and ovarian tumors. However, the overall role of the immune system in cancer initiation and progression is still not well understood [4].

1.2.9 Genetic Instability and Mutation

Almost all the hallmarks we have discussed in this chapter are related to genetic mutation or genetic instability. Genes are almost always organized into circuits or pathways where there are various interrelationships between the expressions of genes. A gene can up- or downregulate the expression of another gene belonging to the same pathway, and as a result the concentrations of various proteins transcribed from the genes can also change. This sort of instability disrupts the normal functioning of a cell, and such genetic instabilities have been detected in most hallmarks of cancer. The alteration of the expression levels of a particular gene varies in cancerous cells; however, the effects of altered expression levels of a single gene could have far-reaching effects. An extreme example of such effect is the triggering of oncogenes or cancer-causing genes, for example, the oncogene Myc can regulate the transcription of thousands of different genes [19]. We will discuss many other gene regulations in relation to the hallmarks of cancer in the subsequent chapters of the book.

Genetic mutation occurs due to copying errors during mitosis and also due to environmental factors like radiation and chemicals. However, normal cellular DNA maintenance machineries are able to correct many of these mutations, and this ensures that the rates of such mutations are kept very low in each cell generation. However, cancer cells can increase the rates of mutation [4, 20]. Such higher rates of mutation are achieved through breaking down of genetic maintenance machinery and increased sensitivity to mutagenic agents. Genetic surveillance mechanisms in normal cells can sense such increased rates of mutation and activate the senescence (cell aging) and apoptosis (cell death) pathways; however, cancer cells can compromise these cellular surveillance mechanisms in several different ways [21]. We should also mention that some epigenetic factors like DNA methylation and histone modifications can alter the concentrations of proteins without modifying the underlying DNA sequence of genes [22]. We discuss these epigenetic mechanisms in details in Chap. 5.

1.2.10 Tumor-Promoting Inflammation

The role of inflammations in tumor initiation and progression was suspected as early as in the nineteenth century [2]. It has been observed for a long time that some tumors harbor an overabundance of cells from both the innate and adaptive arms of the immune system, which protect us by fighting infections. Initially it was thought that the response of the immune system was targeted towards eradicating tumors, and that is definitely one of the reasons why the cells from the immune system infiltrate neoplastic growths. However, it is being understood more and more that the interaction between cancer cells and cells from the immune system is far subtler and cancer cells actively engage the immune system for tumor growth and proliferation [4].

Anti-tumorigenic and pro-tumorigenic mechanisms of the immune system coexist in tumor environments. It is possible that many early-stage neoplastic growths are rejected and cleaned up by the immune system; however, pro-tumorigenic factors of the immune system take over once a tumor is established [23]. Immune system performs many pro-tumorigenic roles similar to healing of inflammations, since healing of inflammation requires building of new tissues and blood vessels similar to a tumor. These pro-tumorigenic roles include supplying bioactive molecules to the tumor microenvironment, including growth factors that sustain proliferative signaling and angiogenesis (growth of new blood and lymphatic vessels), and extracellular matrix-modifying enzymes [4]. Moreover, the pro-tumorigenic actions of the immune system progress rapidly due to feed-forward loops in the signaling pathways. In other words, many such pro-tumorigenic signals result in cancer cells producing bioactive molecules to attract more immune cells to the tumor tissue. For example, the activation of the NF-κB pathway in immune cells induce production of cytokines and that in turn activates the NF-κB pathway in cancer cells to induce chemokines that help in attracting more inflammatory cells into the tumor tissue [23].

1.3 Cancer Diagnosis and Treatment

1.3.1 Risk Factors and Causes of Cancer

A cancerous cell is formed due to accumulation of changes or mutations in its DNA. Therefore, any agent that triggers damage to the genetic material may be a probable cause of cancer. Broadly speaking, the genetic defects that are inherited may be a cause of cancer. Infections with virus, bacteria, etc. may also cause cancer. Environmental factors also play important roles in this respect. Personal traits often leave an impact on the development of carcinogenesis.

Toxic chemicals like benzene, asbestos, nickel, cadmium, vinyl chloride, benzidine, N-nitrosamines, tobacco or cigarette smoke (contains at least 66 known potential carcinogenic chemicals and toxins), aflatoxin, arsenic, and many more chemicals are risk factors for cancer. Ionizing radiations also may contribute to the etiology of cancer. Ultraviolet rays from sunlight; radiation from alpha, beta, and gamma; and X-ray-emitting sources are some examples. Some cancers have been found to have a genetic linkage. Pathogens are also causative factors. Human papillomavirus (HPV), EBV or Epstein-Barr virus, hepatitis viruses B and C, and Kaposi's sarcoma-associated herpes virus (KSHV) are found to be associated with the causation of several cancers. Infection with some bacteria like *Helicobacter pylori* also contributes to cancer development.

1.3.2 Symptoms

It is important to have a knowledge of the signs and symptoms of cancer, but often it is difficult to pinpoint them. This is mainly due to lack of knowledge. Early detection of cancer is the first step of awareness, and it paves a way to cancer control. Earlier the cancer is detected, better is the chance of survival. Therefore, knowing the manifestation of the disease is extremely important and that is possible with prior knowledge of signs and symptoms. There are no unique signs or symptoms of cancer; they depend on the type of cancer, location of the offending growth, and also whether the disease has spread to other parts of the body. In general, there are seven signs of cancer, which should not be ignored. These are (i) a sore that does not heal; (ii) unusual bleeding or discharge; (iii) sudden change in bowel or bladder habits; (iv) indigestion or painful swallowing; (v) lump in the breast, testicles, or any other part of the body; (vi) nagging cough or hoarseness of voice; and (vii) change in size, color, and shape of wart, mole, or sore in the mouth. Other indications are blood with cough, unexplained weight loss or loss of appetite, constant headache, pain in bones, persistent fatigue, nausea or vomiting, low-grade fever without any obvious cause, and repeated infection.

1.3.3 Diagnosis

Identification of the nature of an illness and severity is known as diagnosis. There are certain points which make a diagnosis complete; these are history of an individual's health, physical examination, and tests suggested by the physician. The symptoms often may be non-specific. Clinical diagnosis is based on the signs and symptoms, as understood by a physician. Some of the diagnosis is based on the reports from the laboratory. The parameters tested can give a clue, and the individual is diagnosed on the basis of the test results.

Cancer is not a single disease but an ensemble of diseases having similar behavior. The similarity lies in the fact that cancer cells grow uncontrollably, but each individual cancer is different. Therefore, a single drug cannot cure the disease. This multistage disease therefore needs multimodal treatment. Not only modality, adverse effects, and cost of treatment all need to be taken into consideration. This is why prevention and early detection are important. A better outcome may be expected if detected at an early stage. Therefore, early detection and diagnosis are very important for a good treatment outcome.

If the symptoms are suggestive of cancer, then proper diagnosis is mandatory. Generally, the attending physician considers personal and family history of the patient in the beginning. Physical examination is also vital. Based on these a patient may have to undergo a number of diagnostic tests, which include pathological tests and scans. Certain pathological and biochemical parameters may indicate a sign of cancer. They may help to make a tentative diagnosis, but it may not be possible to make a definitive diagnosis of cancer at this stage. Therefore, these tests are not the only diagnostic factors; they are used as predictive factors. An aberrant value cannot confirm cancer; similarly a negative result may not rule out the possibility of the disease. The doctors have to depend on other diagnostic tools. Biopsy and imaging results are two very important determinants.

1.3.3.1 Biopsy

A part of the tissue from the suspected part of the body is removed. That tissue is examined under a microscope to assess if it is cancerous or not. A tissue may be taken out by surgical procedure, both by excisional and incisional surgery. In excisional biopsy, the entire tumor mass is removed. Often the normal tissues surrounding the growth get removed as well. In incisional biopsy, a part of the tumor is removed and examined. Tissue sample may be taken endoscopically. Sample for biopsy may also be taken through a needle, which is called needle biopsy.

1.3.3.2 Imaging

Imaging methods let the doctor see the presence of a tumor mass. There are several ways to look into the offending growth. These are X-ray, computed tomography (CT), nuclear scan, ultrasound, MRI, and PET scan. X-rays can differentiate soft tissues and bones and use low doses of radiation to create pictures of different parts inside the body employing electromagnetic radiation. Bones, organs, and soft tissues can be imaged, and an estimation of the size of the tumor and its location can be known. Like X-ray, CT scans show the shape, size, and location of a tumor, more clearly than the X-ray. Comparing the CT scan reports before or after treatment, doctors are able to understand the response to the therapy. A dye or any other contrast material is used to increase the contrast. In nuclear scan, a small amount of radioactive substance is used. This is called tracer, which is injected into the vein. The tracer is collected in certain areas and is detected by a scanner, and radioactivity accumulated at different parts is measured. This is also known as radionuclide scan. The injected radioactivity gets eliminated with time.

An ultrasound wave bounces off the tissues like an echo. The echoes are collected by computer to create an image, called sonogram. Ultrasound can detect tumors which are not diagnosed by X-rays but are not as detailed as those obtained by CT or MRI. Ultrasound cannot give a hint if the tumor is cancerous or not. This technique helps to perform a needle biopsy. Rate of blood flow in tumor is more than in normal tissues, and Doppler flow machines can be of help in this regard.

Magnetic resonance imaging (MRI), using strong magnets, can create cross-sectional pictures of internal organs. Cross-sectional view from various angles can be taken. Doctors can plan a treatment out of MRI results. MRI with dye can enhance contrast, and this is very effective in detecting brain and spinal cord tumors. This method can also predict metastasis.

Positron-emission tomography (PET) can give a three-dimensional image of organs and tissues. A small amount of a radioactive substance is injected, which is picked up by the organs and tissues. Cancer cells pick up more radiation, as they use more energy. Sometimes a PET scan is combined with a CT scan. The combination can give an estimate of the stage of cancer.

1.3.4 Treatment

Cancer is not a single disease and its treatment is not unique. Different types of cancer need different combinations of treatments. Treatment regimen depends on the type of cancer, its stage, grade and health condition of the patient, and many other factors. Therefore, an oncologist has to be very judgmental regarding treatment, based on several criteria and conditions.

In some cases, a single treatment is given, whereas, for other cases, a combination treatment is prescribed. The treatment modalities followed to cure cancer are:

(i) *Surgery:* To get rid of the offending growth, surgery is done. The extent of surgery to treat cancer and its successful outcome vary according to the type of cancer and its stage, size, distribution, and location. By surgery it is possible to remove the growth and adjoining tissues. Besides, surgery is also performed to know the location of cancer, if it has spread to other parts of the body. Surgery may be (i) radical surgery (removal of the whole tumor mass), (ii) conserving surgery (e.g., for breast cancer, only the affected part is removed, not the entire breast. If the cancer has advanced much, this is not possible), and (iii) reconstructive surgery (when a part of the body is removed by surgery, e.g., breast or testicle, tissues from patient's own body or silicone material can be used to reconstruct). Other types are surgery for symptomatic relief, recurrent cancer surgery, surgery for metastasis, and surgery for prevention of cancer.

(ii) *Radiation therapy:* Radiation therapy, also known as radiotherapy, is a treatment using ionizing radiation. Ionizing radiation from a linear accelerator can kill the cancer cells, thereby shrinking the tumor. Radiation damages the DNA of cells. Two types of energy, namely, photon or charged particle, can cause damage to the DNA. The damage may be either direct or indirect ionization of the atoms which constitute the DNA. Radiation therapy is often used as an adjuvant to surgery. It is also used with chemotherapy. Radiation therapy is painless but has several side effects. Radiation therapy may be of three types: external beam radiation therapy (EBRT or XRT) or teletherapy, brachytherapy or sealed source radiation therapy, and systemic radioisotope therapy. Radiation is a cause of cancer, and therefore secondary malignancies may occur in some patients as a result of this treatment.

(iii) *Chemotherapy:* It is a mode of treatment using drugs. Surgery and radiation may not access all areas, but chemotherapy can reach every part of the body. These drugs resist the division and growth of cancer cells, thereby killing them. But this treatment modality has a number of adverse effects. Main purpose of chemotherapy is to cure and control the disease. It is also used for palliation. Choice of chemotherapy depends on a number of factors, type and stage of cancer, age and overall health condition of the patient, history of previous cancer treatments, and serious health problems, if any. Chemotherapy may be recommended alone or may be with surgery and radiation. Sometimes chemotherapy is given before surgery or radiation therapy to shrink the tumor. This is called neoadjuvant therapy. When administered after surgery and radiation, it is called adjuvant chemotherapy. In adjuvant chemotherapy residual cancer cells that escape radiation or surgery may be killed.

(iv) *Immunotherapy:* This is also coined as biologic therapy. In this type of therapy, the treatment boosts the immune system, i.e., natural defenses of a body. Immunotherapy stops or slows down the growth of the cancer cells and resists spreading of these cells to other parts of the body, i.e., metastasis. Boosting up the immunity, this type of therapy destroys cancer cells. There are a number of immunotherapies, which are monoclonal antibodies, cancer vaccines, oncolytic virus therapy, and T-cell therapy. Non-specific immunotherapies like interferons and interleukins are also used.

(v) *Targeted therapy:* This is a very important mode of treatment for cancer using drugs. The main types of targeted therapy are small molecules and monoclonal antibodies. This treatment method works by targeting the genes and proteins specific to that cancer.

(vi) *Hormone therapy:* Some of the cancers are dependent on hormones. This therapy slows down or stops the growth of those which utilize hormones for their sustenance. Hormone therapy may treat cancer and ease the symptoms. This modality blocks the ability of the body to produce hormones. Hormone therapy may be used efficiently with other treatment modalities. Before surgery and radiotherapy, this therapy is used to reduce the size of tumors, as an adjuvant therapy. Hormone therapy may kill cancer cells that have returned or may have spread to other parts of the body, i.e., metastasized.

(vii) *Stem cell transplant:* Stem cell transplants are effective in cancer therapy. Blood-forming stem cells get destroyed by chemotherapy drugs and radiation used to treat cancer. Stem cell transplant can restore blood-forming stem cells which are destroyed due to other treatment methods. Peripheral blood, bone marrow, and cord blood can be used to treat cancer. This method of therapy is effective for cancers where blood or immune system is affected. Three main types of blood cells are white blood cells, red blood cells, and platelets. Blood-forming stem cells grow into different types of blood cells, and healthy individuals need all three types. In stem cell transplant, blood-forming stem cells are given, and they are translocated to the bone marrow, where they replace the destroyed cells. These cells can come from the bone marrow, bloodstream, or umbilical cord. Stem cells can be replaced by the individual's own cells, which is called autologous transplant. Stem cells may also come from a different person, which is called allogeneic transplant. Stem cells may be given from an identical twin, which is known as transplant syngeneic.

(viii) *Precision medicine:* In this type of treatment, doctors decide upon the treatment regimen based on the genetic makeup. This is also called personalized medicine. With the help of precision medicine, cancer treatment is tailor made as per the genetic makeup of each individual. Genetic tests guide the treatment regimen. The therapies mentioned above are given alone, or in conjunction to the patients, depending on the type, stage, and other factors. With precision medicine, the doctors will be guided by the genetic predisposition of an individual. This has been discussed in details in Chap. 7.

1.4 Conclusion

We have reviewed the different types of cancers; the factors that can initiate cancers, e.g., genetic imbalance and inflammation; the different hallmarks of cancer; and diagnosis and treatment of cancer in this chapter. The aim of this chapter was not to review these areas thoroughly, rather to give the reader an overview of our current understanding of these areas of research. The reader can explore these topics more thoroughly from the references provided. We will explore all of these areas related to cancer in future chapters more thoroughly in the light of the effects of phytochemicals on cancer cells.

References

1. https://www.cancer.gov/types
2. L.M. Coussens, Z. Werb, Inflammation and cancer. Nature **420**, 860–867 (2010)
3. L.D. Derycke, M.E. Bracke, N-cadherin in the spotlight of cell-cell adhesion, differentiation, embryogenesis, invasion and signaling. Int. J. Dev. Biol. **48**(5–6), 463–476 (2004)
4. D. Hanahan, R.A. Weinberg, Hallmarks of cancer: The next generation. Cell **144**, 646–674 (2011)
5. A.M. Petrilli, C. Fernandez-Valle, Role of Merlin/NF2 inactivation in tumor biology. Oncogene **35**(5), 537–548 (2015)
6. M. Momcilovic, D.B. Shackelford, Targeting LKB1 in cancer – exposing and exploiting vulnerabilities. Br. J. Cancer **113**, 574–584 (2015)
7. A.F. Hezel, N. Bardeesy, LKB1: Linking cell structure and tumor suppression. Oncogene **27**, 6908–6919 (2008)
8. A. Shamas-Din, H. Brahmbhatt, B. Leber, D.W. Andrews, BH3-only proteins: Orchestrators of apoptosis. Biochim. Biophys. Acta **1813**(4), 508–520 (2010)
9. R.J. Youle, A. Strasser, The BCL-2 protein family: Opposing activities that mediate cell death. Nat. Rev. Mol. Cell Biol. **9**, 47–59 (2008)
10. Y. Liu, M. Kulesz-Martin, p53 protein at the hub of cellular DNA damage response pathways through sequence-specific and non-sequence-specific DNA binding. Carcinogenesis **22**(6), 851–860 (2001)
11. J.S. Fridman, S.W. Lowe, Control of apoptosis by p53. Oncogene **22**(56), 9030–9040 (2003)
12. M.E. Peter, A. Hadji, A.E. Murmann, S. Brockway, W. Putzbach, A. Pattanayak, P. Ceppi, The role of CD95 and CD95 ligand in cancer. Cell Death Differ. **22**, 549–559 (2015)
13. E.H. Blackburn, J.W. Szostak, The molecular structure of centromeres and telomeres. Annu. Rev. Biochem. **53**, 163–194 (1984)
14. J.A. Nagy, S.-H. Chang, S.-C. Shih, A.M. Dvorak, H.F. Dvorak, Heterogeneity of the tumor vasculature. Semin. Thromb. Hemost. **36**(3), 321–331 (2012)
15. S. Heerboth, G. Housman, M. Leary, M. Longacre, S. Byler, K. Lapinska, A. Willbanks, S. Sarkar, EMT and tumor metastasis. Clin. Transl. Med. **4**(6) (2015). https://doi.org/10.1186/s40169-015-0048-3
16. N. Hay, Reprogramming glucose metabolism in cancer: Can it be exploited for cancer therapy? Nat. Rev. Cancer **16**, 635–649 (2016)

17. R.G. Jones, C.B. Thompson, Tumor suppressors and cell metabolism: A recipe for cancer growth. Genes Dev. **23**, 537–548 (2009)
18. C.M. Vajdic, M.T. van Leeuwen, Cancer incidence and risk factors after solid organ transplantation. Int. J. Cancer **125**, 1747–1754 (2009)
19. J. van Riggelen, A. Yetil, D.W. Felsher, MYC as a regulator of ribosome biogenesis and protein synthesis. Nat. Rev. Cancer **10**, 301–309 (2010)
20. J.J. Salk, E.J. Fox, L.A. Loeb, Mutational heterogeneity in human cancers : Origin and consequences. Ann. Rev. Pathol. **5**, 51–75 (2010)
21. S. Negrini, V.G. Gorgoulis, T.D. Halazonetis, Genomic instability – an evolving hallmark of cancer. Nat. Rev. Mol. Cell Biol. **11**, 220–228 (2010)
22. R. Jaenisch, A. Bird, Epigenetic regulation of gene expression: How the genome integrates intrinsic and environmental signals. Nat. Genet. Suppl. **33**, 245–254 (2003)
23. S.I. Grivennikov, F.R. Greten, M. Karin, Immunity, inflammation, and cancer. Cell **140**, 883–899 (2010)

Cancer: Genetics and Important Pathways 2

Abstract

We have seen in Chap. 1 that all the hallmarks of cancer are related to molecular mechanisms in cells, driven by expression levels of key genes. Moreover, these genes are many times connected through genetic pathways. We have discussed the apoptosis pathway in Chap. 1. We start this chapter with a review of the fundamental concepts in genetics and molecular biology. We then discuss the genetic landscape of cancer. It is a complex problem to implicate specific genes for particular types of cancer, and our understanding is evolving all the time. We discuss the current understanding of the genetic landscapes for several important types of cancer. Finally we discuss some of the important pathways involved in cancer.

2.1 Introduction

Cancer is predominantly a genetic disease, most human cancers are either due to defects in germline or in somatic cells. These defects can be at the chromosomal level, e.g., chromosomal dislocations, or due to mutations in individual genes. Such defects have important consequences on the metabolic activities of individual proteins transcribed from defective genes. Moreover genes are usually organized as metabolic networks or pathways, and the defective transcription of a gene may have cascading consequences on a pathway. The aim of this chapter is to understand the extent to which these mutations are the cause for cancer initiation, tumorigenesis, and metastasis. We will also discuss some of the most important pathways that have been implicated in different cancers. We start the chapter with some discussion on basic concepts in molecular biology and genetics that will be helpful in following the rest of the chapter.

We refer to the textbook by Alberts et al. [1] for some basic concepts. The conversion from genes to proteins is a three-step process. First a pre-mRNA is

M. Roy, A. Datta, *Cancer Genetics and Therapeutics*,
https://doi.org/10.1007/978-981-13-9471-3_2

formed through transcription of the genetic code from the DNA. Next the intronic (noncoding) sequences of the pre-RNA are spliced out to form mRNA. Finally the mRNA is translated into a protein at the ribosome. For simplicity, we will mention this process as the transcription of a gene into a protein. Proteins have multitude of functions in a cell, including catalyzing metabolic reactions, transporting molecules, responding to intra- and extracellular signals and helping in DNA replication. Almost all of these activities depend on the correct three-dimensional folding of a protein based on its primary amino acid sequence. Any change in the primary amino acid sequence may result in wrong folding, and this may disrupt the normal functioning of a protein.

DNA is first translated into a precursor messenger RNA (pre-mRNA). Next the intronic sequences in the pre-mRNA are spliced out to form the mature mRNA. Then every triplet of bases in the mRNA is translated into an amino acid by transfer RNAs (tRNA) at the ribosome according to the genetic code which has some redundancy, i.e., the same amino acid can be translated from two or more different triplets of bases. However, the translated amino acid may be different when one or more of the bases in the DNA and subsequently in the mRNA change.

Such changes occur due to mutations which can be of several kinds. A single base may be substituted by another base; a single base may be inserted or deleted, in point mutations. In case of substitution, the mutation may be *synonymous*, i.e., the resulting mRNA is translated into the same protein, or *nonsynonymous*, i.e., the translated protein may have a change in its amino acid sequence, and it may be folded differently as a result. In case of insertion or deletion of a base, the reading frame during transcription changes and the resulting protein may have a vastly different primary amino acid sequence. Nonsynonymous mutations may not always change the folded structure or the function of a protein in case an amino acid is replaced by another amino acid with similar properties, e.g., a hydrophobic amino acid is replaced by another hydrophobic amino acid. This type of mutation is also called a *missense* mutation as one amino acid is replaced by another amino acid. A *nonsense* mutation is a point mutation of a single nucleotide such that the mRNA gets a stop codon due to this mutation. As a stop codon indicates the end of transcription, a nonsense mutation results in a vastly different protein.

Another source of change in primary amino acid sequences of a protein is rearrangements of chromosomes, which is a recurrent occurrence in many tumor cells. These rearrangements are usually translocations, i.e., the insertion of a part of a chromosome in a nonhomologous chromosome. These translocations may result in gene deletion and gene fusion. A necessary protein is not transcribed in the first case, and a completely different protein with different functionalities is transcribed in the second case.

The mutations in germline are transferrable from parents to progeny and may result in inherited genes that give predispositions to cancer. A vast majority of mutations in tumorigenesis are due to *somatic* mutations. However, it has been observed that many of these mutations are not associated with tumorigenesis and are called *passenger* mutations. The focus of this chapter is on the mutations that are directly responsible for tumorigenesis and are called *driver* mutations.

Driver mutations give growth advantages to a cell in many different ways. For example, the overexpression of the enzyme telomerase is one such growth advantage. This enzyme is responsible for repairing telomere, the protective base sequence at the end of chromosomes. A few kilobases of telomere are lost in normal cells with every cell division, and at the end chromosomes get fused due to lack of telomere, and cells die due to the activation of the apoptosis pathway. The activation of telomerase is tightly controlled in normal cells; however tumor cells can continue to divide uncontrollably due to overexpression of telomerase and repair of telomere.

Most of the driver mutations in tumors are observed in *oncogenes*. An oncogene is a normal gene that has the potential to impart growth advantages to a cell when mutated. *Proto-oncogenes* are normal genes that impart growth advantages to a cell when mutated. A proto-oncogene becomes an oncogene due to mutation; otherwise proto-oncogenes are normal genes that participate in cellular metabolisms with required levels of expression.

Epigenetics is the study of the changes in gene expression that occur without any change in the primary DNA sequence and hence without any change in the transcribed proteins. Epigenetic causes mostly result in over or under expression of genes and consequently in higher and lower expressions of proteins. One of the most important examples of epigenetic control in cancer is DNA methylation. Methylation of the promoter region of a gene can silence the gene with subsequent effects on other downstream genes in a pathway. Another epigenetic mechanism is *histone modification* which is a covalent posttranslational modification of histone proteins, either through methylation or through acetylation. Histones are responsible for packaging DNA in the cell nucleus, and hence any modification of these proteins affects gene expression as any change in the packaging of DNA changes the availability of DNA for transcriptional machineries. We discuss the role of epigenetic control in cancer in Chap. 5 in details. The interested reader may read that part of Chap. 5 at this point.

2.2 Degree of Gene Mutations in Tumorigenesis

It is important to get an understanding of the degree of mutations in genes and also the timing of these mutations in order to understand the genetic landscape of cancer. Our current knowledge about the mutations in tumor cells has increased tremendously in recent years due to the availability of high-throughput new-generation sequencing technologies. However, there are many problems associated with these technologies in sequencing tumor cells. In particular the sequencing results are not always accurate when large-scale genome rearrangements occur due to translocation of chromosomes and gene fusions. Moreover the heterogeneity of tumor cells from the same tumor as well as tumors due to the same cancer from different patients is well known. Hence it is safe to say that our knowledge of the genetic landscape of cancer is still incomplete. Our review of the genetic landscape of cancer is based on the seminal review article by Vogelstein et al. [2].

Mutations occur in normal cells as well as in neoplastic cells. Hence it is not easy to identify the mutations that are directly related to tumorigenesis, so-called driver mutations. The number of mutations varies widely among different cancer types. For example, the number of mutations in patients with lung tumors or in melanoma can be as high as over 200. This is due to the fact that these patients have been exposed to potent mutagens like cigarette smoke and ultraviolet rays. In fact the number of mutations in lung tumors of patients who are non-smokers is usually much less compared to tumors of smokers. There are two other outliers or extreme cases of mutations. Pediatric tumors have usually far less mutations compared to tumors in adults. Tumors in which the DNA repair mechanisms have been compromised also have a much larger number of mutations.

These trends indicate that the number of somatic mutations increases with age, with exposure to potent mutagens and genetic alterations in DNA repair genes like DNA polymerase. Hence it is better to understand the average number of mutations across all tumor types. The number of mutations in solid tumors from the pancreas, brain, breast, and colon is usually between 33 and 66. Most of these mutations are substitution of a single base by another, for example, C to G substitutions. As discussed before, such substitutions may alter the amino acid in the protein sequence, and this results in wrong folding and malfunctioning of the translated proteins. Such single-base substitutions constitute over 95% of all point mutations; the rest are insertions or deletions of bases. Such mutations change the reading frame, and as a result the primary sequences of the translated proteins change, giving rise to wrongly folded and malfunctioning proteins.

As a majority of mutations are single nucleotide or base substitutions, it is important to further analyze these substitutions. A vast majority of these mutations (over 90%) result in missense changes, coding for a different amino acid. Over 7% result in nonsense changes, and about 2% result in mutations in the untranslated regions (UTR) of the 5’ or 3’ ends.

A single mutation may give a growth advantage for a cell over other normal cells; however, the progression to malignancy (unrestrained growth) requires a collection of mutations in different genes. It has been observed that the number of cells in a tumor with a lower number of mutated genes is usually much less compared to the number of cells with a higher number of mutated genes. This clearly indicates that mutations in more genes give more and more growth advantage to a cell. Though normal cells accumulate mutations, these are mostly passenger mutations that do not give any growth advantage to the cells. The mutations in driver genes on the other hand give these growth advantages. Hence many of the mutations detected in tumor cells are accumulated during the preneoplastic stage of life of the cells (or their parent cells) without giving any growth advantage.

Two questions are relevant in relation to accumulation of somatic mutations: What is the rate of these accumulations, and what is the role of somatic mutations in metastasis? As is well known, the critical stage of any cancer is metastasis, the spread of tumor cells from their native sites to other organs through blood or lymphatic vessels. The first question has been investigated extensively by analyzing samples from different stages of cancer, in particular pancreatic and colorectal cancers. And the conclusion is that it takes decades to accumulate a set of mutations that gives significant growth advantages to cells so that they can progress from benign to neoplastic stage. The second question is not yet settled conclusively. It is not clear whether there are mutations that drive metastasis. Most of the driver mutations detected in metastatic stages of cancers are already present in the original tumor. Though it is possible that we are yet to detect all of the driver mutations, there is another line of reasoning that metastasis is basically stochastic in nature. Even normal cells from different organs proliferate through blood stream to distant sites. So it is not surprising that cancer cells from tumor sites also proliferate. However there is only a small chance that a tumor cell gets the right microenvironment at a remote site to grow into secondary tumors.

2.3 Driver and Passenger Mutations

The identification of driver and passenger mutations is one of the most difficult tasks in determining the genomic landscapes of different cancers. Usually there are passenger as well as driver mutations in driver genes, and as a result it is difficult to determine which mutations are driver mutations imparting a growth advantage. A good example is the APC driver gene and the protein translated from it. The APC protein interacts with the E-cadherin protein that is involved in cell adhesion. A driver mutation in the APC gene may result in colorectal cancer. However the driver mutation in the APC gene is the one that truncates the encoded protein within its N-terminal 1600 amino acids. There are several passenger mutations in the APC gene like missense mutations and C-terminal protein truncating mutation.

Another problem in identifying driver genes and driver mutations is heterogeneity of tumors, as the mutations in different cells within the same tumor and across tumors vary significantly. Many different methods have been used for identifying driver mutations, and some mutations clearly stand out as driver mutations imparting significant growth advantage to a cell. For example, mutations in TP53 (encoding the p53 protein) known as the "guardian of the genome" and KRAS that participates in cell signaling are such driver mutations. These two genes are highly mutated in many cancers and are easy to identify as driver genes. Such highly mutated genes are termed as *mountains*. However, there are usually many *hill*s in most tumors, genes that have relatively few mutations. Hence detection methods that are based on mutation frequencies can only flag the genes for further studies, but cannot identify them definitely as driver genes. Vogelstein et al. [2] used the concept of Mut-driver genes to indicate genes that have a sufficient number of driver gene mutations so that such genes can be unambiguously marked as driver genes.

2.4 Cancer-Specific Genomic Landscapes

Though we have gained overall understanding of the genomic landscape of cancer from the discussion above, it is important to understand that the genomic landscapes for specific cancers vary considerably. Moreover these variations will be important later for understanding the important genetic pathways responsible in some of the major cancers. In this section we review the genomic landscapes of some of the major cancers including breast, lung, liver, melanoma, cervical, and prostate cancers. We will see that many of the driver genes and driver mutations are common across cancers. Hence we discuss the genomic landscape of breast cancer in detail and indicate the driver genes that are different for the other cancers.

2.4.1 Genomic Landscape of Breast Cancer

Breast cancer is one of the most common cancers among women. More than 80% of breast cancers are ER+, i.e., these cancer cells are signaled by the estrogen hormone. The remaining breast cancers are ER-. The genomic landscape differs significantly between these two types of breast cancers and in the cells within the same tumor and across tumors from different patients. We review here the driver genes that are known from recent genome-wide sequencing of breast cancer patients.

Stephens et al. [3] sequenced exons of 21,416 protein-coding genes and 1,664 microRNAs from 100 breast cancer patients; 79 of the patients were ER+ and 21 ER-. microRNAs are noncoding RNAs that have important roles in regulating gene expression by silencing mRNAs by binding to their 3' UTR.

The study by Stephens et al. [3] detected many previously known and confirmed driver genes, previously suspected driver genes, and nine new driver genes. The previously confirmed driver genes include TP53, RB1, PTEN, AKT1, BRCA1, BRCA2, CDH1, GATA3, and PIK3CA; previously suspected driver genes include APC, ARID1A, ARID2, ASXL1, BAP1, KRAS, MAP2K4, MLL2, MLL3, NF1, SETD2, SF3B1, SMAD4, and STK11. The novel driver genes detected are AKT2, ARID1B, CASP8, CDKN1B, MAP3K1, MAP3K13, NCOR1, SMARCD1, and TBX3. We discuss below the different activities of these genes in normal cells and how their mutations and inactivations result in uncontrolled cell growth. An excellent resource for understanding the roles of different genes is the US National Library of Medicine, Genetics Home Reference [4], and NCBI gene database [5]. Our review is based on these two resources. We refer to the paper by Li et al. [32] for a review of the genetic landscape of triple-negative breast cancer, however we omit the details here. Another recent review of somatic mutations in breast cancer is the paper by Nik-Zainal et al. [34].

BRCA1 and BRCA2 genes are known as breast cancer susceptibility genes. These proteins are responsible for DNA repair, and any mutations in these genes result in mutations in other genes due to lack of repairing of the latter mutations.

TP53 is one of the most important tumor suppressor genes and called the "guardian of the genome." The p53 protein transcribed from the T53 gene detects whether the mutations in the DNA are repairable and if not activates the apoptosis pathways. Hence a mutation in the TP53 gene may result in accumulation of mutations in other genes that are responsible for tumorigenesis.

Protein kinases are enzymes that regulate the functions of other proteins through phosphorylation, i.e., acting as a catalyst to attach a phosphate group to other proteins. Phosphorylation is an important posttranslational mechanism for controlling the functions of many proteins, as it activates or deactivates many proteins. Many proteins are phosphorylated temporarily for finer controls of their functions. The phosphate groups are removed later by other enzymes. Mutations in protein kinase genes result in lack of control over expressions of many other genes. The PIK3CA gene is transcribed into the p110α protein that is a catalytic subunit of phosphatidylinositol 3-kinase (PI3K). PIK3CA is strongly suspected to be a driver gene in breast cancer.

Another oncogene AKT1 transcribes a protein AKT1 kinase that is responsible for regulating cell growth and cell division. MAP2K4 is another gene that transcribes the protein kinase 4. This kinase is a direct activator of MAP kinases in response to environmental stress.

STK11 transcribes for a protein serine/threonine kinase 11. STK11 is a tumor suppressor gene that helps in controlling cell growth and proliferation.

PTEN is a cancer suppressor gene that acts as a catalyst for removing phosphate groups from other proteins after they have been phosphorylated. Hence PTEN has a crucial function in posttranscriptional control of other genes. There is some evidence that PTEN plays some role in cell adhesion and movement as well [4].

KRAS is a member of the RAS family of oncogenes, along with HRAS and NRAS. The protein transcribed from KRAS takes part in external cell signaling, i.e., sending signals from the outside of a cell to the nucleus. The signals are either to proliferate or to differentiate. Mutations in KRAS inhibit these normal signaling functions. SMAD4 codes for a protein that is responsible for internal cell signaling, from the cell surface to the nucleus.

The CDH1 gene transcribes a protein epithelial cadherin or E-cadherin. This protein is responsible for cells sticking to one another in the extracellular matrix. E-cadherin is also responsible for signaling cells for controlling cell maturation and movement. E-cadherin is known to be a tumor suppressor protein.

The RB1 gene transcribes a protein pRB which acts as a tumor suppressor by regulating cell growth and cell division. APC and NF1 also code for tumor suppressor proteins that control cell division and proliferation.

Chromatin is the network of DNA and proteins that pack chromosomes. Transcriptional machinery do not get access to genes if the packing is tight, hence the expressions of genes are lower in tightly packed chromatins. Several genes produce proteins that are responsible for chromatin remodeling, that is, they reduce the packing of chromatins, increasing gene expression. Chromatin remodeling proteins are transcribed from the genes ARID1A, AID2, and ASXL1.

Histones are the main protein components of chromatin, and histone methylation and acetylation are two of the most important epigenetic control mechanisms.

SETD2 codes for a protein that is a histone methyltransferase and has regulatory role in gene expression. SETD2 is considered as a tumor suppressor gene.

Transcription factors are proteins that control the rate of transcription of other genes by binding to the DNA sequence close to the gene, for up- or downregulation of the transcription of a gene. Since a hallmark of all cancers is the overexpression or under-expression of different genes, mutated genes of transcription factors have very important roles to play in cancer. GATA3 is an important transcription factor that is known to be mutated in breast tumors. MLL2 and MLL3 code for proteins that are transcription factors.

SF3B1 codes for a protein that is part of a splicing factor complex of proteins. Splicing factors are responsible for splicing out the introns from pre-mRNA to form mature mRNA that is eventually translated into a protein. Hence any mutation in a gene that transcribes a splicing factor may result in aberrant mRNAs and subsequent aberrant proteins.

The new driver genes that Stephens et al. [3] detected are AKT2, ARID1B, CASP8, CDKN1B, MAP3K1, NCOR1, SMARCD1, and TBX3. We now examine the roles of these genes.

AKT2 is an oncogene that codes for a protein kinase that is known to phosphorylate several proteins. This gene is also overexpressed in ovarian and pancreatic cancers [5]. ARID1B codes for a protein that is responsible for chromatin remodeling. As the name suggests, CASP8 codes for a caspase protein that is crucial for the execution phase of the apoptosis pathway. We have discussed the two phases of the apoptosis pathway in Chap. 1. CDN1B is a tumor suppressor gene that codes for a protein responsible for cell cycle regulation. This protein p27 is found in the nucleus and responsible for inhibiting progression of cell division.

MAP3K1 codes for a protein kinase that is activated through autophosphorylation. It binds to other molecules and controls signaling pathways. MAP3K13 is another gene that codes for a similar protein kinase. NCOR1 codes for a chromatin remodeling protein that condenses chromatin and prevents access to certain genes by transcriptional machinery. SMARCD1 is another chromatin remodeling gene. TBX3 encodes a protein that is a transcription factor and inhibits transcription.

2.4.1.1 Summary

The discussion on driver genes for breast cancer gives us some insights into the effects of mutations in these genes. We can see a structure in these mutations in the sense that these genes are involved in some key cellular functions and regulations, and uncontrolled cell growth and proliferation occur when these mutations code for aberrant proteins. For example, these genes are involved in chromatin remodeling, cell signaling, apoptosis pathways, phosphorylation, and intron splicing of mRNAs. Malfunctioning in the proteins encoded by these genes due to mutations in these genes results in proteins that cannot carry out their normal activities, and usually there is a cascading effect of this malfunctioning. Also most genes are organized in pathways that we will discuss later in this chapter, as mutations in genes responsible for similar types of controls indicate the existence of interdependence among these genes.

2.5 Genomic Landscape of Other Cancers

2.5.1 Colorectal Cancer

The reporting of driver genes in colorectal cancer is contradictory. Huang et al. [6] recently conducted a thorough literature review and extensive bioinformatic analysis. They identified six key driver genes, APC, KRAS, BRAF, PIK3CA, SMAD4, and TP53. Out of these only BRAF is a new gene, as we have discussed the other genes in association with breast cancer. BRAF encodes for a protein that is responsible for cell signaling, from outside the cell to the nucleus. This protein is part of an important signaling pathway called RAS/MAPK that is involved in controlling cell growth and proliferation. Huang et al. [6] also found some correlation between metastasis of colorectal cancer with mutations in TP53 and KRAS genes. Patients with synchronous colorectal cancer have two or more primary tumors at the same time. Wang et al. [7] identified APC, KRAS, TP53, and PIK3CA as driver genes for synchronous colorectal cancer. The paper by Haan et al. [29] is another recent review of the genomic landscape of metastatic colorectal cancer.

2.5.2 Liver Cancer

Li et al. [8] conducted a genome-wide analysis for hepatocellular carcinoma (HCC) and found 190,099 somatic mutations in HCC tissues, including 3,750 nonsense, 59,990 missense, and 5,533 insertion/deletions. They confirmed 32 previously reported significantly mutated genes including TP53, CTNNB1, ARID1A, ARID2, AXIN1, and TSC2. They also found 11 novel significantly mutated genes including RNF213, VAV3, TNRC6B, MACC1, LAMA3, and GPAM. We review some of these genes that we have not reviewed before and refer the reader for a full list to their paper and supplementary materials.

CTNNB1 codes for the beta-catenin protein that plays important roles in cell adhesion in tissues and also in cell signaling. AXIN1 codes for a cytoplasmic protein that has an important role in the apoptosis pathway. TSC2 codes for a protein called tuberin. It is a tumor suppressor gene, and the protein works as a tumor suppressor by interacting with many other proteins. The function of the protein encoded by the RNF213 gene is not well understood. Most probably it breaks down or degrades other proteins. VAV3 is a member of the VAV family of genes and encodes a protein that is responsible for cytoskeleton rearrangement and transcriptional control of other proteins. TNRC6B encodes for a protein which interacts with argonaute proteins that participate in gene silencing. Moreover, it plays some role in miRNA-guided posttranslational gene silencing. MACC1 is a key regulator of the hepatocyte growth factor (HCF) pathway that is responsible for cellular growth, cell motility, and metastasis. LAMA3 codes for a part of the laminin 332 protein which is involved in controlling cell adhesion and cell motility. GPAM encodes the protein Glycerol-3-phosphate acyltransferase 1, which is responsible for maintaining cellular triacylglycerol and phospholipid levels. The interested reader may refer to the article by Hama et al. [30] for a review how epigenetic control affects the genomic landscape of liver cancer.

2.5.3 Lung Cancer

About 15% of all lung cancers are small-cell lung cancers, and about 85% are non-small cell lung cancers. Small-cell carcinomas usually have much higher growth rates and earlier development of metastasis. We briefly review the genomic landscapes of both of these cancers.

Semenova et al. [9] note that the study of the genomic landscape of small-cell lung cancer is not as developed as the other cancer types due to the scarcity of patient materials. It is known that this cancer has one of the highest levels of mutation rates due to contact with cigarette smoke, a potent mutagen. Moreover tumor cells in small-cell lung cancer also have one of the highest genetic rearrangements among all cancer types. Some of the highest mutation rates are found not unsurprisingly in the gene TP53, which we have reviewed already. Semenova et al. [9] also mention the tumor suppressor gene RB1 as another gene that has the highest number of mutations. This gene also has been identified as a driver gene in breast cancer, and we have reviewed it earlier. The MYC proto-oncogenes MYC, MYCL, and MYCN have also been found to be significantly upregulated in many small-cell lung cancers. MYC is a proto-oncogene that encodes a protein which has important roles in cell cycle progression and apoptosis. Both MYCL and MYCN are transcription factors. Many small-cell lung cancer cells also lose the functionality of the PTEN gene which is an inhibitor of the PI3K pathway. Substantial upregulation of antiapoptotic proteins, in particular BCL2, has been observed in many cells. Jiang et al. [10] have identified SRSF1 as a key driver gene for small-cell lung cancer. This gene encodes a splicing factor protein that can either activate or repress splicing, depending on its phosphorylation state. Recall that a splicing factor protein is responsible for splicing out the introns from pre-mRNA for forming the mature mRNA.

Govindan et al. [11] conducted a whole-genome and transcriptome sequencing study of tumor cells from non-small scale lung cancer patients. They found significant levels of mutations in some novel genes including DACH1, CFTR, RELN, ABCB5, and HGF. Earlier studies, e.g., [12], have identified possible driver genes including oncogenes EGFR, KRAS, ERBB4, KDR, FGFR4, and NTRK and tumor suppressor genes like TP53, STK11, NF1, RB1, ATM, and APC. We briefly review the genes that we have not discussed before and refer the reader to the papers by Govindan et al. [11] and Ding et al. [12] and the references in these two papers for further details.

DACH1 encodes for a protein that is responsible for transcriptional control of gene expression. It is a chromatin associated protein that also works together with other DNA-binding transcription factors. CFTR encodes a protein that acts as a channel across cell membrane. RELN encodes for the reelin protein that is mainly associated with brain functions. Perhaps this gene has some unknown functionality in other organs. ABCB5 encodes for a protein that is responsible for transmembrane transport. HGF encodes a protein that is responsible for cell growth and cell motility.

We now discuss the driver or potential driver genes known for some time [12]. EGFR codes for a transmembrane protein called epidermal growth factor receptor. This protein attaches to the cell membrane with part of it outside the membrane and the other part inside the membrane. Ligands from outside attach to the protein, and this signal is transmitted inside the cell by other proteins attaching to the internal part of the receptor. This signal is important for cell cycle progression. ERBB4 also encodes for a transmembrane protein and receives signals from external ligands, and it is part of the signaling pathway for mitogenesis and differentiation. KDR is also known as VEGFR. It encodes a protein called vascular endothelial growth factor which is a major growth factor for endothelial cells. FGFR4 encodes a protein called fibroblast growth factor receptor. As the name suggests, it binds to ligands external to the cell and carries the signal inside the cell. It has important functions in cell growth and maturation. NTRK encodes a kinase protein that is involved in regulating other proteins through phosphorylation. The NCF1 gene encodes a protein called neutrophil cystolic factor 1. This protein is a part of a complex NADPH oxydase that is particularly active in phagocyte cells that catch and destroy foreign invaders like bacteria and fungi. ATM encodes for a protein that is found in the nucleus. This protein participates in DNA repair mechanisms. We have reviewed briefly the genomic landscape of lung cancer. More details can be found in the papers we have referred. Also the genomic landscape differs in different ethnic groups. For example, we refer the interested reader to the paper by Nahar et al. [33] for the genomic landscape of Asian EGFR-mutant lung adenocarcinoma.

2.5.4 Genetic Landscape of Melanoma

Melanoma is a tumor of the melanocytes; the pigment melanin, responsible for skin coloring, is made by these cells. Melanoma genomes have one of the highest somatic mutation frequencies among all cancers, and most of these mutations are C-T base change [13] due to the potent mutagen ultraviolet ray.

Hodis et al. [14] analyzed large-scale melanoma exome sequences and detected five well-known cancer genes, BRAF, NRAS, PTEN, TP53, and MAP2K1, and five novel melanoma driver genes PPP6C, RAC1, SNX31, TACC1, and STK19. More recently, Hayward et al. [13] did a whole-genome study and found several significantly mutated genes for various melanoma subtypes, including BRAF, CDKN2A, NRAS, and TP53 in cutaneous melanoma; BRAF, NRAS, and NF1 in acral melanoma; and SF3B1 in mucosal melanoma. We discuss these genes below.

NRAS is an oncogene that encodes a protein called N-Ras which is involved in regulating cell division through signal transduction, relaying signals from outside the cell to the nucleus. MAP2K1 is a protein kinase that acts as an integration point for multiple signals. PPP6C encodes for a catalytic subunit of protein phosphatase which participates in a signaling pathway for cell cycle regulation. The protein encoded by RAC1 is a member of the superfamily of Ras proteins that have many functions including control of cell growth and activation of protein kinases. SNX31 encodes a protein of the sorting nexin family that are responsible for intracellular trafficking of various molecules. TACC1 encodes a protein that has also been implicated in breast cancer, but the function of this protein is not well understood [5]. STK19 encodes for a protein that is a kinase localized in the nucleus. Most probably it acts as a transcriptional regulator. CDKN2A has several transcript variants mainly differing in their first exons. These variants act as inhibitor of CDK4 kinase and also as a stabilizer of the p53 protein. SF3B1 encodes for a protein that is part of a splicing factor protein complex. Recall that splicing factors are responsible for splicing out introns from pre-mRNA to form mature mRNA.

2.5.5 Genomic Landscape of Pancreatic Cancer

Waddell et al. [15] recently conducted a whole-genome analysis of 100 pancreatic ductal adenocarcinomas and detected that chromosomal rearrangements are endemic, affecting several key genes. They detected several genes that were known to be important in pancreatic cancer, namely, TP53, SMAD4, CDKN2A, ARID1A, and ROBO2. They also identified two new candidate driver genes for pancreatic cancer, KDM6A and PREX2. Moreover, they identified mutations in several other genes at low prevalence among patients, including ERBB2, MET, FGFR1, CDK6, PIK3R3, and PIK3CA. Since it is not clear whether these latter six genes can be termed as driver genes, we omit discussing these genes here. We discuss the three genes ROBO2, KDM6A, and PREX2 as we have not discussed these genes before.

The protein encoded by ROBO2 is part of the ROBO family of proteins that act as transmembrane signal receptors. These genes are highly conserved in eukaryotes. KDM6A encodes a protein that is a histone demethylase. Recall that histones are proteins that are responsible for DNA packaging, and hence this gene has important roles in transcriptional control. PLEX2 encodes for a protein that is involved in signal transduction. This gene is also significantly mutated in melanoma [16]. Further details of the genomic landscape of pancreatic cancer can be found in the paper by Hong et al. [31].

2.5.6 Genomic Landscape of Cervical Cancer

Cervical cancer is responsible for about 15% of cancer-related death among women worldwide. Ojesina et al. [17] conducted a whole-exome sequence analysis of 115 cervical carcinoma samples. They detected previously implicated driver mutations in PIK3CA, PTEN, TP53, STK11, and KRAS, as well as novel somatic mutations in MAPK1, HLA-B, EP300, FBXW7, NFE2L2, and ERBB2 genes. MAPK1 encodes for a protein kinase from the MAP kinase family. MAP kinases are known as extracellular signal-regulated kinases and are integration points for several signals and are involved in many important cellular processes including proliferation, differentiation, and transcriptional regulation. HLA-B encodes a protein that is a member of a family called human leukocyte antigen (HLA) complex and helps to distinguish foreign proteins from invaders like bacteria and viruses. EP300 encodes for a protein called p300 that is involved in developmental processes including differentiation. FBXW7 encodes for an F-Box protein that is involved in phosphorylation-dependent ubiquitination. Ubiquitin is a regulatory protein which when attached to another protein (ubiquitination) may mark that target protein in various ways, e.g., for degradation, for changing the cellular localization, and to promote or prevent protein interactions. Ubiquitination is another kind of posttranscriptional control. NFE2L2 encodes for a transcription factor.

2.6 Important Pathways in Cancer

Tumorigenesis is a result of somatic mutations in most cancers; however, cells have DNA repair mechanisms that can repair most of these somatic mutations. Hence progression to tumorigenesis requires accumulation of many mutations in many genes. Moreover most genes or their protein products do not act alone; they are usually organized in networks with interdependence in their expression profiles. For example, overexpression or under-expression of a transcription factor may result in overexpression or under-expression of other downstream genes. Hence it is important to study genetic pathways or organizations of genes in order to get a high-level understanding of tumorigenesis and metastasis.

Several key genetic pathways have been implicated for progression of cancer [18–20]. There are also review articles on specific pathways [21, 22]. We briefly review here some of the most important pathways that have been implicated in tumorigenesis and cancer. These pathways include:

- Apoptosis pathway – We have discussed this pathway in details in Chap. 1;
- TGFβ signaling pathway;
- NF-κB signaling pathway;
- Wnt signaling pathway;
- MAPK/ERK signaling pathway;
- PI3K/AKT/mTOR signaling pathway;

2.6.1 TGFβ Signaling Pathway

The transforming growth factorβ signaling pathway has several different variations and myriads of functions in normal cell activities. The effect of TGFβ is different depending on the cellular context, and the understanding of these contexts is not yet complete [22]. Here we discuss only the canonical Smad-mediated pathway through which TGFβ signaling works. We refer the reader to the review papers [19, 20, 22, 23] for further details and other noncanonical pathways. We illustrate this pathway in Fig. 2.1. TGFβ is an extracellular signaling ligand that sends a signal

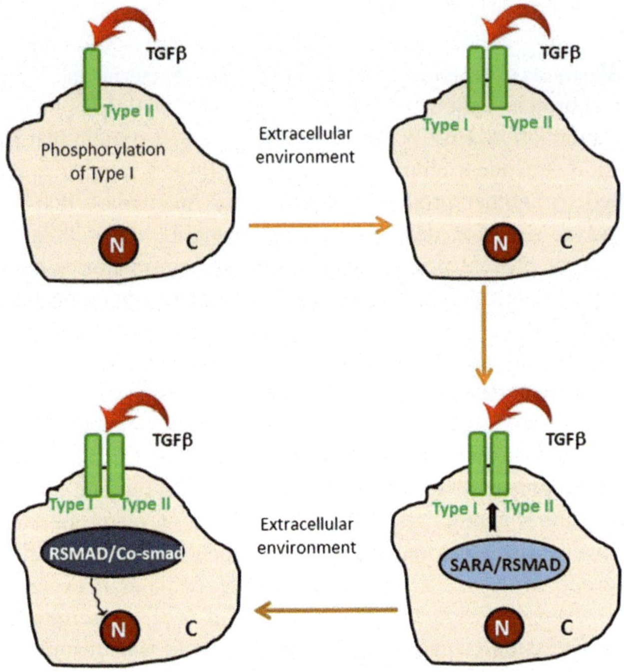

Fig. 2.1 An illustration for the TGFβ pathway. "N" and "C" denote the nucleus and the cytoplasm, respectively

to the nucleus of cells so that transcription factors can increase or decrease the transcriptions of certain genes. It is known that TGFβ can affect the transcription of hundreds of genes [22], and these genes have important roles in many diseases including cancer, heart disease, and diabetes. The effect of TGFβ signaling in normal cells is to stop the cell cycle for preventing proliferation, promote apoptosis, and induce cell differentiation.

Extracellular TGFβ complexes bind to the serine/threonine kinase TGFβ receptors on the cell membrane. TGFβ receptors are divided into two structurally similar receptor families, type I and type II, both of which act in sequence in the pathway. First TGFβ binds to a type II receptor that then recruits a type I receptor and phosphorylates it. This promotes the kinase activity of the type I receptor and starts the signaling. The main agents in signaling in the canonical TGFβ pathway are the Smad family of proteins. Some members of the Smad family work as signal carriers (R-Smad), and some act as signal inhibitors (I-Smad).

We first discuss the R-Smads; these include the proteins Smad1, Smad2, Smad3, Smad5, and Smad8. Another protein called SARA (Smad anchor for receptor activation) recruits the R-Smad proteins to the type I TGFβ receptor. The type I receptor phosphorylates the R-Smad, and after that R-Smad forms a heteromeric complex with co-Smad like Smad4. This complex is then transported to the nucleus by a mechanism involving the cytoplasmic protein importin. This complex then acts as a transcription factor in the nucleus in association with other DNA-binding proteins. The job of the I-Smad proteins like Smad6 and Smad7 is to prevent the phosphorylation of the R-Smad by the type I receptor. The I-Smad proteins work as negative regulators against rampant TGFβ signaling in normal cells.

The TGFβ signaling pathway plays very complex roles in cancer [20]. This signaling is primarily a tumor suppressor in normal cells as it controls cell cycle progression and induces apoptosis. The roles of TGFβ signaling as a tumor suppressor has been verified in many studies by observing the mutations of TGFβ type II receptor, Smad2, and Smad4 in colorectal, gastric, pancreatic, and ovarian cancers. TGFβ signaling switches from tumor suppressor to tumor promoter once the tumor suppressing role of TGFβ signaling has been overcome by cancer cells. Two of the most important tumor promoting activities of TGFβ are inducing epithelial to mesenchymal transition and inducing angiogenesis.

2.6.2 NF-κB Signaling Pathway

While TGFβ signals come from outside the cell, NF-κB signaling is intracellular. NF-κB is a protein complex that is found in almost all cell types. This complex is responsible for DNA transcription, cytokine (small proteins that are important for many signaling pathways) production, and the survival of cells. This pathway is activated due to different stimuli like stress, heavy metals, reactive oxygen species, ultraviolet radiation, and different kinds of antigens due to bacteria and viruses.

NF-κB is a transcription factor that does not require production in response to stress and hence can act very rapidly. NF-κB signaling is very strictly regulated in normal cells, and this regulation is indispensible for correct cellular functions. We discuss the canonical NF-κB signaling pathway in brief and refer the reader to the review papers [19, 24] for more details. We illustrate this pathway in Fig. 2.2.

The NF-κB is a family of transcription factors including NF-κB1/p105, NF-κB2/p100, RelA/p65, RelB, and c-Rel. All NF-κB have an N-terminus homology domain called Rel. NF-κB1 and NF-κB2 are initially produced as the proteins p105 and p100, respectively, which then mature into the NF-κB1 and NF-κB2 proteins. All NF-κB proteins exist in the cytoplasm in a sequestered or inactive state. This sequestration is done by iκB (inhibitor of κB) proteins. NF-κB is activated due to signals external to the cell as mentioned before. The activation starts when the iκB proteins are degraded by a kinase called iκB kinase due to these external signals. iκB kinase phosphorylates the iκB proteins. This causes the iκB proteins to be degraded due to ubiquitination by proteasome. This causes the NF-κB proteins to be free, and they are transported to the nucleus. They then act as transcription factors in association with other proteins to enhance the transcription of specific genes that can mitigate the effects of the external signals. NF-κB acts as a transcription factor for iκB indicating how tightly controlled is the NF-κB pathway in normal cells, as the iκB proteins inhibit the pathway by sequestering NF-κB as explained before.

The NF-κB pathway acts both as a tumor suppressor and a tumor promoter. Its action as a tumor promoter has been observed in many cancers. One important

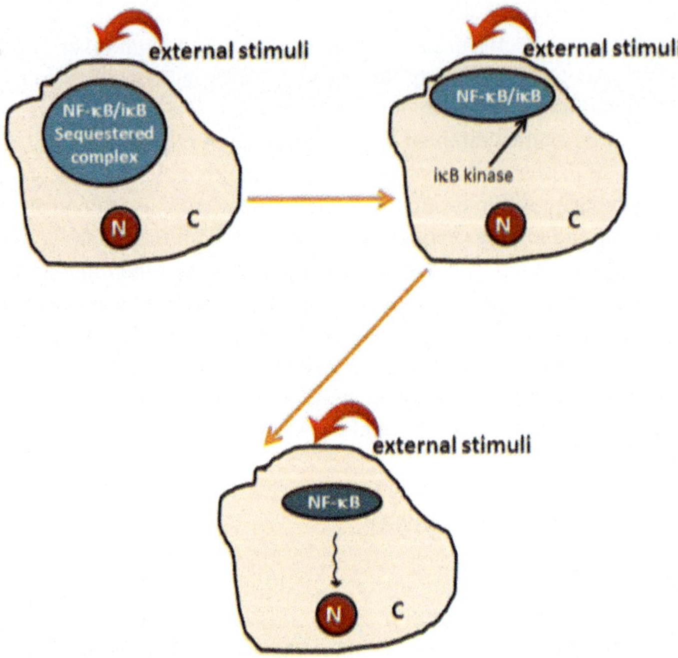

Fig. 2.2 An illustration for the NF-κB pathway

example is human B-cell malignancies [19] due to mutations and mis-regulation of the Rel gene, encoding the c-Rel protein. The Rel gene is overexpressed substantially in Hodgkin's lymphoma, non-Hodgkin's B-cell lymphoma, and T-cell lymphomas [19]. Most probably the overexpression of the Rel gene overcomes the inhibitory effects of iκB proteins, resulting in overexpression of NF-κB target genes and proliferation of B cells. NF-κB also acts as a tumor promoter in several other inflammation-related cancers. Overexpression of iκB reduced tumor growth considerably in liver cancers, indicating that the blocking of the NF-κB pathway is an inhibitor for tumor growth.

2.6.3 Wnt Signaling Pathway

The Wnt signaling pathway is actually a collection of complex cell signaling pathways that carry extracellular signals first to the cytoplasm, and then certain transcription factors accumulate in the nucleus due to these signaling. The end result of such signaling is the regulation of certain genes.

The history of the Wnt signaling pathway is quite interesting. It was first identified in carcinogenesis, through the identification of a proto-oncogene int-1. However, it was clear soon that the same gene is responsible for development and

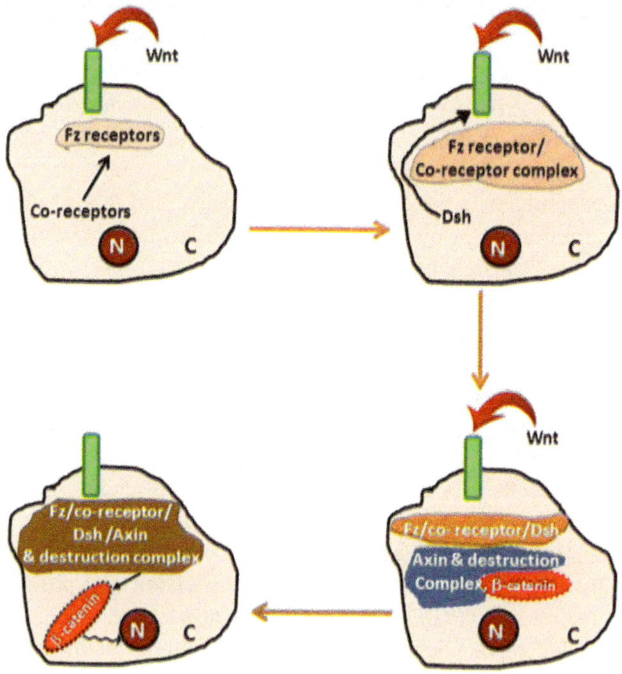

Fig. 2.3 An illustration for the Wnt pathway

body axis patterning in *D. melanogaster*, the fruit fly. Hence this gene was renamed as Wnt or "wingless/integrated," and it was soon found that this gene is highly conserved in many species, from fruit fly to humans and has been implicated in prostate and breast cancers. We only discuss the canonical Wnt signaling pathway in this brief review and refer to the paper by Zhan et al. [25] for further details.

The end result of the Wnt signaling pathway is the accumulation of certain transcription factors in the nucleus due to signals external to a cell. The difference between the canonical and noncanonical pathways is that the canonical pathway involves the protein β-catenin as the transcription co-activator, while noncanonical pathways do not involve this protein. We illustrate the Wnt pathway in Fig. 2.3.

The first step in this signaling pathway is the release of the Wnt ligand from the secreting cells. This is a complex step with a variety of mechanisms in Wnt signaling, and we refer the reader to the paper by Zhan et al. [25] for the details. Once the Wnt protein is released in the intercellular environment, it binds to the Frizzled (Fz) family receptor proteins on the surfaces of other cells. The Frizzled family receptors take help of co-receptors, and several co-receptors have been identified, most prominent among those are LRP-5/6 (lipoprotein receptor-related proteins).

The next major participant in this signaling pathway is the Dishevelled (Dsh) protein, which resides in the cytoplasm. The signal is transmitted to the Dsh protein through direct contact with the Fz receptor. The most important part of the canonical pathway is the accumulation of β-catenin in the cytoplasm. Usually β-catenin is degraded through ubiquitination by a destruction complex comprising several proteins including Axin, APC, and PP2A. However this destruction complex is inactivated after Wnt is bound to the Fz receptor and the destruction complex moves to the plasma membrane. As β-catenin is released from the destruction complex, it accumulates in the cytoplasm. It is transported to the nucleus, where it acts as a transcription factor by recruiting other transcriptional coactivators. Several components of the Wnt signaling pathway have been implicated in colorectal cancer and pancreatic ductal adenocarcinoma [25].

2.6.4 MAPK/ERK Signaling Pathway

MAPK/ERK is also an external signal-initiated complex pathway. Similar to the Wnt signaling pathway, the purpose of this pathway is to send a signal from outside a cell to the nucleus of the cell. We present a very simple view of the pathway here and refer the reader to the paper by Zhang and Liu [26] for details. The signaling molecule is usually a mitogen (that signals cell division and proliferation), e.g., epidermal growth factor (EGF). This pathway is also called the RAS-RAF-MEK-ERK pathway due to the other proteins involved in the pathway. The pathway involves several proteins, RTK and RAS (a small G protein), and three protein kinases RAF, MEK, and ERK. We illustrate this pathway in Fig. 2.4.

The external signal is provided by EGF. The tyrosine kinase receptor EGFR (other possible receptors are Trk A/B and FGFR) binds to EGF, and the pathway is activated after that. EGFR is phosphorylated, and a complex of a docking protein GRB2 and SOS (a guanine nucleotide exchange factor) binds to the phosphorylated residue of EGFR. SOS is activated at this point, and it helps in removal of GDP (guanosine diphosphate) from a member of the Ras family. Ras binds to GTP (guanosine triphosphate) at this point and becomes active.

The next phase of the pathway starts after this through the activation of RAF-MEK-ERK chain. Ras activates the RAF kinase, which in turn phosphorylates MEK (mitogen-activated protein kinase), and MEK phosphorylates and activates MAPK (another mitogen-activated protein kinase). MAPK was originally known as *extracellular signal-regulated kinase* or ERK.

MAPK can phosphorylate several proteins which are transcription factors. These transcription factors in turn regulate the transcription of genes that control cell cycle; this is the culmination of this pathway.

The RAS family of proteins including HRAS, KRAS, and NRAS are frequently mutated in human cancers. For example, NRAS is mutated in melanoma. The RAF family protein BRAF is also frequently mutated in melanoma. Mutated BRAF is a good therapeutic target.

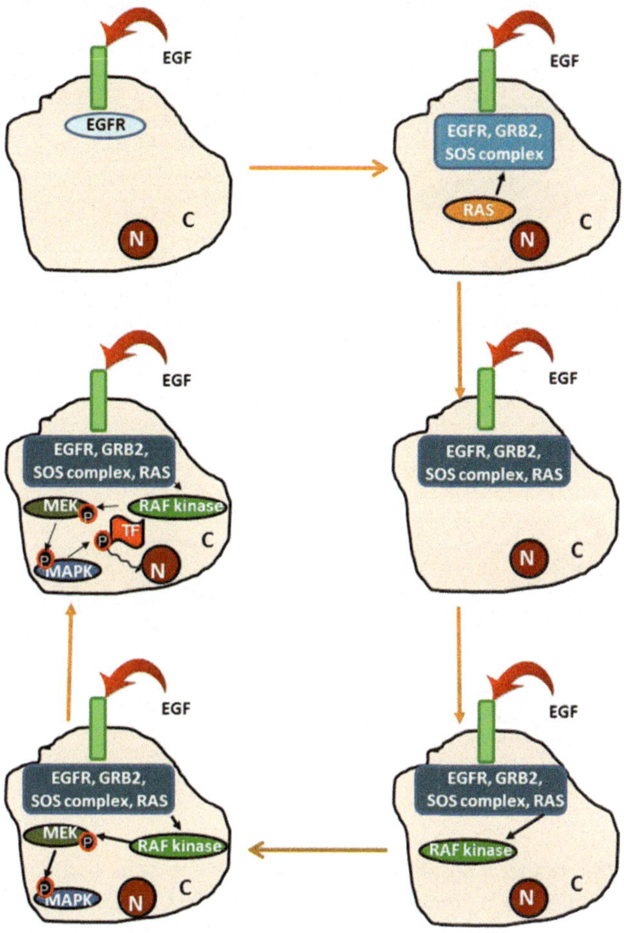

Fig. 2.4 An illustration for the MAPK/ERK pathway

2.6.5 PI3K/AKT/mTOR Signaling Pathway

PI3K/AKT/mTOR is an intracellular signaling pathway; in other words it is not activated from outside the cell through some extracellular ligand. This pathway regulates cell cycle and is responsible for cell proliferation. This pathway is often overactive in several cancers, reducing apoptosis and increasing cell proliferation. It is a complex pathway, and the components of this pathway are still under active research. We discuss this pathway briefly, referring the reader to the articles by Hemmings and Restuccia [27] and Yu and Cui [28] for further details. The pathway is illustrated in Fig. 2.5.

The PI3K is a family of intracellular lipid kinases that phosphorylate PtdIns (phosphatidylinositides), also known as PIP2. Class I PI3K lipid kinase converts

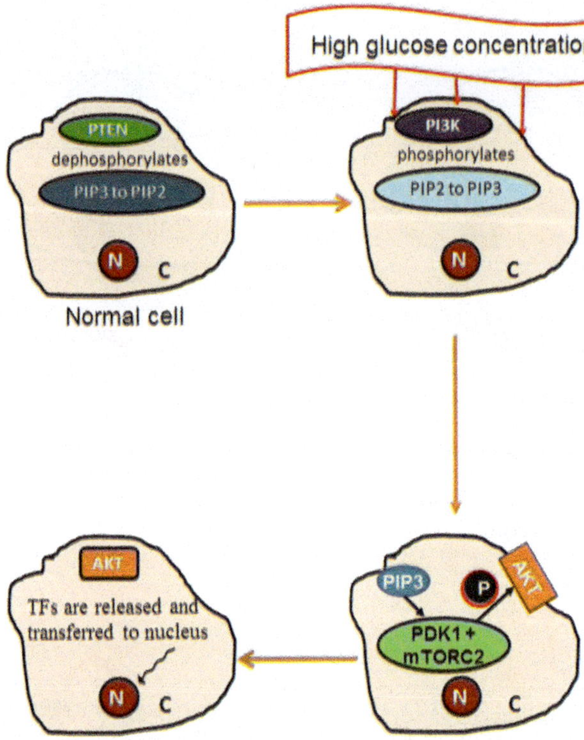

Fig. 2.5 An illustration for the PI3K/AKT pathway

PIP2 to PIP3, another phosphatidylinositide. PIP3 recruits and activates kinases that have the pleckstrin homology domain (PH), e.g., PDK1. Phosphatase and tensin homolog (PTEN) is an inhibitor of the PI3K and controls the signaling duration of PIP3. The serine/threonine kinase AKT also has a PH domain and is recruited to the plasma membrane along with PDK1. AKT is activated by the phosphorylation of its amino acid residues by PDK1 and mTORC2. AKT can phosphorylate many target proteins once it is activated, e.g., GSK3 (glycogen synthase kinase 3), TSC2 (tuberous sclerosis 2), caspase 9, and AKT1S1. AKT has a very broad range of downstream effects in cell proliferation, apoptosis, and angiogenesis. Hence an overactivation of this pathway can inhibit apoptosis and enhance cellular proliferation.

The PI3K/AKT/mTOR pathway has been implicated in ovarian cancer, in several types of breast cancers, and in urothelial cancer. In fact the aberrations in this pathway are the most common forms of abnormality detected in several breast cancers.

2.7 Conclusion

It is important to understand cancer genetics for effective prevention and cure of cancer. We have first reviewed the fundamentals of molecular biology and genetics and then discussed the genetic landscape of cancer. Mutations in driver genes are the key drivers in tumorigenesis, and hence it is important to understand these mutations. However, the driver genes involved in various cancers are not the same; hence we have reviewed the genetic landscapes of individual cancers for most of the major cancers. We have discussed the apoptosis pathway in Chap. 1 and discussed some of the other most important genetic pathways in this chapter. Many of the therapeutic targets are genes associated with these pathways, and hence understanding these pathways is important for understanding the actions of drugs as well as phytochemicals in the future chapters.

References

1. B. Alberts, A.D. Johnson, J. Lewis, D. Morgan, M. Raff, K. Roberts, P. Walter, *Molecular Biology of the Cell*, 6th edn. (Garland Science, New York, 2015)
2. B. Vogelstein, N. Papadopoulos, V.E. Velculescu, S. Zhou, L.A. Diaz Jr., K.W. Kinzler, Cancer genome landscapes. Science **339**, 1546–1558 (2013)
3. P.J. Stephens et al., The landscape of cancer genes and mutational processes in breast cancer. Nature **486**(7403), 400–404 (2012). https://doi.org/10.1038/nature11017
4. US National Library of Medicine, Genetics Home Reference, https://ghr.nlm.nih.gov/gene
5. NCBI gene resources, https://www.ncbi.nlm.nih.gov/gene
6. D. Huang et al., Mutations of key driver genes in colorectal cancer progression and metastasis. Cancer Metastasis Rev. **37**, 173–187 (2018)
7. X. Wang et al., The molecular landscape of synchronous colorectal cancer reveals genetic heterogeneity. Carcinogenesis **39**(5), 708–718 (2018)
8. X. Li et al., Genomic analysis of liver cancer unveils novel driver genes and distinct prognostic features. Theranostics **8**(6), 1740–1751 (2018)
9. E.A. Semenova, R. Nagel, A. Berns, Origins, genetic landscape, and emerging therapies for small cell lung cancer. Genes Dev. **29**, 1447–1462 (2015)
10. L. Jiang et al., Genomic landscape survey identifies SRSF1 as a key oncodriver in small cell lung cancer. PLOS Genet. (2016). https://doi.org/10.1371/journal.pgen.1005895
11. R. Govindan et al., Genomic landscape of non-small cell lung cancer in smokers and never smokers. Cell **150**(6), 1121–1134 (2012). https://doi.org/10.1016/j.cell.2012.08.024
12. L. Ding et al., Somatic mutations affect key pathways in lung adenocarcinoma. Nature **455**, 1069–1075 (2008)
13. N. Hayward et al., Whole-genome landscapes of major melanoma subtypes. Nature **545**, 175–180 (2017). https://doi.org/10.1038/nature22071
14. E. Hodis et al., A landscape of driver mutations in melanoma. Cell **150**(2), 251–263 (2012). https://doi.org/10.1016/j.cell.2012.06.024
15. N. Waddell et al., Whole genomes redefine the mutational landscape of pancreatic cancer. Nature **518**(7540), 495–501 (2015). https://doi.org/10.1038/nature14169
16. M.F. Berger et al., Melanoma genome sequencing reveals frequent PLEX2 mutations. Nature **485**, 502–206 (2012). https://doi.org/10.1038/nature11071
17. A. Ojesina et al., Landscape of genomic alterations in cervical carcinomas. Nature **506**(7488), 371–375 (2014). https://doi.org/10.1038/nature12881
18. B. Vogelstein, K.W. Kinzler, Nat. Med. **10**, 789–799 (2004)

19. O. Dreesen, A.H. Brivanlou, Signaling pathways in cancer and embryonic stem cells. Stem Cell Rev. **3**(1), 7–17 (2007)
20. A. Weiss, L. Attisano, The TGFbeta superfamily signaling pathway. WIREs Dev. Biol. **2**, 47–63 (2013). https://doi.org/10.1002/wdev.86
21. R. Sever, J.S. Brugge, Signal transduction in cancer. Cold Spring Harb. Perspect. Med. **5**, 1006098 (2015). https://doi.org/10.1101/cshperspect.a006098
22. J. Messagué, TGFβ signalling in context. Nat. Rev. Mol. Cell Biol. **13**, 616–630 (2012)
23. F. Verrecchia, A. Mauviel, Transforming growth factor-β signaling through the Smad pathway: role in extracellular matrix gene expression and regulation. J. Invest. Darmatol. **118**(2), 211–215 (2002)
24. Y. Xia, S. Shen, I.M. Verma, NF-κB, and active player in human cancers. Cancer Immunol. Res. **2**(9), 823–830 (2014)
25. T. Żhan, N. Rindtorff, M. Boutros, Wnt signaling in cancer. Oncogene **36**, 1461–1473 (2017)
26. W. Zhang, H.T. Liu, MAPK signal pathways in the regulation of cell proliferation in mammalian cells. Cell Res. **12**(1), 9–18 (2002)
27. B.A. Hemmings, D.F. Restuccia, PI3K-PKB/AKT pathway. Cold Spring Harb. Perspect. Biol. **4**, a011189 (2012)
28. J.S.L. Yu, W. Cui, Proliferation, survival and metabolism: the role of PI3K/AKT/mTOR signaling in pluripotency and cell fate determination. Development **143**, 3050–3060 (2016)
29. J.Ć. Haan et al., Genomic landscape of metastatic colorectal cancer. Nat. Commun. **5**, 5457 (2014). https://doi.org/10.1038/ncomms6457
30. N. Hama et al., Epigenetic landscape influences the liver cancer genomic architecture. Nat. Commun. **9**, 1643 (2018). https://doi.org/10.1038/s41467-018-03999-y
31. S.M. Hong et al., Molecular signatures of pancreatic cancer. Arch. Pathol. Lab. Med. **135**(6), 716–727 (2011)
32. S. Li, H.M. Lu, M.H. Black, The current genetic landscape of triple-negative breast cancer. J. Lab. Precis. Med. **3**, 94 (2018)
33. R. Nahar et al., Elucidating the genomic architecture of Asian EGFR-mutant lung adenocarcinoma through multi-region exome sequencing. Nat. Commun. **9**, 216 (2018). https://doi.org/10.1038/s41467-017-02584-z
34. S. Nik-Zainal et al., Landscape of somatic mutations in 560 breast cancer whole-genome sequences. Nature **534**, 47–54 (2016). https://doi.org/10.1038/nature17676

Fundamentals of Phytochemicals

3

Abstract

The beneficial effects of phytochemicals in health and well-being are known for thousands of years. Many of the fruits and vegetables in our diet contain many phytochemicals that are important for chemoprevention. We discuss the details of phytochemicals and their chemical structures and classifications in this chapter. Though there is no universally accepted classification of phytochemicals, we have used their chemical structures for classifying them in different groups. Further, we identify and discuss different phytochemicals in each group, and their specific beneficial effects in treatment of different diseases, with an emphasis on cancer. Finally, we briefly discuss phytochemicals in some common fruits and vegetables and their beneficial effects in cancer prevention and treatment.

3.1 Introduction

The term "phytochemical," also coined as phytonutrient, has gained enormous importance in recent times, though many of these compounds are age-old and had found use in several ailments in ancient times. Phytochemicals are non-nutritive plant chemicals that have protective or disease-preventive properties. Though they are not essential nutrients and are not required by the human body for sustaining life, regular consumption of plant products shows enormous health benefits, owing to the presence of phytochemical and antioxidants such as flavonoids, proanthocyanins, phenols, anthocyanin, flavonols, carotenoid, vitamin C, and β-carotene [36, 54].

Best health care can be achieved if emphasis is on prevention of the disease. Thomas Edison rightly said, "The doctor of the future will give no medicine, but will interest his patients in the care of the human frame, in diet and in the cause and prevention of disease." Similar thoughts were in the mind of Benjamin Franklin who said "An ounce of prevention is worth a pound of cure" and Desiderius Erasmus who said the well-known phrase, "Prevention is better than cure." Lifestyle has a great

M. Roy, A. Datta, *Cancer Genetics and Therapeutics*,
https://doi.org/10.1007/978-981-13-9471-3_3

impact on human health, and many cancers can be prevented by lifestyle choices. Therefore, the motto of cancer control should be to prevent the preventable and treat the treatable. It was first advocated by Wattenberg that a platter full of fruits and vegetables may impart protection against several diseases including cancer.

Phytochemicals are responsible for the color, odor, and flavor of fruits and vegetables. A correlation between regular and generous intake of plant-derived foods and beverages and good health has been well known for centuries. Phytochemicals are naturally present in the diets in most Southeast Asian countries, as plenty of fruits, vegetables, spices, and whole grains are integral parts of the diet in these countries [84]. Maharshi Charaka, known as the "Father of Indian medicine," mentioned the usefulness of a plethora of plants in his book, *Charaka Samhita*. This book has been regarded in the western world as a good source of alternate medicine. Use of plants as a means to treat ailments had been prevalent in Sumerian and Akkadian civilizations. Also, Chinese and Egyptian medicine systems depend on plant sources. The origin of the word *phyto* is Greek, which means plant; hence phytochemicals are the compounds that are abundant in plant-based foods.

Phytochemicals provide protection against several noncommunicable diseases (NCDs). They include chronic diseases like diabetes, cardiovascular and neurodegenerative diseases, cancer, etc. Two-thirds of death worldwide is attributable to inadequate consumption of fruits, vegetables, spices, etc. [65]. In order to reduce the risk of noncommunicable diseases, World Health Organization (WHO) has suggested daily intake of five to eight portions of fruits and vegetables [32]. The active ingredients of the phytochemicals are metabolic products of the plants – either primary or secondary. Generally, these are not produced as a by-product of primary metabolism, hence do not play an integral role in survival of the plant; rather they take an active part in secondary metabolism. Depending on the biosynthetic origins, secondary metabolites may be of four groups: terpenoids, alkaloids, phenolics, and sulfur compounds.

3.2 Roles of Phytochemicals in Cancer and Other Diseases

Phytochemicals have the unique property that they differentially target cancer cells via various signaling pathways, sparing the normal cells; therefore, they are safe for human consumption. Phytochemicals are rich in several properties, which have been explored since ages. They are antioxidant, anticarcinogenic, antimicrobial, antidiabetic, and anti-inflammatory. They beneficially target several signaling pathways that lead to disease processes. An atom or a group of atoms, having unpaired electrons, are highly reactive. Generation of reactive oxygen species (ROS) is common during normal physiological processes. ROS may be singlet oxygen, superoxide, peroxyl radicals, hydroxyl radicals, and peroxynitrite that may culminate into oxidative stress leading to cellular damage [70]. Free radicals are by-products of metabolism, which are capable of causing cells to lose their structure and function and eventually destroying them. An excessive concentration of ROS causes damage to several biomolecules, like DNA, proteins, and lipids. Cellular DNA repair machineries can cope with a low level of DNA damage but can be overwhelmed if DNA damage is excessive. An excessive level of DNA damage can lead to carcinogenesis. Phytochemicals protect cells against the damage caused by reactive oxygen species (ROS) as antioxidants. Antioxidants reduce oxidative damage. Antioxidants are capable of stabilizing or deactivating free radicals before they attack cells. Antioxidants help to achieve chemoprevention by stopping or reversing carcinogenesis. We will discuss in Chap. 5 the molecular mechanisms related to the actions of phytochemicals on ROS.

Inadequate consumption of plant-based diet increases the risk of cancer to a larger extent [98]. Therefore, phytochemicals may pave a way to cancer control [99, 100]. As a result of imparting protection against damage due to free radicals, they prevent diseases, such as atherosclerosis, cardiac and cerebral ischemia, neurodegenerative diseases, and cancer [52, 113]. Apart from acting as antioxidants, these plant-derived products activate detoxifying enzymes and immune and hormonal response. Cancer is not a single disease but rather a group of diseases comprising abnormal cell proliferation and may spread to distal parts of the body. Development of cancer is also complicated and is called carcinogenesis, which is a multistep process. These phytonutrients act at all the steps, preventing the onset and progression of the disease process. They regulate cell cycle machinery and hence check proliferation of cancer cells, by targeting Akt, NF-κB, MAPK, and Bcl-2 mediated apoptotic pathways. We have discussed these pathways in details in Chap. 2. Phytochemicals also interfere with carcinogen metabolism, hindering the formation of active carcinogens from procarcinogens and also inducing phase II detoxification enzymes [73], which has been discussed in detail in Chap. 5. Apart from regulating cell proliferation, phytochemicals are modulators of cell differentiation and inflammation. As antimicrobial agents, they check the growth of fungi, yeast, and bacteria, many of which may cause cancer. Ulceration in stomach is often caused by *H. pylori*, which is a cause of stomach cancer. Phytochemicals act against formation of ulcer. Diabetes and inflammation are critically linked to cancer, and therefore, owing to their antidiabetic and anti-inflammatory activities, they put a check on carcinogenesis. Several transcription factors are also modulated by these plant-derived products.

Different types of phytochemicals work differently. Some are antioxidants and protect the cells against oxidative damage. Thus, chances of development of cancer can be reduced. Some may have hormonal action and affect metabolic pathways, which are linked to the development of cancer and heart diseases. Some act by stimulation of enzymes. There are some phytochemicals that interfere with DNA replication; therefore, proliferation of cancer cells can be controlled by these phytochemicals. Some may possess antibacterial effect and prevent adhesion of pathogens to the cell wall. Some phytochemicals have been found to lower LDL cholesterol, preventing blood clotting.

3.3 Sources of Phytochemicals

Foods containing phytochemicals are already part of our daily diet. The easiest way to get phytochemicals is to eat a diverse range of fruits and vegetables; a daily intake of five to nine servings is best to have a good outcome. We first discuss a few terminologies:

Designer food – Processed foods that are supplemented with food ingredients naturally rich in disease-preventing substances

Functional food – Any modified food or food ingredient that may provide a health benefit beyond the traditional nutrients it contains

Nutraceutical – Specific chemical compounds in food, including vitamins and additives that may aid in preventing disease

Pharmafood – Food or nutrient that claims medical or health benefits, including the prevention and treatment of disease

Phytochemicals are of various colors; therefore, we can have sound health by eating a right mix of food that is enriched with phytochemicals. Moreover, colorful foods are not only a feast for the eye but are key to sound health. Richness of color is directly proportional to the phytochemical content. Figure 3.1 represents different types of phytochemicals in different color category. But it is important to know what is behind these colors. The blue and the purple color of the fruits and vegetables that we eat are mainly due to anthocyanins, which are good antioxidants. Presence of chlorophyll imparts the green color. Isothiocyanates and indoles are generally green and are rich in anticancer properties. Isothiocyanates induce enzymes that are capable of removing carcinogenic compounds. Green vegetables are excellent sources of vitamin K, folic acid, potassium, etc.

A greenish yellow coloration is often due to presence of lutein, which helps to protect against age-related macular degeneration. These foods are also rich in vitamin C. The foods that are yellowish orange in color contain beta-carotene in abundance. Therefore, these foods are good source of antioxidants. Beta-cryptoxanthin, beta-carotene, and alpha-carotene are carotenoids, which may be converted in the body to vitamin A, a nutrient integral for vision and immune function, as well as skin and bone health. They also have high level of vitamin C.

From scarlet to crimson, red phytochemicals are attractive due to their vibrant color. The red color of certain fruits and vegetables is attributed to a carotenoid, lycopene. Lycopene is a good antioxidant which prevents cancer. Red-colored phytochemicals are also good source of flavonoids, which are anti-inflammatories and antioxidants. Cranberries are marked for their antioxidant potential, and their red color is not due to lycopene but due to anthocyanins. Flavonoids are an important class of phytochemicals, which are colorless. They aid in counteracting free radical formation. There are a large number of flavonoids (flavonols, flavones, flavanones, flavan-3-ols, anthocyanins) which will be discussed later.

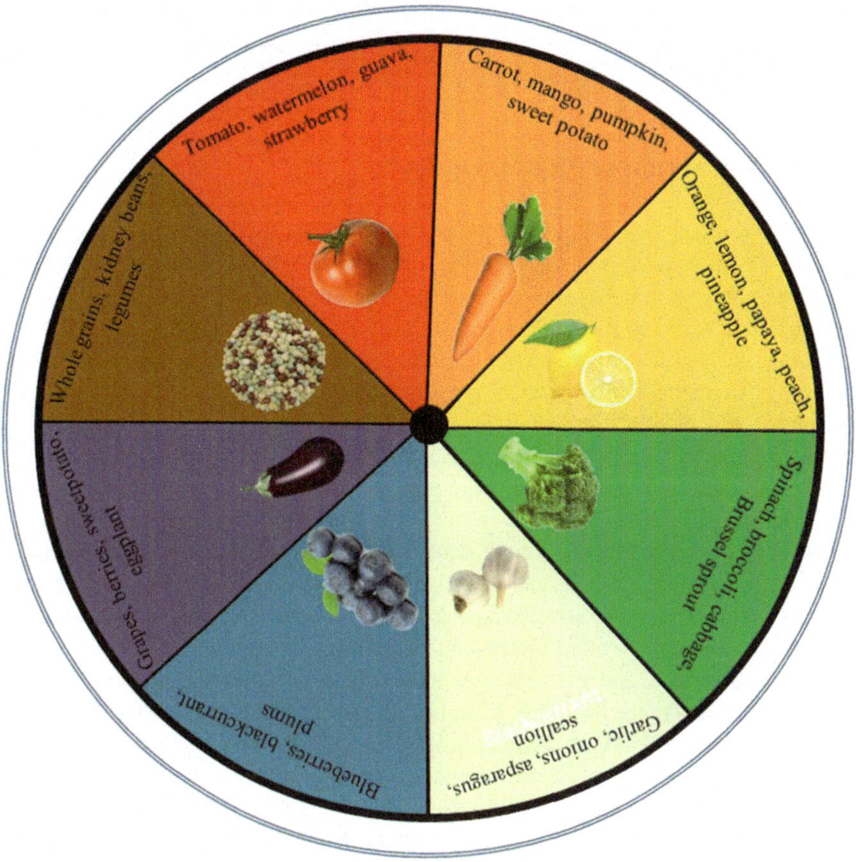

Fig. 3.1 The color wheel of phytochemicals

Many drugs and remedies of diseases have been isolated from plants from the dawn of civilization. These medicines from natural sources had great impacts on human health, in prevention of diseases, as well as in treatment. It is important to know the principle of action of these phytochemicals. They owe their activity to certain chemical groups, commonly referred to as active ingredients. Structural modifications of these active compounds are used in many modern-day drugs. These plant molecules and the resulting drugs generally target similar genes and metabolic pathways. Many drugs show a number of toxicities and side effects. Hence, plant products along with drugs may work in synergism and give a better outcome [3, 13, 17, 43].

In addition to curing diseases, phytochemicals like alkaloids, glycosides, flavonoids, volatile oils, tannins, resins, etc. have been used in a wide range of commercial and industrial applications such as flavors, aromas and fragrances, enzymes, preservatives, cosmetics, bio-based fuels and plastics, natural pigments, and bioactive compounds. The research on phytochemicals and use of phytochemicals are on the rise because of the harmful side effects of the synthetic compounds [83].

3.4 Extraction of Phytochemicals

Identifying the active compounds in phytochemicals, their extraction, and quantitation is an important step in using phytochemicals in medicine. Therefore, understanding the processes of extraction, isolation, and purification is important. Extraction of active ingredient may be achieved by using solvents and processes like microwave-assisted extraction (MAE) and ultrasonic-assisted extraction (UAE). In order to extract the phytochemical from leaves and seeds, solvents like methanol, hexane, ethyl alcohol, etc. may be used. Multiple solvents are often used for extracting phytochemicals, and the choice of solvent depends on the polarity of the solute. Best outcome is achieved if the polarity of the solvent matches with that of the solute.

MAE is often used instead of other methods for the extraction of antioxidants, as time required for this process and the volume of the solvents have been minimized in MAE [110]. Microwaves are electromagnetic radiations, and the MAE technique is efficient in extracting active ingredients of plants with minimum loss of material. MAE utilizes heat, and extraction of antioxidants is attained with low volumes of extraction solvents [8]. This method is also very sensitive as yield of target molecules is high. UAE is commonly used in food processing technology and is the easiest method of extraction of phytochemicals. Ultrasound disrupts plant cell walls, thereby allowing the solvent to penetrate into the cells, leading to higher yields of phytochemicals. UAE works at low temperature; therefore, degradation of the extracted compound is minimized or reduced [30].

Selection and collection of plant materials are crucial to isolate and characterize active phytochemicals, and the next important step is to ascertain possible active molecules. Most importantly, while searching for the active plant molecule, proper methods like in vitro assays are desirable. Such analyses can shed light on the bioactivity of the compound with precision, for example, whether it is an antioxidant, or antibacterial, or if it is cytotoxic [72]. Accuracy, simplicity, and time costs are very important in these analyses. Several solvents may be tested for the plant extracts thus collected to check the maximum bioactivity.

Column chromatography is used for isolation and purification, but high-pressure liquid chromatography (HPLC) is a quicker way to purify phytochemicals. A better technology may be high-resolution mass spectrometry to ascertain the mass of active compounds in medicinal plants. UHPLC-Orbitrap is a modern method that can allow rapid fingerprinting and metabolomics analysis [25]. Once the product is extracted, it needs to be purified before use. Purification can be achieved by silica gel column chromatography and thin-layer chromatography (TLC). For separation of phytochemicals, silica, alumina, cellulose, and polyamide are commonly used.

For identification of purified compounds, several techniques may be employed, for example, spectroscopic techniques like UV-visible, infrared (IR), nuclear magnetic resonance (NMR), and mass spectroscopy can identify the purified compounds [85]. Extraction entails the separation of active ingredients of plant materials using selective solvents in a liquid or powder form. The steps followed for this purpose are decoctions, infusions, fluid extracts, tinctures, semisolid extracts, or powdered extracts [92]. During the extraction step, active parts get separated and solvents get diffused. The unwanted materials may be eliminated with the aid of a suitable solvent. The resultant tincture may be used as medicine, but exact dosage cannot be ascertained at this stage. To prescribe a drug, tablet, or capsule form is most suitable. The tincture obtained is usually a mixture of a number of metabolites. The extraction procedure begins with maceration, which comprises breaking and softening of the plant material using a liquid. This is followed by infusion, percolation, decoction, and extraction. Extraction procedures may vary, and there are different parameters that influence the quality of extraction, e.g., the plant part taken and choice of solvents. Selection of a solvent is very important, as it should not have toxicity and should not interfere with the extraction procedure; it should evaporate easily at low temperature and should be efficient for absorption [112]. The choice of solvents also depends on the quantity and type of phytochemical to be extracted [75].

Some common solvents are tabulated in Table 3.1 [22]

Table 3.1 Common solvents for extraction of phytochemicals

Solvent	Specificity	Extracted material
Acetone	Hydrophilic and lipophilic components, mixes with water easily, volatile and minimum toxicity	Phenol, flavonoids
Alcohol	Efficient with cell walls and seeds. Aromatic organic compounds are obtained through ethanol or methanol extraction	Polyphenols, flavonol, terpenoids, sterols, alkaloids, tannins, anthocyanins, terpenoids, saponins, lactones, etc.
Chloroform	Dried barks are extracted using hexane, chloroform, and methanol. Most active material is obtained with chloroform extract	Terpenoids, flavonoids
Dichloromethanol		Terpenoids
Ether		Alkaloids, terpenoids, coumarins, fatty acids
Water	Universal solvent	Anthocyanins, starches, tannins, saponins, terpenoids, polypeptides, lectins

3.5 Classification of Phytochemicals

Broadly, phytochemical compounds have primary and secondary constituents. Primary constituents comprise chlorophylls, proteins, amino acids, and common sugars, whereas terpenoids, alkaloids, and phenolic compounds are secondary constituents [103, 116]. Based on the chemical structure and properties, phytochemicals can be classified into various groups. These are phenolics, carotenoids, alkaloids, N-containing compounds, and organosulfur compounds [11]. These groups are illustrated in Fig. 3.2.

3.5.1 Phenolics

Phenolics contain a hydroxyl group –OH, and an aromatic hydrocarbon group is bonded to this hydroxyl group. These are mostly found in abundance in fruits, vegetables, nuts, and beverages of plant origin. These compounds are different from alcohols, as the hydroxyl group is bonded with an unsaturated carbon atom. Phenolic compounds are produced from the plants as secondary metabolites [24, 28]. Phenolics elicit both antioxidant and prooxidant activities. Phenolic compounds are scavengers of endogenous reactive oxygen species (ROS). Free radical scavenging potential and the capacity to modulate intracellular pro- and antioxidant enzymes contribute to their role as antioxidants [38, 97, 130]. The prooxidant activity of these compounds also contributes to the anticarcinogenic potential [2, 29, 37, 90, 95], mainly through interaction of transition metal ions, generally found in living systems [39]. Phenolic compounds may be further divided into phenolic acids, tannins, stilbenes, coumarins, and flavonoids. These are described in details in Table 3.2.

Fig. 3.2 The classification of phytochemicals

Table 3.2 Classification of phenolics with structure

Types of phenolics	Brief description	Structure	Natural source
Phenolic acids	Non-flavonoid compounds, these phenolic acids are aromatic carboxylic acids. They contain a phenol ring and one or more carboxylic acid group that imparts functionality. Their basic backbone is either hydroxybenzoic (C1–C6) or hydroxycinnamic (C3–C6) acid and all phenolic acids are their derivatives	$C_7H_6O_3$ 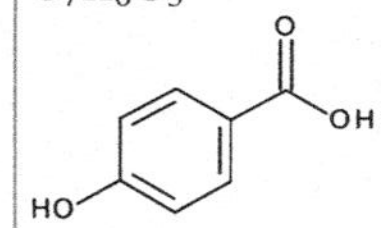 $C_9H_8O_3$ 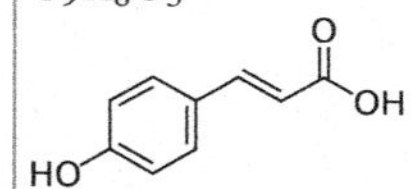	Raspberry, grapes, strawberries, walnut, cranberries, black currants
Tannins	These water-soluble polyphenols are of low nutritional value but are anticarcinogenic. Tannins tend to interact with aqueous solutions of proteins and other macromolecules forming insoluble precipitates, but they differ in their affinity towards proteins. According to the structure and properties, tannins may be categorized into hydrolysable (HT) and condensed tannins (CT) or proanthocyanidins and phlorotannins	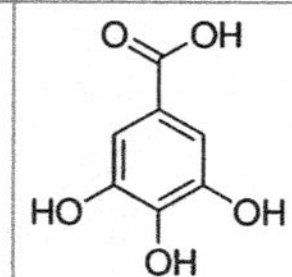	Grapes, tea, coffee, berries, pomegranates, barley, nuts, chocolates (cocoa), squash, legumes (e.g. chickpeas and beans), cinnamon, vanilla, cloves, thyme
Stilbenes	Stilbenes are non-flavonoids, with a polyphenolic structure. They contain a 1,2-diphenylethylene nucleus. Typical characteristic feature of stilbenes is the presence of an ethylene moiety with one phenyl group, on either side of C-C double bond. Two possible isomers are found for stilbenes: trans, denoted by E, and cis, denoted by Z. The cis forms are less stable	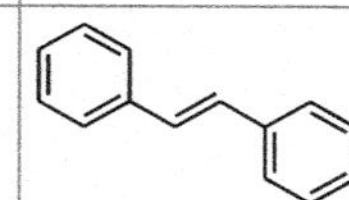 $C_6H_5CH{=}CHC_6H_5$	Grape skin, peanuts, blue berries, cranberries
Coumarins	Generally, these are colorless substances, consisting of fused benzene and α-pyrone rings. They elicit some biological activity but had limited approval due to toxicity to the liver or kidney. Coumarins are anti-inflammatory, antitumor, antibacterial, antifungal, anticoagulant, and antihypertensive among other properties	$C_9H_6O_2$ 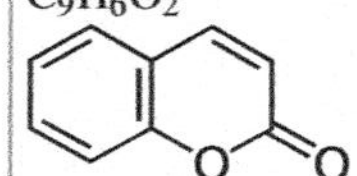	Strawberries, black currants, apricots, and cherries, vanilla grass, citrus fruits, green tea, tonka bean, sweet woodruff, mullein, sweet grass, cassia cinnamon, sweet clover
Flavonoids	Flavonoids are extensively found in many fruits and vegetables Flavonoids are based upon a 15-carbon skeleton consisting of two benzene rings, linked via a heterocyclic pyran ring. They can be subdivided into anthocyanidins, flavanols, flavones, flavonols, flavonones, and isoflavones. They aid in weight management, cardiovascular disease, diabetes, cancer prevention, nerve disorder, fight against infectious diseases	C_6-C_3-C_6 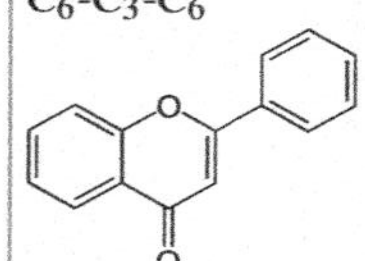	Tea, citrus fruit, berries, apples, legumes, soy products, broccoli, Brussels sprout, tropical fruits, garlic, spinach, lettuce, tomatoes, beans, almonds, peaches, pears, strawberries, celery, watermelon, etc.

3.5.1.1 Phenolic Acids

Phenolic acids comprise aromatic acids, containing a phenolic ring and carboxylic acid. They work very well against oxidative damage and therefore aid in control of coronary disease, stroke, and cancer. They have a variety of pharmacological attributes which include their anti-inflammatory, antioxidant, and antimutagenic and anticarcinogenic potential. They are found in plants in the form of esters or glycosides, which exist in conjugation with other phytochemicals like flavonoids, alcohols, sterols, etc. Most of them are conjugated through ester, ether, or acetal bonds to other molecules. The skeletons of all phenolic acids are the same; they only vary in the number and position of hydroxyl groups.

Of the two basic backbones, hydroxybenzoic acids are derived from p-hydroxybenzoic acid, salicylic acid, 2,3-dihydroxybenzoic acid, 2,5-dihydroxybenzoic acid, 3,4-dihydroxybenzoic acid, 3,5-dihydroxybenzoic acid, gallic acid, and vanillic acid, whereas the hydroxycinnamic acids include caffeic acid, ferulic acid, sinapic acid, etc. and their esterified forms. The formation of hydroxycinnamic acids requires hydroxylation and methylation. Phenolic acids are readily absorbed in human body. Due to reactivity of phenol moiety, the phenolic acids are good antioxidants. They are good radical scavengers and that is attributed to donation of hydrogen atoms [61, 127]. Phenolic acids also elicit antibacterial properties, e.g., ferulic, isovanillic, p-hydroxycinnamic, p-hydroxybenzoic, caffeic, etc. Caffeic acid along with its esterified derivatives shows antitumor potential in colon carcinoma [80]. They also inhibit allergic manifestations, e.g., gallic acid. Gallic acid is also an anti-inflammatory agent. Caffeic acid and some of its derivatives show antiviral potential; they inhibit human immunodeficiency virus type 1 (HIV-1) integrase that catalyzes the integration of HIV-1 DNA into the host chromatin [27]. Above all, these phenolic acids possess antiproliferative activity, thereby inhibiting carcinogenesis. Therefore, phenolic acids are stud with therapeutic potential. Notable among the phenolic acids are ferulic acid, salicylic acid, cinnamic acids, hydroxybenzoic acid, gentisic acid, gallic acid, vanillic acid, sinapic acid, digallic acid, diferulic acid, rosmarinic acid, etc. [34].

3.5.1.2 Tannins

Tannins are soluble in water and are a class of astringent that bind with proteins and precipitate or shrink them. They are bitter plant products. They contain many hydroxyl and carboxyl groups to form complexes with macromolecules. In order to function as a protein binder, tannin molecule requires a minimum of 12 hydroxyl groups and 5 phenyl groups. By virtue of its free radical scavenging activity, tannins act as antioxidants. They also chelate transition metals, inhibit prooxidant enzymes, and check lipid peroxidation [74, 76, 115]. They induce programmed cell death and inhibit the mutagenic potential of carcinogens, thereby acting as an anticancer agent [76]. Tannins disrupt cell walls and membranes and complex with metal ions. They are anti-inflammatory, antiseptic, and antifungal and aid in wound healing [7, 76, 79]. Molecular weight of tannins varies from 500 to over 3000 [9] for gallic acid esters and 20,000 for proanthocyanins.

Tannins may be classified into three types, hydrolysable tannins, phlorotannins/proanthocyanidins, and condensed tannins. Hydrolysable tannins and proanthocyanins complex with proteins in a similar fashion. Condensed tannins are found in foods; whereas hydrolysable tannins are not common in foods, the only exception is pomegranate. The main sources of hydrolysable tannins, phlorotannins, and condensed tannins are plants, brown algae, and tree heartwood, respectively. There is another class of tannins, called pseudotannins [55]. These are low molecular weight compounds. They differ from hydrolysable and condensed tannins in the property that unlike them, pseudotannins do not change color in Goldbeater's skin test. Some of the pseudotannins are gallic acid from rhubarb, flavan-3-ols from tea and cocoa, chlorogenic acid from coffee, and ipecacuanhic acid from *Carapichea ipecacuanha.* Tea hinders the absorption of iron, due to the fact that tea, being a tannin, complexes with iron and its absorption is inhibited. However, this problem does not arise if tea and iron-containing foods are taken separately. This iron-complexing property of tannins could be exploited in male patients with hemochromatosis, which is now recognized as a relatively common disorder. Tannins can also react with thiamine and decrease its absorption [117] .

3.5.1.3 Stilbenes

Isomeric molecules with similar molecular formulae and sequences, varying in the three-dimensional orientations, are termed stereoisomers. Stilbenes possess two stereoisomers, the trans isomer E-stilbene and the cis isomer Z-stilbene. Stability of Z-stilbene is less than E-stilbene, and Z-stilbene is soluble in organic solvents. Photochemical conversion of the trans form to cis form is possible. The hydroxylated derivatives of stilbene are stilbenoids, with a chemical structure of C_6-C_2-C_6. Stilbenoids are found in plants and bacteria. Resveratrol is a rich source of stilbenoids [109].

3.5.1.4 Coumarins

An aromatic compound, coumarin, belongs to the benzopyrone class. It has a sweet smell, similar to that of freshly mown hay. Coumarin is bitter in taste and a suppressor of appetite. It is often used as an enhancer of aroma. Coumarin is synthesized through hydroxylation and glycolysis of cinnamic acid. Coumarin and some of its derivatives belong to the class phenylpropanoids. The derivatives are 7-hydroxycoumarin or umbelliferone, 6,7-dihydroxycoumarin or aesculetin, 7-methoxycoumarin or herniarin, etc. The gene UGT1A8 encodes an enzyme with glucuronidase activity that binds with coumarin in humans [93]. Compounds derived from coumarins are coumarinoids, brodifacoum, warfarin, auraptene, bromadiolone, etc.

3.5.2 Flavonoids

Flavonoids have a 15-carbon phenylpropanoid core (C_6-C_3-C_6 system), arranged into 2 aromatic rings connected by a heterocyclic pyran ring. They have various modes of action, some that are highly implicated in cancer including (i) free radical scavenging, (ii) antioxidant, (iii) anti-inflammatory, (iv) regulator of enzymes, including cyclooxygenase, lipo-oxygenase, and xanthine-oxidase, (v) inhibitors of butyryl choline esterase and acetyl choline esterase, (vi) antimutagenic, (vii) inhibitors of NF-κB. Depending on the oxidation status and saturation of the heterocyclic ring, flavonoids may be divided into several groups. Of the 12 major subclasses, 6 are found in our diet. These are:

- Anthocyanidins
- Flavan-3-ols
- Flavonols
- Flavones
- Flavanones
- Isoflavones

Anthocyanidins These plant pigments are bioactive flavonoids that aid in good health. At acidic pH, anthocyanidins possess color, whereas in alkaline pH, they are colorless. Malvidin, pelargondin, peoidin, cyanidin, apigeninidin, delphinidin, diosmetinidin, guibourtinidin, fisetinidin, luteolinidin, robinetinidin, tricetinidin, capensinidin, petunidin, europinidin, aurentinidin, columnidin, hirsutinidin, and rosinidin all belong to the anthocyanidin group. They possess antioxidant, antimicrobial, and anticancer potential and are effective against various diseases particularly noncommunicable ones. Of these, the health-promoting effects are attributed to the antioxidant potential, via free radical scavenging pathway, cyclooxygenase pathway, mitogen-activated protein kinase pathway, and inflammatory cytokine signaling. Anthocyanidins are found in abundance in berries (including redberry, blueberry, and purple berries), plums, grapes (red and purple), and pomegranates. They also aid in obesity control [57].

Flavan-3-ols These are 3-hydroxy derivatives of flavanones, where the hydroxyl group is always bound to position 3 of the C ring. There are three types of flavanols: monomers (catechins), dimers, and polymers. (+)-catechin, (+)-gallocatechin, (−)-epicatechin (EC), (−)-epigallocatechin (EGC), (−)-epicatechin 3-gallate (ECG), (−)-epigallocatechin 3-gallate (EGCG), theaflavin, theaflavin 3-gallate, theaflavin 3′-gallate, theaflavin 3,3′ digallate, and thearubigins are commonly known flavanols. They do not possess any double bond between positions 2 and 3. Flavan-3-ols have good anti-inflammatory potential. Flavan-3-ols are abundant in bananas, apples, tea (catechin), cocoa, grapes, fava beans, blueberries, peaches, and pears.

Flavonols Having a keto group, they are the building block of proanthocyanins. They possess a hydroxyl group in position 3 of the C ring. Flavonols have different extent of glycosylation, methylation, and hydroxylation patterns; thus, they are the largest subgroup of flavonoids. The most explored flavonols are quercetin, kaempferol, myricetin, rutin, morin, fisetin, galangin, rhamnetin, robinin, and spirenoside. They are good anti-inflammatory and antioxidant agents and interfere with important signaling pathways, leading to disease prevention. They are found in plenty in onions, leeks, lettuce, Brussels sprouts, kale, broccoli, tea, berries, beans, apples, etc. [48].

Flavones Flavones are characterized by the presence of a double bond between positions 2 and 3 of the C ring. There is a keto group at position 4. Generally, a hydroxyl group is found at position 5 of the A ring. Hydroxylation at other positions may differ. Common compounds under this group are apigenin, tangeretin, baicalein, chrysin, diosmetin, diosmin, luteolin, etc. Flavones cause DNA damage and cell cycle arrest at G_1 or G_2, leading to induction of apoptosis. These are found in celery, parsley, broccoli, thyme, dandelion, and chamomile tea and in various herbs like oregano, rosemary, etc.

Flavanones These are generally found in citrus fruits. The bitter taste of citrus fruits is attributed to this class of compounds. They gained importance due to their role as free radical scavenger. They are structurally similar to flavones, with a difference that the double bond between positions 2 and 3 is saturated in flavanones. Eriodictyol, hesperidin, hesperetin, likvirtin, naringin, naringenin, and pinocembrin are common flavanones. They are abundantly found in oranges, lemons, grapes, grapefruits, etc.

Isoflavones Isoflavonoids are a large and very distinctive subgroup of flavonoids. They act like the hormone estrogen; therefore, they are called phytoestrogens. This subgroup includes genistein, glycitein, irilone, luteone, prunetin, pratensein, and daidzein. They differ from flavones due to the location of the phenyl group. They are found in soya beans and leguminous plants like peas, peanuts, chickpeas and navy beans, alfalfa, etc.

3.5.3 Carotenoid

Carotenoids play a pivotal role in photosynthesis and therefore are intricately associated with plant life. They are efficient in quenching singlet oxygen and trapping peroxyl radicals. Carotenoids (also known as terpenoids) are responsible for the characteristic red, orange, yellow, and purple tinge of carrots, corn, canaries, etc., though anthocyanins also play a role for the vibrant coloration. The beautiful hue during autumn fall is contributed by the carotenoids; the yellow-orange color comes when green chlorophyll is degraded for the season. We consume a reasonably good portion of carotenoids through our daily diet.

Generally, carotenoids are in trans form but may be transformed to the cis form during processing. Carotenoids may be divided into two classes: the one containing oxygen is known as xanthophyll (e.g., lutein, zeaxanthin, etc.), and the other that does not contain oxygen is called carotene. Carotenes contain unsaturated hydrocarbons like α-carotene, β-carotene, lycopene, etc. Carotenoids are sometimes converted to retinoids, with provitamin activity but with much higher antioxidant efficiency. They can regulate several enzymatic activities and boost up immune functions and are anticarcinogenic.

The basic structure of carotenoids, containing a system of conjugated double bonds, is a repetition of branched five-carbon units, and these molecules are called isoprenoids. These double bonds are responsible for the absorption of light. Most carotenoids have a symmetrical tetraterpene skeleton, by end-to-end linkage of two C20 moieties, and the linear C40 hydrocarbon backbone is susceptible to diverse structural modifications. Some of the widely studied carotenoids are α-carotene, β-carotene, β-cryptoxanthin, lutein, zeaxanthin, and lycopene. β-Carotene, which gets converted to vitamin A, is the most abundant form in fruits and vegetables. β-Carotene becomes vitamin A at a much higher rate than α-carotene or β-cryptoxanthin. α-Carotene and β-cryptoxanthin on the other hand are mainly available as retinyl ester (20–75%) through cleavage of the β-carotene molecule. The majority of conversion to vitamin A takes place in the intestinal mucosa. Typical structures of carotenes are:

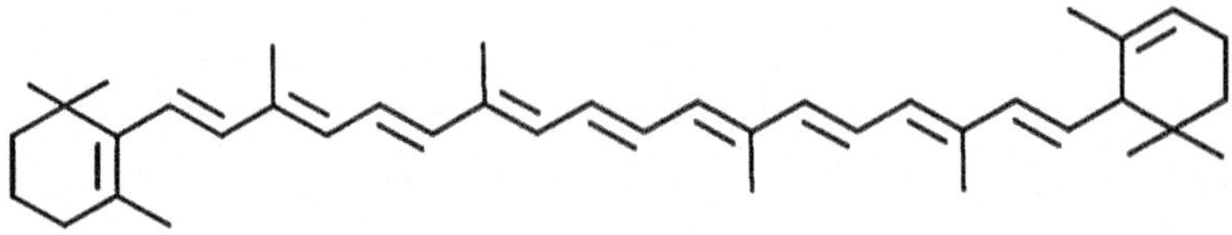

α-Carotene

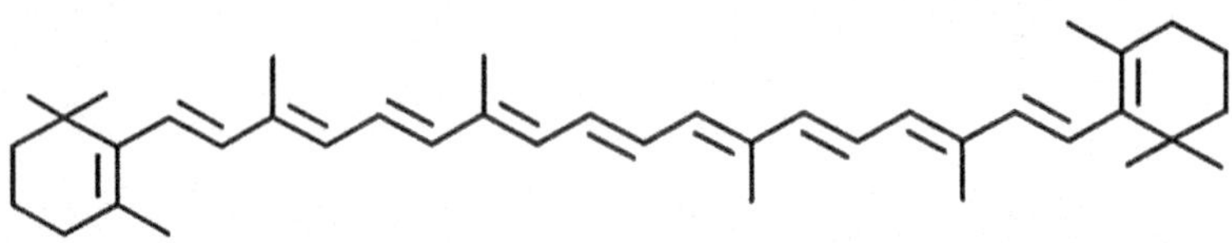

β-Carotene

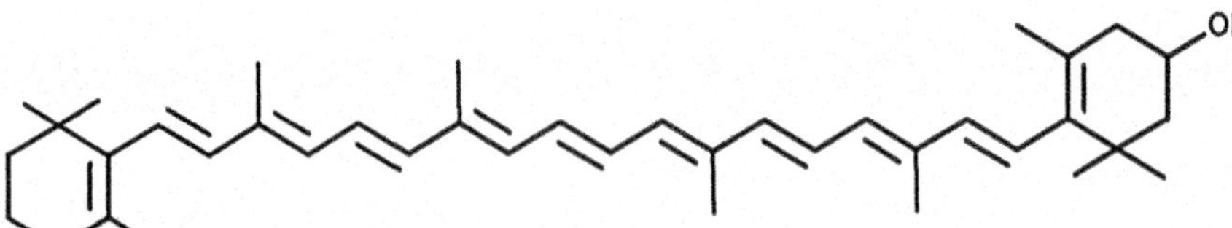

β-Cryptoxanthin

Typical structures of xanthophylls are:

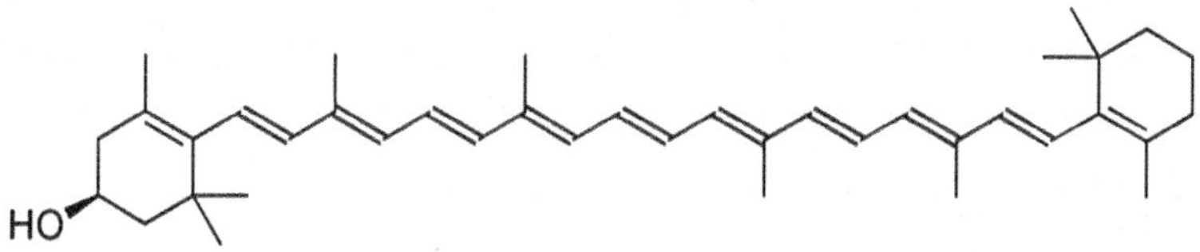

Carotenoids are highly lipophilic molecules and are located inside cell membranes. The hydrocarbons reside in the inner part of lipid bilayer, whereas those with attached oxygen are aligned almost perpendicularly to the membrane surface.

Carotenoids are found in beet root, apricots, carrots, pumpkin, sweet potato, grapefruit, tomatoes, watermelon, mango, papaya, peaches, prunes, squash, oranges, etc.

3.5.4 Alkaloid

*Alkaloid*s are nitrogen containing naturally occurring cyclic organic molecules and having diverse physiological effects. They are generally organic bases and react with acids to give salt, and their solution is alkaline. These bioactive compounds act as reservoir to supply nitrogen, as they have nitrogen atom in the heterocyclic ring. Alkaloids, having aromatic planar structure like piperine, etc., show their action by intercalating with the DNA strands [19]. Some of the alkaloids elicit microbial activity through white blood cells that help to get rid of unwanted and harmful microorganisms and cell debris [79]. The alkaloids also affect the central nervous system via stimulation of neurotransmitters [78]. They also have analgesic potential and work directly on the brain, e.g., morphine, codeine, etc., isolated from opium.

The first alkaloid isolated was morphine. Generally, alkaloids are found as salts of organic acids, e.g., glycosides of sugar. They are extracted with alcohol and generally soluble in water. Alkaloids find use in medicine since ages, particularly for the treatment of fever, snakebite, and mental disorder. Even a minuscule amount of alkaloid leaves a therapeutic impact. These optically active, colorless substances are often bitter in taste. Alkaloids may be classified into true, proto-, and pseudo-alkaloids. True alkaloids are obtained from amino acids, and they have a heterocyclic ring with nitrogen. Atropine, nicotine, morphine, etc. are examples of true alkaloids. Besides, there are certain true alkaloids which may contain terpene or peptide fragments. Examples of terpene and peptide fragment-containing alkaloids are evonine and ergotamine, respectively. Proto-alkaloids, which are not a part of the heterocyclic ring, also contain nitrogen atoms derived from amino acids. Typical examples include adrenaline, ephedrine, nescaline, etc. Pseudo-alkaloids, on the contrary, are not derived from amino acids. They include terpene-like, steroid-like, and purine-like alkaloids. Caffeine, theobromine, theacrine, theophylline, coniine, capsaicin, solanidin, etc. belong to this class. There are two more classes of alkaloids, polyamine alkaloids and peptide or cyclopeptide alkaloids.

As per pharmacognosy, alkaloids may be classified on the basis of presence of nitrogen within or outside the ring; these are heterocyclic and non-heterocyclic alkaloids:

(i) *Heterocyclic (typical):*

- *Pyridine*: Pyridine alkaloids are alkaloids containing a pyridine ring along with a pyrrolidine ring or a piperidine ring. Examples include nicotine and anabasine which are found in plants of the genus *Nicotiana* including tobacco. Nicotine by interaction with acetylcholine receptors increases the concentration of several neurotransmitters, e.g., dopamine. Anabasine is found in tobacco plant and functionally similar to nicotine. The liver enzyme cytochrome P451 metabolizes nicotine to form cotinine [12]. Anticancer activities of several pyrimidine derivatives have been investigated on several cancer cell lines, and the viability of the cells has been found to diminish. Apart from anticancer activities, they have antiviral, antidiabetic, and antimicrobial properties as well [4]. Some of the derivatives of pyridine show inhibitory activity against Herpes simplex type I, HIV, etc. Some of the derivatives elicit their anticancer potential by inhibition of PI3Kp110a, which is a potential target for anticancer drugs. Pyridine and its different derivatives showed cytotoxicity against several human cancer cell lines [6]. They are found to inhibit tyrosine kinase and (CDK) cyclin-dependent kinases which control proliferation of cancer cells [102].
- *Piperidine*: These are alkaloids with a double bond between the nitrogen and the alpha carbon of the piperidine ring. Piperidine is a product from black pepper, and its spicy taste is attributed to this compound [104]. Hemlock, the famous poison, is the source of coniine [44]. Piperidine originates from lysine, and it is used to treat bronchitis and gonorrhea and is an insecticide as well. Absorption of selenium, which is important for glutathione functioning, is greatly influenced by piperidine. It also enhances the uptake of the B vitamins and hence may be used to treat systemic vitamin B deficiency. So far as drug discovery is concerned, piperidine is valuable in designing anticancer drugs [33]. Piperidine is an inhibitor of stress protein Hsp70, which is implicated in cancer, imparting resistance to chemotherapeutic drugs. Targeting Hsp70 reverses drug resistance in lapatinib-resistant breast cancer cells [128]. It acts as a chemoenhancer in ovarian cancer as well [118].
- *Pyrrole*: Pyrroles are heterocyclic aromatic organic compounds with a five-membered ring. These are colorless but develop color on exposure to air. Pyrrolizidine alkaloids normally need metabolic activation to show their effect [15]. Pyrrolic alkaloids have an allylic structure which imparts high reactivity of these compounds. These are therefore reactive alkylating agents which bind to the DNA, proteins, etc.
- *Pyrrolidine*: They have five-member N-containing rings and are derived from ornithine (or arginine in some cases) and lysine. Pyrrolidine alkaloids are found in a number of families but are best known from the

Solanaceae, Convolvulaceae, and Erythroxylaceae families. A pyrrolidine alkaloid isolated from cocoa leaves is hygrine, a thick yellow oil with an unpleasant taste and odor that has many uses in the pharmaceutical industry. Nicotine is another example of a natural compound containing pyrrolidine. Pyrrolidine derivatives reduce cancer-related weakness and wasting of muscle and adipose tissue [18].

- *Tropane*: They contain a tropane ring. These alkaloids are found in plants belonging to the genus *Solanum*. Development of chemotherapeutic drug resistance is a major problem in cancer therapy. Tropane alkaloids have been found to overcome multidrug resistance in cancer. These modulators may increase the efficacy of the medicine vinblastine many times and aid in reversal of drug resistance.
- *Quinoline*: It is a heterocyclic aromatic organic compound. Quinine and camptothecin are some quinoline alkaloids that are rearranged products of indole. These compounds elicit promising anticancer activities like inhibition of cell growth, induction of programmed cell death, prevention of metastatic events like inhibition of angiogenesis, cell migration, etc.
- *Quinolizidine*: Many of the alkaloids are derivatives of this nitrogen-containing heterocyclic compound. Many are toxic, though not all. Toxic ones are tetracyclic with a pyridone nucleus. These alkaloids include lupinine, epilupinine, and lusitanine [16].
- *Aporphine*: These are reduced isoquinoline. They form the core of a class of quilonine alkaloids. These alkaloids have affinity towards dopaminergic, adrenergic, and serotonergic receptors; hence, they are good platforms for designing of central nervous system (CNS) ligands [41].
- *Isoquinoline*: Alkaloids containing isoquinoline ring are known as isoquinolines, which are generally used for manufacturing the drug vasodilator. These narcotic alkaloids (like papaverine, etc.) are mainly found in poppy, aiding in pain relief and sleep. These are also highly addictive alkaloids, e.g., morphine and codeine. These alkaloids inhibit the enzyme phosphodiesterase, thereby degrading cyclic AMP, which is a derivative of ATP and important in many biological processes. Tumor promoter 12-O-tetradecanoylphorbol-13-acetate (TPA)-induced activation of Epstein-Barr virus early antigen (EBV-EA) has been inhibited by several isoquinolines, like 1-benzylisoquinolines, and protoberberines in cancer cells. Isoquinoline alkaloids are also potential cancer chemopreventive agents [20].
- *Indole*: Indoles have a heterobicyclic configuration, having a six-membered ring fused to a five-membered pyrrole ring. Indoles are natural compounds that can be found in numerous types of plants. They are found in cruciferous vegetables, consumption of which reduces the chances of colon, breast, and prostate cancers. Indole-based alkaloids are important in cancer chemotherapy, particularly as an adjuvant with Vinca alkaloids.
- *Purine*: These water-soluble compounds are heterocyclic aromatic organic in nature, consisting of a pyrimidine ring fused to an imidazole ring. These alkaloids have both pharmacological and therapeutic activities. They have effects

on central nervous system (CNS), the kidney, blood vessels, smooth muscles, the liver, and other organs. They inhibit the activity of phosphodiesterase, thereby increasing the production of adrenaline. Methylated derivatives of 2,6 dioxy-purine or xanthine, including caffeine, theobromine, and theophylline, are of pharmacological importance.

- *Terpenoids*: Terpenoid alkaloids are a number of compounds derived from pathways that lead to various groups of terpenes such as monoterpenoids, sesquiterpenoids, diterpenoids, sesterterpenoids, triterpenoids, tetraterpenoids, and steroids. Isoprenoids are terpenoids, derived from five-carbon isoprene units. These structures vary from one another in functional groups, as well as in their basic carbon skeletons. These lipids can be found in all classes of living organisms and are one of the largest group of natural products. Monoterpenoid alkaloids (e.g., gentianine), sesquiterpenoid alkaloids (e.g., dendrobine), and diterpenoids (e.g., aconitine) are found in Gentianaceae, Orchidaceae, and Ranunculaceae, respectively. Terpenoids show anti-inflammatory, antibacterial, anti-protozoan, antifungal, antiviral, and anti-allergen activities, in addition to acting as an immune booster and antineoplastic agent [114, 116]. Terpenoids find use as chemo- and radiosensitizers, thereby enhancing efficacy of treatment with minimal side effects. Besides, they show chemopreventive and therapeutic effects against colon, breast, prostate, and skin cancers [111].
- *Pyrrolizidine alkaloids*: One nitrogen atom at the bridgehead between two fused five-member carbon rings is the typical structure of pyrrolizidine alkaloids. They play a vital role in several diseases, including cancers, pulmonary hypertension, cirrhosis of liver, etc. [26]. These alkaloids elicit their toxic effects due to biotransformation into strong reactive pyrroles by oxidases present in the liver, which alkylate nucleic acids and proteins [21]. Pyrrolizidine alkaloidosis is caused by chronic poisoning due to ingestion of plants containing pyrrolizidine alkaloids. Often cereal and forage crops can be polluted with pyrrolizidine-containing seeds. Some plants like ragwort (*Senecio jacobaea*), woolly groundsels (*Senecio redellii*, *Senecio longilobus*), rattleweed (*Crotalaria retusa*), and seeds of yellow tarweed (*Amsinckia intermedia*) are common sources of pyrrolizidine alkaloids.
- *Imidazole*: A white or colorless solid, soluble in water, it is an aromatic heterocycle, classified as a diazole having nonadjacent nitrogen atoms. Many antifungal drugs and some of the antibiotics and sedatives contain this alkaloid [87].
- *Indolizidine*: Indolizidine alkaloids contain the indolizidine nucleus with two different cycles. These alkaloids are active biotoxins.
- *Steroidal alkaloids*: These are alkaloids with organic ring backbones, featuring nitrogen-based functional groups. They possess a tetracyclic cyclopentanophenanthrene backbone, which is responsible for the steroidal character. These alkaloids interact with a number of cellular targets, and due to this, they may cease proliferation of cancer cell [120]. Two types of steroidal alkaloids are *Solanum* alkaloids and *Veratrum* alkaloids. These alkaloids

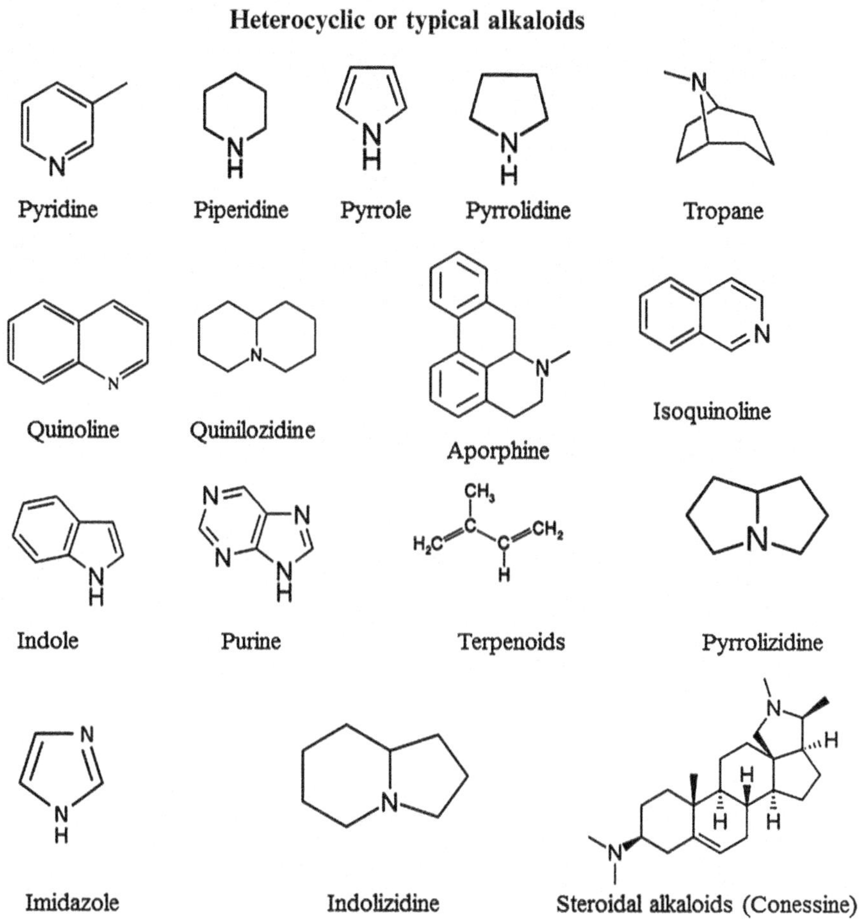

Fig. 3.3 Classification of heterocyclic alkaloids

inhibit acetylcholinesterase. They may be used to treat brain, lung, mammary gland, pancreas, and skin cancers, due to their ability to disrupt the hedgehog signaling pathway.

The structures are given in Fig. 3.3.

(ii) *Non-heterocyclic alkaloids or atypical alkaloids*

These are less abundant in nature. These alkaloids have a nitrogen atom, which is not a part of the ring system. Typical examples are hordenine, mescaline, ephedrine, colchicine, erythromycin, jurubin, pachysandrine A, taxol, and others. Of these, hordenine belongs to the phenethylamine class and is obtained from barley. It has antibacterial properties. Mescaline is known for its hallucinogenic effects. Ephedrine is an adrenergic agent which is used to treat asthma and hay fever. Colchicine, a spindle poison is often used to treat gout. Erythromycin is an antibiotic. Jurubin is a steroid with a 3-amino group. Taxol is an anticancer agent. The structures are given in Fig. 3.4 below.

(iii) *Other alkaloids*

- *Phenanthrene alkaloids*: These are polycyclic aromatic hydrocarbons consisting of three fused benzene rings. These are opium alkaloids derived from poppy seeds and serve as effective pain relievers and narcotics. Morphine, the principal alkaloid though an efficient pain reliever, is also addictive, unlike these alkaloids. Derivatives of phenanthrene are capable

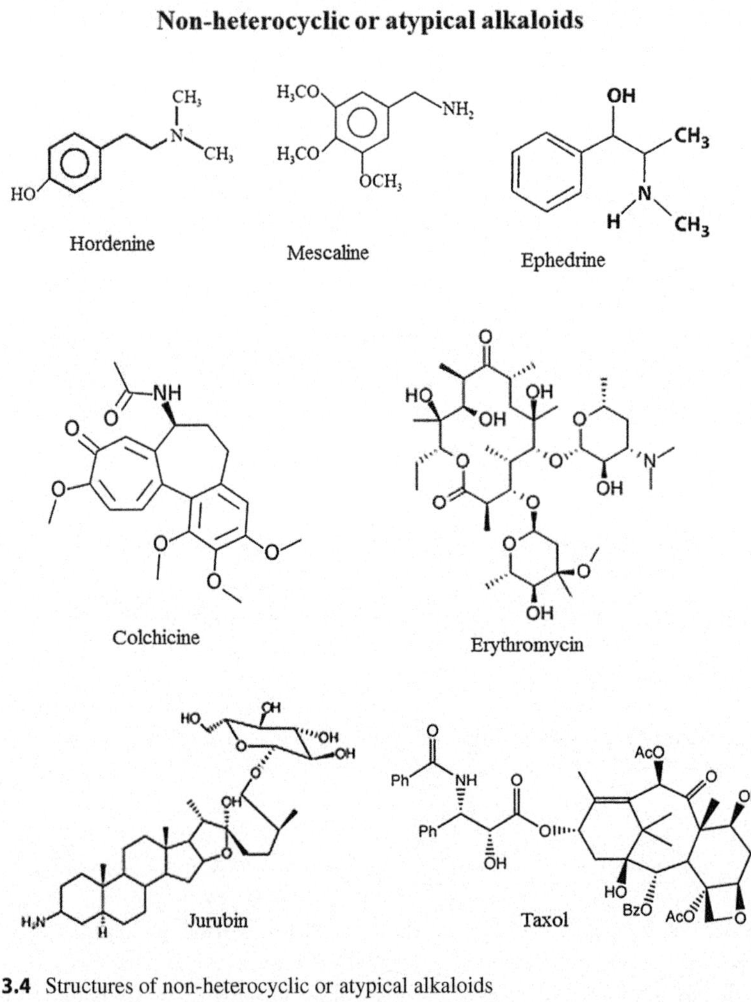

Fig. 3.4 Structures of non-heterocyclic or atypical alkaloids

of killing cancer cells. Phenanthrenes are found in plants, particularly in the Orchid family, of which *Dendrobium*, *Bulbophyllum*, *Eria*, *Maxillaria*, *Bletilla*, *Coelogyne*, *Cymbidium*, *Ephemerantha*, and *Epidendrum* are some examples. Phenanthrene derivatives have been found to induce G2/M cell cycle arrest and apoptosis via interruption of the proteins involved in cell cycle progression. Inactivation of Akt by phenanthrene disrupts the NF-κB signaling pathway. Hence, they may be a potential candidate for anticancer therapy [66].

- *Phenylethylamine*: These are neuromodulators and may also act as neurotransmitters. These are primary amines, where an ethyl group connects the amino group and the benzene ring. These are used in the treatment of asthma. Source of phenylethylamine is ephedra, e.g., ephedrine, pseudoephedrine, norephedrine, catecholamine (dopamine), etc. Phenethylamine (β-phenylethylamine or 2-phenylethylamine) is a monoamine. Chocolates, after microbial fermentation, contain this alkaloid in abundance. It is also found in mistletoe (*Viscum album*) and in *Brassica* species like cabbage, cauliflower, and kale.
- *Tropolone*: This seven-membered aromatic ring compound, a cyclic ketone, is structurally similar to various natural antifungal compounds. It serves as a ligand precursor due to its unusual electronic structure. Tropolone derivatives are inhibitors of histone deacetylase (HDAC), which plays a leading role in carcinogenesis. Specific HDAC inhibitors are used in case of hematological malignancies [81]. However, many inhibitors are toxic with various side effects; hence, tropolone may be used as an inhibitor as it is of natural origin [42].
- *Steroidal*: They have a tetracyclic cyclopentanophenanthrene backbone, which imparts a steroidal feature. They may be of two types, *Solanum* and *Veratrum* alkaloids. Solanidine, a steroidal alkaloid, and chaconine are found in potatoes. *Solanum* alkaloids are also found in Jerusalem cherries and tomatoes. *Veratrum* may be isolated from the roots and rhizomes of the plant of the genus *Veratrum*, e.g., veratridine, cyclopamine, and jervine [46]. They have antimicrobial, anti-inflammatory, and antiestrogenic properties. They modulate cytokines like interleukins 2 and 8 and JNK, among others.

3.5.5 Nitrogen-Containing Compounds

Many plant-derived foods contain nitrogen compounds. These may be of organic or inorganic forms. Organic forms are amines, amides, amino acids, pyrimidines, proteins, nucleic acids, etc., while inorganic forms are nitrates and ammonia. Pyrazines are also nitrogen-containing compounds. These are cyclic nitrogen-containing compounds which contribute flavor to many natural foods. Grapes and wine contain many nitrogen compounds [49]. Many of the alkaloids contain nitrogen atoms and have already been covered in Sect. 3.5.4.

3.5.6 Organosulfur Compounds

Sulfur-containing organic compounds are organosulfur compounds. The element sulfur is necessary for the growth and sustenance of life. Plants are capable of utilizing inorganic sulfur and reducing them to amino acids containing sulfur. Unfortunately, mammals fail to do that, and therefore they are dependent on sulfur-containing amino acids from plants. Pharmacological attributes of these organosulfur compounds are common in nature. Organosulfur compounds occur in the form of certain essential amino acids, e.g., cysteine, cystine, and methionine, which are components of proteins, of tripeptide glutathione, and of enzymes, coenzymes, vitamins, and hormones. Many of the organosulfur compounds have a foul smell.

Life-saving antibiotics, penicillin and cephalosporin, are organosulfur compounds, which are derived from fungi. Another fungi-derived organosulfur compound is gliotoxin, which may have some antiviral properties. Some of the organosulfur compounds impart flavor to wine, cheese, chocolate, coffee, etc., and they are volatile in nature [89]. They have medicinal properties as well, some prevent platelet aggregation, and some show anticancer properties. Some other organosulfur compounds with a characteristic smell are allicin in garlic and lanthionine in shiitake mushroom. Onions are also an abundant source of bivalent organosulfur compounds. Organosulfur compounds are classified as bivalent, tri-coordinated, and tetra coordinated. Not only allicin but also garlic contains a number of other organosulfur compounds like allin, ajoene, s-allylcysteine, etc., and they inhibit cancer as a blocking agent and prevent activation of carcinogens. Some vegetables that contain organosulfur compounds are cruciferous vegetables like cabbage and Brussels sprouts. Phenethyl isothiocyanates are potential blockers of carcinogenesis. Some other bivalent organosulfur compounds that are of pharmacological importance are cephalosporins, captopril, D-penicillamine, ranitidine, etc.

Organosulfur compounds are good antioxidants as well. At the same time, they show some non-antioxidant effects, i.e., they have anti-inflammatory, antiaging, antiplatelet, and immunomodulatory properties. Owing to these properties, these compounds gained in importance in various pathological conditions. They are used for the treatment of diabetes, cardiovascular diseases, neurodegenerative disorders, and cancer.

3.5.7 Benefits of Dietary Phytochemicals

We will discuss the details of chemoprevention and sensitization of phytochemicals in Chap. 5. Here we summarize some of the effects due to phytochemicals present in our daily dietary intake.

Bioactive compounds are stud with a plethora of health benefits. Many of these that we consume in our daily diet have potential anticancer activities. Curcumin, well known for its anticancer properties, is a main ingredient in Indian recipes. The most popular global beverage, next to water, is tea. Green tea already has gained importance as a cancer-fighting plant product, but black tea is also almost equally beneficial. There is a long list of phytochemicals showing promising potential in cancer prevention and treatment. Genistein from soybean, isothiocyanates from cruciferous vegetables, resveratrol from grapes, gingerol from ginger, and lycopene from tomato are a few in this long list.

Curcumin The diferuloylmethane belongs to the ginger family. Curcumin preferentially kills cancer cells, sparing the normal healthy cells. Curcumin is nontoxic up to a dose of 12 g/day, and it has anti-inflammatory, antioxidant, and anticancer properties. Curcumin modulates the transcription factor NF-κB, which is involved in various cellular responses to stress, cytokines, free radicals, etc. and plays a vital role in immune response. Aberrant regulation of NF-κB is intricately linked to cancer. mTOR, the mammalian target of rapamycin, behaves like a serine/threonine protein kinase regulating cell growth, cell proliferation, cell motility, cell survival, and protein synthesis. The intracellular signaling pathway PI3K/AKT/mTOR regulates the cell cycle and hence has a direct relation to cellular quiescence, proliferation, and cancer. Overactivation of mTOR signaling significantly contributes to the initiation and development of tumors, and mTOR activity was found to be deregulated in many types of cancer [125]. Curcumin inhibits mTOR [10] and modulates growth of tumor cells through regulation of a number of signaling pathways including cell proliferation pathway (cyclin D1, c-myc), cell survival pathway (Bcl-2, Bcl-x, cFLIP, XIAP, c-IAP1), caspase activation pathway (caspase-8, 3, 9), tumor suppressor pathway (p53, p21), death receptor pathway (DR4, DR5), mitochondrial pathways, and protein kinase pathway (JNK, Akt, and AMPK) [91].

Cyanidin Berries, like grapes, cranberry, blackberry, etc., apples, plums, and red cabbage contain cyanidin, which, by virtue of its antioxidant and radical scavenging effects, aid in cancer control. This compound inhibits cancer cell proliferation and iNOS and COX-2 gene expressions [129]. Cyanidin regulates COX-2 expression by suppressing NF-κB and AP-1. These are regulated by the MAPK/ERK pathway, and cyanidin regulates this protein, as well as targets other important genes like MKK-4, MEK1, and Raf-1, ultimately modulating COX-2 expression [58].

Catechin Tea is a popular beverage throughout the world. Among different types of tea, green and black are most popular. Green tea, containing catechin compounds, is known for its health benefits. EGCG is notable among these catechins. It binds to Bcl-xl, an anti-apoptotic protein, and inhibits its action; hence, apoptosis is facilitated [63]. Green tea is reported to inhibit mitogen-activated protein kinases (MAPK), cyclin-dependent kinases, NF-κB, topoisomerase I, and metastatic proteins, matrix metalloproteinases [62]. Black tea has been reported to be at par with green tea, so far as antioxidant potential is concerned [107, 108]. Polyphenolic contents from tea are notable antioxidants and show antimutagenic and antigenotoxic potential and may aid in prevention of cancer [96, 101]. Chemopreventive potential of the active ingredients of tea is supposedly mediated by the Nrf2 pathway [106].

We discuss the details of this pathway in Chap. 5. Intake of black tea has been found to prevent development of cancer of the oral cavity and skin induced by inorganic arsenic in animal model [31, 94].

Genistein Soy foods are an alternative to meat as they are rich in proteins; these are quite popular and healthy too. Plant proteins containing fibers, potassium, magnesium, and vitamins are found in soy products. Genistein, the main ingredient of soy foods, is an isoflavone, an antioxidant. These antioxidants possess antiangiogenic effects and elicit anticancer properties by modulating several enzymes related to cell division and cell survival. It is a tyrosine kinase inhibitor, which inhibits topoisomerase II [67, 69]. Most of the breast cancers are estrogen receptor positive. Isoflavones may act like estrogen, binding of which helps in proliferation of mammary cells. Genistein competes with estrogen to bind to estrogen receptor (ER). Genistein is probably involved in the JNK pathway in inducing AP-1 activity [35].

Sulforaphane We consume a lot of organosulfur compounds; as mentioned before, these compounds are found in abundance in cruciferous vegetables such as broccoli, cabbage, etc.

Sulforaphane is produced when the enzyme myrosinase transforms glucosinolate into sulforaphane, and this happens mainly due to chewing, when the plant part is crushed. Sulforaphane enhances the expression of tumor suppressor proteins and inhibits the epigenetic marker histone deacetylase (HDAC) activity [23]. sulforaphane inhibits Akt, ERK signaling, COX-2, and cyclin D1 protein expression [105]. Sulforaphane has been found to inhibit Keap1, thereby retarding the degradation of Nrf2 and increasing the expression of Nrf2 [53].

Sulforaphane has an inhibitory action on NF-κB. The expression of phosphorylated c-Jun N terminal kinase (p-JNK), phosphorylated extracellular signal-regulated kinases (p-ERK), and phosphorylated-Akt (p-Akt) has been found to be downregulated by sulforaphane [47].

Fisetin Fisetin, abundantly found in apple, grape, onion, cucumber, etc., is a flavone used as a coloring agent. It has antioxidant, anti-inflammatory, and anticarcinogenic potential. It regulates protein kinases [121]. Fisetin also modulates PI3K/Akt and mTOR signaling [56]. Fisetin by G1-phase cell cycle arrest reduces cell viability. Fisetin efficiently diminishes the NF-κB and activator protein AP-1. The development of drug resistance and the mainstay of prognosis may be overcome by fisetin [5]. It has an inhibitory effect on the deadliest stage of cancer, metastasis, as adhesion, migration, and invasion of cancer cells are inhibited [64].

Lycopene A bright red carotene, lycopene, is found generally in red fruits and vegetables, including in ripe tomatoes, red carrots, and watermelons. Some red fruits like strawberry and cherries do not contain lycopene; some foods like asparagus and parsley, though not red in color, contain lycopene [40]. It shows antioxidant and chemoprotective activities. DNA damage due to oxidative stress has been found to be reduced by lycopene [86]. Supplementation of this product aids in prostate cancer by modulation of both nuclear factor-2 (erythroid derived-2)-mediated oxidative stress response and male androgen levels in healthy prostate tissues of prostate cancer patients [68].

PEITC PEITC showed a strong potency against melanoma. It has shown its efficacy in several cancers like breast [71], lung [126], cervical [119], prostate [124], sarcoma [122], and myeloma cells [50]. PEITC activates caspase-3, culminating in cell cycle arrest at G2/M phase [126].

Resveratrol It is a natural phenol found in red grapes, peanuts, etc. It is a potent chemopreventive agent as it prevents initiation process via induction of phase II drug-metabolizing enzymes. The promotion stage of carcinogenesis can be hindered by inhibition of cyclooxygenase and hydroperoxidase functions. Promotion stage is also inhibited by resveratrol [51]. Inhibition of metastasis is achieved by inhibition of hypoxia inducible factor-1α and MMP-9 expression [123].

Indole-3-carbinol (I3C) Sources of I3C are *Brassica* vegetables. Diindolylmethane (DIM) is a derivative of I3C. I3C shows its action through modulation of the receptor tyrosine kinase/PI3K/Akt signaling pathway [88].

I3C and DIM have promising anticancer activities against hormone-responsive cancers like breast, prostate, and ovarian cancers [1]. In a recent study, it is concluded that DIM rather than I3C is the active agent in cell culture studies [14].

Allium Vegetables Garlic, onion, shallot, leek, chive, etc. belonging to this class have medicinal potentials. These vegetables are popular in cuisines worldwide and are valued for their potential medicinal properties. Increased consumption of allium vegetables decreases the risk of cancer [77]. Current knowledge of the mechanisms by which these compounds cause apoptosis indicates that they target various apoptotic signaling molecules from initiation to execution. Molecules affected include the map kinases JNK, ERK1/2, and p38; p53, NF-κB, and bcl-2/bax family constituents; and caspases, but not all of the signaling molecules were influenced by each *Allium* compound [45]. These compounds lead to cell cycle arrest and reduce Cdk1/cyclin B kinase activity or activation of ERK1/2 [45, 59].

Gingerol Ginger has been known since ages for its medicinal value. This nontoxic plant product has a broad spectrum of pharmacological attributes, e.g., it has antitumor, antioxidant, and anti-inflammatory properties [60]. Ginger has been found to decrease iNOS and TNF-α expression via suppression of IκBα phosphorylation and NF-κB nuclear translocation [82]. It prevents invasion through regulation of MMP-9 (Weng et al. [2010]).

3.6 Conclusion

Agents that are harmless, with known mechanism of action, and efficacious are always important for better outcomes in cancer prevention and treatment. Plant-derived molecules are unique as they are capable of increasing the efficacy of drugs with minimal toxicity and without side effects. Moreover, phytochemicals also help in cancer prevention. There are myriads of phytochemicals, and there is no universally accepted classification of them. We have used a structure-based classification and reviewed all well-known phytochemicals in this chapter. We have also discussed their roles in cancer therapy briefly, as we will discuss this in details in the later chapters of the book.

References

1. A. Acharya, I. Das, S. Singh, T. Saha, Chemopreventive properties of indole-3-carbinol, diindolylmethane and other constituents of cardamom against carcinogenesis. Recent Pat. Food Nutr. Agric. **2**(2), 166–177 (2010)
2. A. Ahmad, S.F. Asad, S. Singh, S.M. Hadi, DNA breakage by resveratrol and Cu(II): Reaction mechanism and bacteriophage inactivation. Cancer Lett. **154**, 29–37 (2000)
3. AICR, *World Cancer Research Fund, American Institute for Cancer Research. Food, Nutrition, and Physical Activity, and the Prevention of Cancer: A Global Perspective* (AICR, Washington, DC, 2007)
4. S.H. Allen, A.B. Johns, S.K. Gudmundsson, A.G. Freeman, F.L. Boyd, H.C. Sexton, W.D. Selleseth, K. Creech, R.K. Moniri, Synthesis of C-6 substituted pyrazolo[1,5-a]pyridines with potent activity against herpesviruses. Bioorg. Med. Chem. **14**, 944–954 (2006)
5. A. Angelini, C. Di Ilio, M.L. Castellani, P. Conti, F. Cuccurullo, Modulation of multidrug resistance p-glycoprotein activity by flavonoids and honokiol in human doxorubicin-resistant sarcoma cells (MES-SA/DX-5): Implications for natural sedatives as chemosensitizing agents in cancer therapy. J. Biol. Regul. Homeost. Agents **24**(2), 197–205 (2010)
6. A.M. Attia, A.A. El-Shehawy, A convenient method for the synthesis of 2-(beta-D-glycopyranosylthio) pyridines. Nucleosides Nucleotides Nucleic Acids **22**(9), 1737–1746 (2003)
7. O.A. Awoyinka, I.O. Balogun, A.A. Ogunnowo, Phytochemical screening and in-vitro bioactivity of *Cnidoscolus aconitifolius* (Euphorbiaceae). J. Med. Plants Res. **1**, 063–065 (2007)
8. T.S. Ballard, P. Mallikarjunan, K. Zhou, S. O'Keefe, Microwave-assisted extraction of phenolic antioxidant compounds from peanut skins. Food Chem. **120**, 1185–1192 (2010)
9. E.C. Bate-Smith, T. Swain, Flavonoid compounds, in *Comparative Biochemistry*, ed. by M. Florkin, H. S. Mason, III edn., (Academic, New York, 1962), pp. 75–809
10. C.S. Beevers, L. Chen, L. Liu, Y. Luo, N.J. Webster, S. Huang, Curcumin disrupts the mammalian target of rapamycin-raptor complex. Cancer Res. **69**(3), 1000–1008 (2009)
11. Bellik Y, Boukraâ L, Alzahrani H A, Bakhotmah B A, Abdellah F, . Hammoudi S M and Iguer-Ouada M, 2013. Molecular mechanism underlying anti-inflammatory and Anti-allergic activities of phytochemicals: An update, Molecules, 18, 322–353
12. N. Benowitz, J. Hukkanen, J. Peyton III, Nicotine chemistry, Metabolism, Kinetics and Biomarkers. Handb. Exp. Pharmacol. **192**, 29–60 (2009)
13. H. Boeing, A. Bechthold, A. Bub, S. Ellinger, D. Haller, A. Kroke, E. Leschik-Bonnet, M.J. Müller, H. Oberritter, M. Schulze, P. Stehle, B. Watzl, Critical review: Vegetables and fruit in the prevention of chronic diseases. Eur. J. Nutr. **51**(6), 637–663 (2012)

14. H.L. Bradlow, M.A. Zeligs, Diindolylmethane (DIM) spontaneously forms from indole-3-carbinol (I3C) during cell culture experiments. In Vivo **24**(4), 387–391 (2010)
15. M.X. Chen, L. Li, D.F. Zhong, S.J. Shen, J. Zheng, X.Y. Chen, 9-Glutathionyl-6,7-dihydro-1-hydroxymethyl-5H-pyrrolizine is the major pyrrolic glutathione conjugate of retronecine-type pyrrolizidine alkaloids in liver microsomes and in rats. Chem. Res. Toxicol. **29**(2), 180–189 (2016)
16. Y.D. Cho, R.O. Martin, Resolution and unambiguous identification of microgram amounts of 22 lupin alkaloids by sequential use of thin-layer and gas-liquid chromatography and mass spectrometry. Anal. Biochem. **44**, 49–57 (1971)
17. A.J. Cooper, N.G. Forouhi, Z. Ye, B. Buijsse, L. Arriola, B. Balkau, et al., Fruit and vegetable intake and type 2 diabetes: EPIC-InterAct prospective study and meta-analysis. Eur. J. Clin. Nutr. **66**(10), 1082–1092 (2012)
18. L. Cornara, M. Biagi, J. Xiao, B. Burlando, Therapeutic properties of bioactive compounds from different honeybee products. Front. Pharmacol. **8**, 412 (2017)
19. M.M. Cowan, Plant products as antimicrobial agents. Clin. Microbiol. Rev. **12**(4), 564–582 (1999)
20. W. Cui, K. Iwasa, H. Tokuda, A. Kashihara, Y. Mitani, T. Hasegawa, Y. Nishiyama, M. Moriyasu, H. Nishino, M. Hanaoka, C. Mukai, K. Takeda, Potential cancer chemopreventive activity of simple isoquinolines, 1-benzylisoquinolines, and protoberberines. Phytochemistry **67**(1), 70–79 (2006)
21. T.P.T. Cushnie, B. Cushnie, A.J. Lamb, Alkaloids: An overview of their antibacterial, antibiotic-enhancing and antivirulence activities. Int. J. Antimicrob. **44**, 377–386 (2014)
22. K. Das, R.K.S. Tiwari, D.K. Shrivastava, Techniques for evaluation of medicinal plant products as antimicrobial agent: Current methods and future trends. J. Med. Plants Res. **4**(2), 104–111 (2010)
23. R.H. Dashwood, E. Ho, Dietary histone deacetylase inhibitors: From cells to mice to man. Semin. Cancer Biol. **17**(5), 363–369 (2007)
24. G. Duthie, A. Crozier, Plant-derived phenolic antioxidants. Curr. Opin. Lipidol. **11**(1), 43–47 (2000)
25. C. Echiburu-Chau, L. Pastén, C. Parra, J. Bórquez, A. Mocan, M.J. Simirgiotis, High resolution UHPLC-MS characterization and isolation of main compounds from the antioxidant medicinal plant Parastrephia lucida (Meyen). Saudi Pharm. J. **25**(7), 1032–1039 (2017)
26. J.A. Edgar, R.J. Molyneux, S.M. Colegate, Pyrrolizidine alkaloids: Potential role in the etiology of cancers, pulmonary hypertension, congenital anomalies, and liver disease. Chem. Res. Toxicol. **28**(1), 4–20 (2014)
27. M. Falsaperlaa, G. Morgiab, A. Tartaronec, R. Arditoc, G. Romano, Support ellagic acid therapy in patients with hormone refractory prostate cancer (HRPC) on standard chemotherapy using vinorelbine and estramustine. Phosphate. Eur Urol **47**, 449 (2005)
28. C.G. Fraga, M. Galleano, S.V. Verstraeten, P.I. Oteiza, Basic biochemical mechanisms behind the health benefits of polyphenols. Mol. Asp. Med. **31**(6), 435–445 (2010)
29. G. Galati, P.J. O'Brien, Potential toxicity of flavonoids and other dietary phenolics: significance for their chemopreventive and anticancer properties. Free Radic. Biol. Med. **37**(3), 287–303 (2004)
30. P. Garcia-Salas, A. Morales-Soto, A. Segura-Carretero, A. Fernandez-Gutierrez, Phenolic-compound-extraction systems for fruit and vegetable samples. Molecules **15**, 8813–8826 (2010)
31. A. Ghosh, A. Mukherjee, S. Mukherjee, M. Roy, *Role of Black Tea in Prevention of Skin Carcinogenesis* (LAP LAMBERT Academic Publishing, Germany, 2019)
32. Global Action Plan for the Prevention and Control of NCDs 2013–2020 (World Health Organization, Geneva, 2013); Global Status Report on Noncommunicable Diseases 2014. (World Health Organization, Geneva, 2014)
33. P. Goel, O. Alam, M.J. Naim, F. Nawaz, M. Iqbal, M.I. Alam, Recent advancement of piperidine moiety in treatment of cancer- A review. Eur. J. Med. Chem. **157**, 480–502 (2018)

34. M. Goleniowski, M. Bonfill, R. Cusido, P. Javier, Zphenolic acids, in *Natural Products*, ed. by K. G. Ramawat, J. M. Me'rillon, (Springer-Verlag, Berlin/Heidelberg, 2013)
35. A. Gopalakrishnan, C.J. Xu, S.S. Nair, C. Chen, V. Hebbar, A.N. Kong, Modulation of activator protein-1 (AP-1) and MAPK pathway by flavonoids in human prostate cancer PC3 cells. Arch. Pharm. Res. **29**(8), 633–644 (2006)
36. N. Gungor, M. Sengul, Antioxidant activity, total phenolic content and selected physicochemical properties of white mulberry (Morus Alba L.). Fruits. Int. J. Food Prop. **11**(1), 44–52 (2008)
37. S.M. Hadi, S.F. Asad, S. Singh, A. Ahmad, Putative mechanism for anticancer and apoptosis inducing properties of plant-derived polyphenolic compounds. IUBMB Life **50**, 167–171 (2000)
38. B. Halliwell, Are polyphenols antioxidants or pro-oxidants? What do we learn from cell culture and in vivo studies? Arch. Biochem. Biophys. **476**(2), 107–112 (2008)
39. B. Halliwell, The wanderings of a free radical. Free Radic. Biol. Med. **46**(5), 531–542 (2009)
40. USDA Nutrient Database 2014. *Foods highest in lycopene, Nutrition Data, USDA Nutrient Database, version SR-21*. nutritiondata.com. Conde Nast. Retrieved 19 Aug 2014 (2014)
41. N. Kapadia, W. Harding, Aporphine alkaloids as ligands for serotonin receptors. Med. Chem. (Los Angeles) **6**, 241–249 (2009)
42. S.L. Haney, C. Allen, M.L. Varney, K.M. Dykstra, E.R. Falcone, S.H. Colligan, Q. Hu, A.M. Aldridge, D.L. Wright, A.J. Wiemer, S.A. Holstein, Novel tropolones induce the unfolded protein response pathway and apoptosis in multiple myeloma cells. Oncotarget **8**(44), 76085–76098 (2017)
43. L. Hartley, E. Igbinedion, J. Holmes, N. Flowers, M. Thorogood, A. Clarke, S. Stranges, L. Hooper, K. Rees, Increased consumption of fruit and vegetables for the primary prevention of cardiovascular diseases. Cochrane Database Syst. Rev. **6**, CD009874 (2013)
44. T.A. Henry, *The Plant Alkaloids*, 4th edn. (The Blakiston Company, Philadelphia, 1949)
45. A. Herman-Antosiewicz, S.V. Singh, Signal transduction pathways leading to cell cycle arrest and apoptosis induction in cancer cells by Allium vegetable-derived organosulfur compounds: A review. Mutat. Res. **555**, 121–131 (2004)
46. A. Hollman, Veratrum alkaloids. Br. Heart J. **65**(5), 286 (1991)
47. R. Hu, T.O. Khor, G. Shen, W.S. Jeong, V. Hebbar, C. Chen, C. Xu, B. Reddy, K. Chada, A.N. Kong, Cancer chemoprevention of intestinal polyposis in ApcMin/+ mice by sulforaphane, a natural product derived from cruciferous vegetable. Carcinogenesis **27**(10), 2038–2046 (2006)
48. T. Iwashina, Flavonoid properties of five families newly incorporated into the order Caryophyllales. Bull. Natl. Mus. Nat. Sci. **39**, 25–51 (2013)
49. R.S. Jackson, 6 – chemical constituents of grapes and wine. Wine Sci. 347–426 (2014)
50. J. Jakubikova, D. Cervi, M. Ooi, K. Kim, S. Nahar, S. Klippel, D. Cholujova, M. Leiba, J.F. Daley, J. Delmore, J. Negri, S. Blotta, D. McMillin, T. Hideshima, P. Richardson, J. Sedlak, K. Anderson, C. Mitsiades, Anti-tumor activity and signaling events triggered by the isothiocyanates, sulforaphane and PEITC in multiple myeloma. Haematologica **96**(8), 1170–1179 (2011)
51. M. Jang, L. Cai, G.O. Udeani, K.V. Slowing, C.F. Thomas, C.W. Beecher, H.H. Fong, N.R. Farnsworth, A.D. Kinghorn, R.G. Mehta, R.C. Moon, J.M. Pezzuto, Cancer chemopreventive activity of resveratrol, a natural product derived from grapes. Science **275**(5297), 218–220 (1997)
52. M.A. Jayasri, L. Mathew, A. Radha, A report on the antioxidant activities of leaves and rhizomes of Costus pictus D. Don. Int. J. Int. Bio. **5**(1), 20–26 (2009)
53. W.S. Jeong, Y.S. Keum, C. Chen, M.R. Jain, G. Shen, J.H. Kim, W. Li, A.N. Kong, Di_erential expression and stability of endogenous nuclear factor E2-related factor 2 (Nrf2) by natural chemopreventive compounds in HepG2 human hepatoma cells. J. Biochem. Mol. Biol. **38**(2), 167–176 (2005 Mar 31)
54. M.P. Kahkonen, A.I. Hopia, H.J. Vuorela, J.P. Rauha, K. Pihlaja, T.S. Kujala, M. Heinonen, Antioxidant activity of plant extracts containing phenolic compounds. J. Agric. Food Chem. **47**, 3954–3962 (1999)

55. A. Kar, *Pharmacognosy and Pharmacobiotechnology*. New Age International. pp. 44–. ISBN 978-81-224-1501-8. Archived from the original on 2 June 2013. Retrieved 31 Jan 2011 (2003)
56. N. Khan, F. Afaq, F.H. Khusro, V.M. Adhami, Y. Suh, H. Mukhtar, Dual inhibition of PI3K/AKT and mTOR signaling in human non-small cell lung cancer cells by a dietary flavonoid fisetin. Int. J. Cancer **130**(7), 1695–1705 (2012)
57. H.E. Khoo, A. Azlan, S.T. Tang, S.M. Lim, Anthocyanidins and anthocyanins: Colored pigments as food, pharmaceutical ingredients, and the potential health benefits. Food Nutr. Res. **61**(1), 1361779 (2017)
58. J.E. Kim, J.Y. Kwon, S.K. Seo, J.E. Son, S.K. Jung, S.Y. Min, M.K. Hwang, Y.S. Heo, K.W. Lee, H.J. Lee, Cyanidin suppresses ultraviolet B-induced COX-2 expression in epidermal cells by targeting MKK4, MEK1, and Raf-1. Biochem. Pharmacol. **79**(10), 1473–1482 (2010)
59. L.M. Knowles, J.A. Milner, Possible mechanism by which allyl sulfides suppress neoplastic cell proliferation. J. Nutr. **131**, 1061S–1066S (2001)
60. M. Kumara, M.R. Shylajab, P.A. Nazeemc, T. Babu, 6-Gingerol is the most potent Anticancerous compound in ginger (Zingiber officinale Rosc.). J. Dev. Drugs **6**, 1 (2017)
61. J.L. Lamaison, C. Petitjeanfreytet, Medicinal Lamiaceae with antioxidant activity, potential sources of rosmarinic acid. Pharm. Acta Helvetiae **66**, 185 (1996)
62. J.D. Lambert, C.S. Yang, Mechanisms of cancer prevention by tea constituents. J. Nutr. **133**(10), 3262S–3267S (2003)
63. M. Leone, D. Zhai, S. Sareth, S. Kitada, J.C. Reed, M. Pellecchia, Cancer prevention by tea polyphenols is linked to their direct inhibition of antiapoptotic Bcl-2-family proteins. Cancer Res. **63**(23), 8118–8121 (2003)
64. Y.C. Liao, Y.W. Shih, C.H. Chao, X.Y. Lee, T.A. Chiang, Involvement of the ERK signaling pathway in fisetin reduces invasion and migration in the human lung cancer cell line A549. J. Agric. Food Chem. **57**(19), 8933–8941 (2009)
65. S.S. Lim, T. Vos, A.D. Flaxman, G. Danaei, K. Shibuya, H. Adair-Rohani, M. Amann, H.R. Anderson, K.G. Andrews, M. Aryee, C. Atkinson, L.J. Bacchus, A.N. Bahalim, K. Balakrishnan, J. Balmes, S. Barker-Collo, A. Baxter, M.L. Bell, J.D. Blore, F. Blyth, C. Bonner, G. Borges, R. Bourne, M. Boussinesq, M. Brauer, P. Brooks, N.G. Bruce, B. Brunekreef, C. Bryan-Hancock, C. Bucello, R. Buchbinder, F. Bull, R.T. Burnett, T.E. Byers, B. Calabria, J. Carapetis, E. Carnahan, Z. Chafe, F. Charlson, H. Chen, J.S. Chen, A.T. Cheng, J.C. Child, A. Cohen, K.E. Colson, B.C. Cowie, S. Darby, S. Darling, A. Davis, L. Degenhardt, F. Dentener, D.C. Des Jarlais, K. Devries, M. Dherani, E.L. Ding, E.R. Dorsey, T. Driscoll, K. Edmond, S.E. Ali, R.E. Engell, P.J. Erwin, S. Fahimi, G. Falder, F. Farzadfar, A. Ferrari, M.M. Finucane, S. Flaxman, F.G. Fowkes, G. Freedman, M.K. Freeman, E. Gakidou, S. Ghosh, E. Giovannucci, G. Gmel, K. Graham, R. Grainger, B. Grant, D. Gunnell, H.R. Gutierrez, W. Hall, H.W. Hoek, A. Hogan, H.D. Hosgood 3rd, D. Hoy, H. Hu, B.J. Hubbell, S.J. Hutchings, S.E. Ibeanusi, G.L. Jacklyn, R. Jasrasaria, J.B. Jonas, H. Kan, J.A. Kanis, N. Kassebaum, N. Kawakami, Y.H. Khang, S. Khatibzadeh, J.P. Khoo, C. Kok, F. Laden, R. Lalloo, Q. Lan, T. Lathlean, J.L. Leasher, J. Leigh, Y. Li, J.K. Lin, S.E. Lipshultz, S. London, R. Lozano, Y. Lu, J. Mak, R. Malekzadeh, L. Mallinger, W. Marcenes, L. March, R. Marks, R. Martin, P. McGale, J. McGrath, S. Mehta, G.A. Mensah, T.R. Merriman, R. Micha, C. Michaud, V. Mishra, K. Mohd Hanafiah, A.A. Mokdad, L. Morawska, D. Mozaffarian, T. Murphy, M. Naghavi, B. Neal, P.K. Nelson, J.M. Nolla, R. Norman, C. Olives, S.B. Omer, J. Orchard, R. Osborne, B. Ostro, A. Page, K.D. Pandey, C.D. Parry, E. Passmore, J. Patra, N. Pearce, P.M. Pelizzari, M. Petzold, M.R. Phillips, D. Pope, C.A. Pope 3rd, J. Powles, M. Rao, H. Razavi, E.A. Rehfuess, J.T. Rehm, B. Ritz, F.P. Rivara, T. Roberts, C. Robinson, J.A. Rodriguez-Portales, I. Romieu, R. Room, L.C. Rosenfeld, A. Roy, L. Rushton, J.A. Salomon, U. Sampson, L. Sanchez-Riera, E. Sanman, A. Sapkota, S. Seedat, P. Shi, K. Shield, R. Shivakoti, G.M. Singh, D.A. Sleet, E. Smith, K.R. Smith, N.J. Stapelberg, K. Steenland, H. Stöckl, L.J. Stovner, K. Straif, L. Straney, G.D. Thurston, J.H. Tran, R. Van Dingenen, A. van Donkelaar, J.L. Veerman, L. Vijayakumar, R. Weintraub, M.M. Weissman, R.A. White,

H. Whiteford, S.T. Wiersma, J.D. Wilkinson, H.C. Williams, W. Williams, N. Wilson, A.D. Woolf, P. Yip, J.M. Zielinski, A.D. Lopez, C.J. Murray, M. Ezzati, M.A. AlMazroa, Z.A. Memish, A comparative risk assessment of burden of disease and injury attributable to 67 risk factors and risk factor clusters in 21 regions, 1990–2010: a systematic analysis for the Global Burden of Disease Study 2010. Lancet **380**(9859), 2224–2260 (2012)

66. J.C. Lin, S. Yang, T. Hong, S. Yu, Q. Shi, L. Wei, H.-Y. Chen, P.-C. Yang, K.-H. Lee, Phenanthrene-based Tylophorine-1 (PBT-1) inhibits lung Cancer cell growth through the Akt and NF-κB pathways. J. Med. Chem. **52**(7), 1903–1911 (2009)
67. M. Lopez-Lazaro, E. Willmore, C.A. Austin, Cells lacking DNA topoisomerase II beta are resistant to genistein. J. Nat. Prod. **70**(5), 763–767 (2007)
68. M.J. Magbanua, R. Roy, E.V. Sosa, V. Weinberg, S. Federman, M.D. Mattie, M. Hughes-Fulford, J. Simko, K. Shinohara, C.M. Haqq, P.R. Carroll, J.M. Chan, Gene expression and biological pathways in tissue of men with prostate cancer in a randomized clinical trial of lycopene and fish oil supplementation. PLoS One **6**(9), e24004 (2011)
69. J. Markovits, C. Linassier, P. Fosse, J. Couprie, J. Pierre, A. Jacquemin-Sablon, J.M. Saucier, J.B. Le Pecq, A.K. Larsen, Inhibitory effects of the tyrosine kinase inhibitor genistein on mammalian DNA topoisomerase II. Cancer Res. **49**(18), 5111–5117 (1989)
70. M.P. Mattson, A. Cheng, Neurohormetic phytochemicals: Low-dose toxins that induce adaptive neuronal stress responses. Trends Neurosci. **29**, 632–639 (2006)
71. Y.J. Moon, D.A. Brazeau, M.E. Morris, Dietary phenethyl isothiocyanate alters gene expression in human breast cancer cells. Evid. Based Complement Alternat. Med. **2011**, pii 462525 (2011)
72. N. Mulinacci, D. Prucher, M. Peruzzi, A. Romani, P. Pinelli, C. Giaccherini, F.F. Vincieri, Commercial and laboratory extracts from artichoke leaves: Estimation of caffeoyl esters and flavonoidic compounds content. J. Pharm. Biomed. Anal. **34**, 349–357 (2004)
73. R. Munday, C.M. Munday, Induction of phase II detoxification enzymes in rats by plant-derived isothiocyanates: Comparison of allyl isothiocyanate with sulforaphane and related compounds. J. Agric. Food Chem. **52**(7), 1867–1871 (2004)
74. M.C. Navarro, M.P. Montilla, M.M. Cabo, M. Galisteo, A. Caceres, C. Morales, I. Berger, Antibacterial, antiprotozoal and antioxidant activity of five plants used in izabal for infectious diseases. J. Phytother. Res. **17**, 325–329 (2003)
75. N.S. Ncube, A.J. Afolayan, A.I. Okoh, Assessment techniques of antimicrobial properties of natural compounds of plant origin: Current methods and future trends. Afr. J. Biotechnol. **7**(12), 1797–1806 (2008)
76. S.N. Ngoci, C.M. Mwendia, C.G. Mwaniki, Phytochemical and cytotoxicity testing of Indigofera lupatana baker F. J. Anim. Plant Sci. **11**(1), 1364–1373 (2011)
77. H.L. Nicastro, S.A. Ross, J.A. Milner, Garlic and onions: Their cancer prevention properties. Cancer Prev. Res. (Phila.) **8**(3), 181–189 (2015)
78. S. Njeru, J. Matasyoh, G.M. Charles, M. Mwendia, G. Kobia, A review of some phytochemicals commonly found in medicinal plants. Int J Med. Plants. Photon **105**, 135–140 (2013)
79. K.O. Ogunwenmo, O.A. Idowu, C. Innocent, E.B. Esan, O.A. Oyelana, Cultivars of Codiaeum variegatum (L.) Blume (Euphorbiaceae) show variability in phytochemical and cytological characteristics. J. Biotechnol. **20**, 2400–2405 (2007)
80. M.R. Olthof, P.C.H. Hollman, M.B. Katan, Chlorogenic acid and caffeic acid are absorbed in humans. J. Nutr. **131**, 66 (2001)
81. S.N. Ononye, M.D. VanHeyst, E.Z. Oblak, W. Zhou, M. Ammar, A.C. Anderson, D.L. Wright, Tropolones as lead-like natural products: The development of potent and selective histone deacetylase inhibitors. ACS Med. Chem. Lett. **4**(8), 757–761 (2013)
82. A.A. Oyagbemi, A.B. Saba, O.I. Azeez, Molecular targets of [6]-gingerol: Its potential roles in cancer chemoprevention. Biofactors **36**(3), 169–178 (2010)
83. V. Paritala, K.K. Chiruvella, I.H.A. Ripain, A. Mohammed, A recent review on phytochemical constituents and medicinal properties of kesum (Polygonum minus Huds.). Asian Pac. J. Trop. Biomed. **4**(6), 430–435 (2014)

84. V. Pasricha, G. Satpathy, R.K. Gupta, Phytochemical & Antioxidant activity of underutilized legume *Vicia faba* seeds and formulation of its fortified biscuits. J. Pharmacogn. Phytochem. **3**(2), 75–80 (2014)
85. I.E. Popova, C. Hall, A. Kubátová, Determination of lignans in flaxseed using liquid chromatography with time-of-flight mass spectrometry. J. Chromatogr. A **1216**, 217–229 (2009)
86. M. Porrini, P. Riso, A. Brusamolino, C. Berti, S. Guarnieri, F. Visioli, Daily intake of a formulated tomato drink affects carotenoid plasma and lymphocyte concentrations and improves cellular antioxidant protection. Br. J. Nutr. **93**(1), 93–99 (2005)
87. A.F. Pozharskii, A.T. Soldatenkov, A.R. Katritzky, *Heterocycles in Life and Society*, 1st edn. (Wiley, Chichester/New York/Weinheim/Brisbane/Singapore/Toronto, 1997)., 301 p
88. X. Qian, T. Melkamu, P. Upadhyaya, F. Kassie, Indole-3-carbinol inhibited tobacco smoke carcinogen-induced lung adenocarcinoma in A/J mice when administered during the postinitiation or progression phase of lung tumorigenesis. Cancer Lett. **311**(1), 57–65 (2011)
89. M.C. Qian, X. Fan, K. Mahattanatawee, eds. Volatile sulfur compounds in food. ACS symposium series 1068. American Chemical Society (2011)
90. A. Rahman, S.M. Shahabuddin Hadi, J.H. Parish, Complexes involving quercetin, DNA and Cu(II). Carcinogenesis **11**, 2001–2003 (1990)
91. J. Ravindran, S. Prasad, B.B. Aggarwal, 2009. Curcumin and cancer cells: How many ways can curry kill tumor cells selectively? AAPS J. **11**(3), 495–510 (2009)
92. J.P. Remington, *The Science and Practice of Pharmacy*, 21st edn. (Lippincott Williams & Wilkins, Philadelphia, 2005), pp. 773–774
93. J.K. Ritter, F. Chen, Y.Y. Sheen, H.M. Tran, S. Kimura, M.T. Yeatman, I.S. Owens, A novel complex locus UGT1 encodes human bilirubin, phenol, and other UDP-glucuronosyltransferase isozymes with identical carboxyl termini. J. Biol. Chem. **267**(5), 3257–3261 (1992)
94. M. Roy, S. Mukherjee, Black tea extract prevents 4-nitroquinoline 1-oxide induced oral tumorigenesis in mice by targeting protein tyrosine kinases and associated biological response. Clin. Oncol. Res. **2**(1), 2–7 (2019)., available online at www.sciencerepository.org
95. M. Roy, S. Chakraborty, M. Siddiqi, R.K. Bhattacharya, Induction of apoptosis in tumor cells by natural phenolic compounds. Asian Pac. J. Cancer Prev. **3**(1), 61–67 (2002)
96. M. Roy, S. Chakraborty, D. Sinha, R.K. Bhattacharya, M. Siddiqi, Anticlastogenic, antigenotoxic and apoptotic activity of epigallocatechin gallate, a green tea polyphenol. Mutat. Res. **523–524**, 33–41 (2003)
97. M. Roy, S. Chakraborty, D. Sinha, T. Kundu, R.K. Bhattacharya, Polyphenol and cancer. Sci. Cult. **70**(3–4), 136–141 (2004)
98. M. Roy, A. Mukherjee, S. Mukherjee, J. Biswas, Phytonutrients from fruits and vegetables in breast Cancer control, in *Anticancer Properties of Fruits and Vegetables: A Scientific Review*, ed. by D. A. B. Kunnumakkara, (World Scientific Publishing, Singapore, 2014), pp. 75–101, Chapter 3
99. M. Roy, A. Mukherjee, S. Mukherjee, J. Biswas, Phytonutrients from fruits and vegetables in breast cancer control. Ind. Perfumer **60**(3) (2016)
100. M. Roy, A. Mukherjee, S. Mukherjee, J. Biswas, Nutraceuticals in leukemia. J. Ayurvedic Herbal Med. **3**(1), 41–47 (2017)
101. M. Roy, M. Siddiqi, R.K. Bhattacharya, Cellular and molecular response to tea polyphenols characteristic of cancer chemoprevention. Asian Pacific J. Cancer Prev. **2**, 109–116 (2001)
102. F. Ruebsam, D. Murphy, C.V. Tran, L.-S. Li, J. Zhao, P. Dragovich, H. McGuire, A.X. Xiang, S. Zhongxiang, B.K. Ayida, K.J. Blazel, H.K. Sun, Y. Zhou, H. Qing, C.R. Kissinger, S.E. Webber, R.E. Showalter, A.M. Shah, M. Tsan, L. Kirkovsky, Discovery of tricyclic 5,6-dihydro-1H-pyridin-2-ones as novel, potent, and orally bioavailable inhibitors of HCV NS5B polymerase. Bioorg. Med. Chem. Lett. **19**, 6404–6412 (2009)
103. M. Saxena, J. Saxena, R. Nema, D. Singh, A. Gupta, Phytochemistry of medicinal plants. J. Pharmacogn. Phytochem. **1**(6), 168–182 (2013)

104. A. Senning, *Elsevier's Dictionary of Chemoetymology* (Elsevier, Amsterdam, 2006). ISBN 0-444-52239-5
105. G. Shen, T.O. Khor, R. Hu, S. Yu, S. Nair, C.T. Ho, B.S. Reddy, M.T. Huang, H.L. Newmark, A.N. Kong, Chemoprevention of familial adenomatous polyposis by natural dietary compounds sulforaphane and dibenzoylmethane alone and in combination in ApcMin/+ mouse. Cancer Res. **67**(20), 9937–9944 (2007)
106. G. Shen, C. Xu, R. Hu, M.R. Jain, S. Nair, W. Lin, C.S. Yang, J.Y. Chan, A.N. Kong, Comparison of (-)-epigallocatechin-3-gallate elicited liver and small intestine gene expression profiles between C57BL/6J mice and C57BL/6J/Nrf2 (−/−) mice. Pharm. Res. **22**(11), 1805–1820 (2005)
107. D. Sinha, M. Roy, Antagonistic role of tea against sodium arsenite induced oxidative DNA damage and inhibition of DNA repair in Swiss albino mice. J. Environ. Pathol. Toxicol. Oncol. **30**(4), 1–11 (2011)
108. D. Sinha, S. Roy, M. Roy, Antioxidant potential of tea reduces arsenite induced oxidative stress in Swiss albino mice. J. Food Chem. Toxicol. **48**(4), 1032–1039 (2010)
109. V.S. Sobolev, B.W. Horn, T.L. Potter, S.T. Deyrup, G.B. Gloer, Production of Stilbenoids and phenolic acids by the peanut plant at Early Stages of growth. J. Agric. Food Chem. **54**(10), 3505–3511 (2006)
110. S. Suzara, D.A. Costa, Y. Gariepyb, S.C.S. Rochaa, V. Raghavanb, Spilanthol extraction using microwave: Calibration curve for gas chromatography. Chem. Eng. Trans. **32**, 1783–1788 (2013)
111. R.J. Thoppil, A. Bishayee, Terpenoids as potential chemopreventive and therapeutic agents in liver cancer. World J. Hepatol. **3**(9), 228–249 (2011)
112. P. Tiwari, B. Kumar, M. Kaur, G. Kaur, H. Kaur, Phytochemical screening and extraction: A review. Internationale Pharmaceutica Sciencia **1**(1), 98–106 (2011). Jan–March 2011 | Vol. 1 | Issue 1 Available online http://www.ipharmsciencia.com ©2011 IPS]
113. S.N. Uddin, M.A. Akond, S. Mubassara, M.N. Yesmin, Antioxidant and antibacterial activities of Trema Cannabina. Middle East J. Sci. Res. **3**, 105–108 (2008)
114. R.J.S. Vega, N.C. Xolalpa, A.J.A. Castro, C.P. González, J.P. Ramos, S.P. Gutiérrez, Terpenes from natural products with potential anti-inflammatory activity, in *Terpenes and Terpenoids*, ed. by S. Perveen, A. A. Taweel, (Intech Open, 2018). https://doi.org/10.5772/intechopen.73215
115. K. Vit, K. Katerina, R. Zuzana, K. Kamil, J. Daniel, J. Ludek, O. Lubomir, Mini review: Condensed and hydrolysable tannins as antioxidants influencing the health. J. Med. Chem. **8**, 436–447 (2008)
116. A. Wadood, M. Ghufran, S. Jamal, M. Naeem, A. Khan, R. Ghaffar, Phytochemical analysis of medicinal plants occurring in local area of Mardan. Biochem Anal. Biochem. **2**(144), 2–4 (2013)
117. L.S. Wang, G. Zhou, B. Zhu, J. Wu, J.G. Wang, A.M. Abd El-Aty, T. Li, J. Liu, T.L. Yang, D. Wang, X.Y. Zhong, H.H. Zhou, St John's wort induces both cytochrome P450 3A4-catalyzed sulfoxidation and 2C19-dependent hydroxylation of omeprazole. Clin. Pharmacol. Ther. **75**(3), 191–197 (2004)
118. M. Wang, K. Li, Z. Zou, L. Li, L. Zhu, Q. Wang, W. Gao, Y. Wang, W. Huang, R. Liu, K. Yao, Q. Liu, Piperidine nitroxide Tempol enhances cisplatin-induced apoptosis in ovarian cancer cells. Oncol. Lett. **16**(4), 4847–4854 (2018)
119. X. Wang, S. Govind, S.P. Sajankila, L. Mi, R. Roy, F.L. Chung, Phenethyl isothiocyanate sensitizes human cervical cancer cells to apoptosis induced by cisplatin. Mol. Nutr. Food Res. **55**(10), 1572–1581 (2011)
120. C. Wiart, *Lead Compounds from Medicinal Plants for the Treatment of Cancer* (Elsevier, Boston, 2013)
121. R.J. Williams, J.P. Spencer, C. Rice-Evans, Flavonoids: Antioxidants or signaling molecules? Free Radic. Biol. Med. **36**(7), 838–849 (2004)
122. C.L. Wu, A.C. Huang, J.S. Yang, C.L. Liao, H.F. Lu, S.T. Chou, C.Y. Ma, T.C. Hsia, Y.C. Ko, J.G. Chung, Benzyl isothiocyanate (BITC) and phenethyl isothiocyanate (PEITC)-mediated

generation of reactive oxygen species causes cell cycle arrest and induces apoptosis via activation of caspase-3, mitochondria dysfunction and nitric oxide (NO) in human osteogenic sarcoma U-2 OS cells. J. Orthop. Res. **29**(8), 1199–1209 (2011)

123. H. Wu, X. Liang, Y. Fang, X. Qin, Y. Zhang, J. Liu, Resveratrol inhibits hypoxia-induced metastasis potential enhancement by restricting hypoxia-induced factor-1 alpha expression in colon carcinoma cells. Biomed. Pharmacother. **62**(9), 613–621 (2008)
124. D. Xiao, A.A. Powolny, M.B. Moura, E.E. Kelley, A. Bommareddy, S.H. Kim, E.R. Hahm, D. Normolle, B. Van Houten, S.V. Singh, Phenethyl isothiocyanate inhibits oxidative phosphorylation to trigger reactive oxygen species-mediated death of human prostate cancer cells. J. Biol. Chem. **285**(34), 26558–26569 (2010)
125. K. Xu, P. Liu, W. Wei, mTOR signaling in tumorigenesis. Biochim. Biophys. Acta **1846**(2), 638–654 (2014)
126. H. Yan, Y. Zhu, B. Liu, H. Wu, Y. Li, X. Wu, Q. Zhou, K. Xu, Mitogen-activated protein kinase mediates the apoptosis of highly metastatic human non-small cell lung cancer cells induced by isothiocyanates. Br. J. Nutr. **106**(12), 1779–1791 (2011)
127. N. Yanishlieva, M. Gordon, Antioxidants in food. J. Agric. Food Chem. **52**, 2391 (2001)
128. Y. Zeng, R. Cao, T. Zhang, S. Li, W. Zhong, Design and synthesis of piperidine derivatives as novel human heat shock protein 70 inhibitors for the treatment of drug-resistant tumors. Eur. J. Med. Chem. **97**, 19–31 (2015)
129. Y. Zhao, X. Hu, X. Zuo, M. Wang, Chemopreventive effects of some popular phytochemicals on human colon cancer: A review. Food Funct. **9**, 4548–4568 (2018)
130. L. Ziberna, M. Lunder, M. Spela, V. Andreja, T. Federica, P. Sabina, D. Gorazd, Acute cardioprotective and cardiotoxic effects of bilberry anthocyanins in ischemia–reperfusion injury: Beyond concentration-dependent antioxidant activity. Cardiovasc. Toxicol. **10**, 283–294 (2010)
131. C.J. Weng, C.F. Fu, H.W. Huang, C.T. Ho, G.C. Yen, Anti-invasion e_ects of 6-shogaol and 6-gingerol, two active components in ginger, on human hepatocarcinoma cells. Mol. Nutr. Food Res. **54**(11), 1618–1627 (2010 Nov). https://doi.org/10.1002/mnfr.201000108

Drugs and Phytochemicals

4

Abstract

Plant-derived substances, or phytochemicals, have been used for treating many diseases from ancient times. In this chapter we give a broad overview of the phytochemicals that are used as drugs, with a focus on the drugs that are used in chemotherapy. We start with a classification of anticancer drugs and their use in chemotherapy. We then discuss the mechanisms through which these drugs kill cancer cells. We then give an overview of how phytochemicals are used for treating many different diseases. Finally, we discuss the use of phytochemicals that have shown promise as anticancer drugs and agents.

4.1 Introduction

The aim of this chapter is to give an introduction to chemotherapeutic drugs and their classification, with an emphasis on plant-derived drugs from phytochemicals. An understanding of chemotherapeutic drugs will help in understanding how these drugs target cancer cells and the genetic pathways that are targeted by these drugs. Though the purpose of chemotherapy is to kill the neoplastic cells, these drugs also kill normal cells that are essential for correct functioning of healthy tissues. Killing of normal cells causes many unwanted side effects. One of the challenges in chemotherapy is the choice of appropriate drugs that can reduce these side effects. We will see in Chap. 5 that phytochemicals play important roles in chemosensitization, for increasing the efficacy of chemotherapeutic drugs. Chapter 6 will cover the details how phytochemicals can help in overcoming drug resistance in cancer and inhibit metastasis, the most advanced stage of cancer.

Several treatment strategies are followed for controlling and eliminating neoplastic growths or tumors. These treatments comprise surgery, where the tumor is removed by instrumental techniques, radiotherapy, a therapy involving ionizing radiation and chemotherapy, a treatment strategy employing one or more anticancer

M. Roy, A. Datta, *Cancer Genetics and Therapeutics*,
https://doi.org/10.1007/978-981-13-9471-3_4

drugs. The treatment regimen depends on several factors, like stage of cancer and origin and site of tumor among others. Our focus in this book is chemotherapy. We first discuss the overall classification of drugs used in chemotherapy. We will discuss plant-derived drugs in details later.

4.2 Classification of Chemotherapy Drugs

We have seen in Chap. 1 that an important hallmark of cancer is uncontrolled growth. We have also seen in Chap. 2 that this uncontrolled growth is achieved by cancer cells by manipulating the expression levels of key genes, which in turn results in the malfunctioning of the normal genetic pathways. Moreover cancer cells can signal other cells that in turn become neoplastic. It is natural that the goal of chemotherapy is to arrest this uncontrolled growth by killing cancer cells. This is achieved mainly through interventions in different phases of the cell cycle, from damaging the DNA to intervention in key pathways. Cancer drugs are classified according to their function, chemical structures, and how they are related to other drugs. We classify cancer drugs according to the classification adopted by the American Cancer Society (ACS) [10]. There are mainly six types of cancer drugs according to this classification, alkylating agents, antitumor antibiotics, antimetabolites, mitotic inhibitors, topoisomerase inhibitors, and corticosteroids.

4.2.1 Alkylating Agents

Alkylating agents include altretamine, busulfan, carboplatin, carmustine, chlorambucil, cisplatin, cyclophosphamide, dacarbazine, lomustine, melphalan, oxaliplatin, temozolomide, and thiotepa. Alkylating agents can intervene in all the phases of the cell cycle and mainly act through damaging DNA. They are used for treating several different cancers including multiple myeloma, Hodgkin's lymphoma, leukemia, and sarcoma.

4.2.2 Antimetabolites

Antimetabolites include 5-fluorouracil (5-FU), 6-mercaptopurine (6-MP), capecitabine, cytarabine, floxuridine, fludarabine, gemcitabine, hydroxyurea methotrexate, and pemetrexed. The antimetabolites have very similar structure as the bases used by cells to build DNA (adenine, guanine, cytosine, thymine) and RNA (adenine, guanine, cytosine, uracil). Antimetabolites can act as decoys for these bases, and as a result both DNA and RNA are damaged due to wrong molecules as components, resulting in cell cycle arrest. Antimetabolites are used for treating several cancers including cancer of the intestinal tract, breast and ovary, and leukemia.

4.2.3 Antitumor Antibiotics

Antitumor antibiotics are divided into two major groups, anthracyclines, including daunorubicin, doxorubicin, epirubicin, and idarubicin, and non-anthracyclines including actinomycin-D, bleomycin, mitomycin-C, and mitoxantrone. Their role is not to treat infection, like normal antibiotics. Their mode of action is to damage DNA so that the normal cell cycle is disrupted. One of the ways they damage DNA is through DNA intercalation, i.e., by inserting other molecules in between the bases of DNA. They are used as chemotherapeutic drugs in various cancers.

4.2.4 Topoisomerase Inhibitors

They are divided into two groups, topoisomerase I inhibitors, including topotecan and irinotecan, and topoisomerase II inhibitors, including etoposide, teniposide, and mitoxantrone. Topoisomerases I and II are two important enzymes that help in normal unwinding of the DNA double strand during transcription or cell division. Inhibiting these two enzymes may result in single- or double-strand DNA breaks, thus arresting cell cycle progression. These drugs are used for treating lung, gastrointestinal, ovarian cancers, as well as leukemia.

4.2.5 Mitotic Inhibitors

These chemotherapeutic drugs are derived from plants and include docetaxel, estramustine, ixabepilone, paclitaxel, vinblastine, vincristine, and vinorelbine. Their mode of action is to block cell division at the mitosis phase. They prevent microtubule functions to achieve this. Microtubules are important cellular structures that are required for cell division. Mitotic inhibitors prevent microtubule formation, thus disrupting mitosis, resulting in cell cycle arrest.

4.2.6 Corticosteroids

Corticosteroids are hormones and hormonelike drugs that include prednisone, methylprednisolone, and dexamethasone. They are used for treating many kinds of cancers.

We discuss the details of actions of these drugs and their use in chemotherapy below.

4.3 Mode of Action of Chemotherapeutic Drugs

Therapeutic drugs used in treatment of cancer employ the cytotoxic and cytostatic potential of the active compounds. Chemotherapy drugs may be classical which directly target the DNA or contemporary that target the aberrantly expressed proteins in the cancer cells [42]. An alkylating agent employed in cancer treatment attaches an alkyl group to the DNA, particularly to the guanine base of DNA, at position 7 of the purine ring. Alkylating agents are toxic to the cancer cells, but they are also toxic to the normal cells. Some of the alkylating agents conventionally used as chemotherapeutic drugs like doxorubicin, carboplatin, cisplatin, oxaliplatin, cyclophosphamide, melphalan, etc. bind covalently with DNA, cross-link DNA, and generate strand breaks.

Many chemotherapeutic drugs act via interference with the topoisomerases, enzymes that participate in the winding of DNA, over or under winding depending on the requirement. Topoisomerase inhibitors like doxorubicin and irinotecan act by disrupting DNA replication. Mitosis is a process of dividing a cell and its nucleus into two cells which each has their own nucleus. Cell division produces genetically identical cells, and the number of chromosomes in the daughter cell is the same as the mother cell.

Microtubules are one of the major components of the cytoskeleton that aid in structural support, intracellular transport, and DNA segregation. α- and β-tubulin polymerize into dynamic microtubules. Certain drugs bind with the tubulin, hampering the microtubule dynamics and hence the DNA segregation and cell division. Certain chemotherapeutic drugs like paclitaxel, docetaxel, vinblastine, vincristine, etc. work on tubulins.

Folic acid analogue methotrexate is an antimetabolite, which inhibits dihydrofolate reductase (DHFR) [27, 66]. Methotrexate and another antimetabolite 5-fluorouracil regulate DNA synthesis, and by doing so they kill cancer cells.

There are certain other drugs that target markers that are aberrantly expressed in cancer. Vascular endothelial growth factor (VEGF) receptor, a member of growth factors, is a signaling protein produced by cells that stimulates the formation of blood vessels. These signaling proteins are therefore implicated in both vasculogenesis and angiogenesis. A tumor mass needs sufficient blood supply for growth, and tumors with high level of VEGF can maintain their growth and can easily metastasize. Certain drugs like aflibercept, bevacizumab, ranibizumab, and pegaptanib may restrict the spread of the disease by inhibiting VEGF. All these drugs are used to treat macular degeneration as well. Bevacizumab is used to treat metastatic cancers [96]. Rituximab acts on CD20 in lymphoma, imatinib acts on Bcr/Abl, gefitinib targets epithelial growth factor receptor, and bortezomib is a proteasome inhibitor [53].

Certain cancers have some genetic abnormalities, e.g., in chronic myeloid leukemia or CML cells; the Philadelphia chromosome is responsible for the pathogenesis. It is a fusion gene formed by a reciprocal translocation between chromosome 9 and chromosome 22; this is indicated as t(9;22). ABL1, a tyrosine kinase, is responsible for cellular differentiation as well as regulation of the cell cycle in normal cells. In presence of fusion products, tyrosine kinase gets activated, leading to uncontrolled proliferation of cells. Therefore, a tyrosine kinase inhibitor is a drug of choice for treatment of CML, and ABL tyrosine kinase inhibitor imatinib is commonly used. Often the CML cells get resistant to imatinib, and it is necessary to use the second-generation tyrosine kinase inhibitors like dasatinib, nilotinib, and bosutinib as these are better than imatinib.

Imatinib is the first-generation ABL tyrosine kinase inhibitor. The use of CML includes use in newly diagnosed adult and pediatric patients and in patients after the failure of interferon-alpha therapy. Though treatment using imatinib is effective and lasting for many patients, 30–40% of patients may eventually need further treatment [74]. The most probable cause may be point mutations in the tyrosine kinase domain of the BCR-ABL fusion gene. These point mutations make the protein insensitive to inhibitory effects of imatinib.

This type of disease progression necessitated the development of the second-line tyrosine kinase inhibitors (TKI): dasatinib, nilotinib, and bosutinib. Dasatinib and bosutinib are also inhibitors of SRC. In epithelial cancers, epidermal growth factor receptor (EGFR), which is a member of the ErbB family of proteins, is a vital factor, overexpression of which supports tumor growth, invasion, and metastasis [56]. In normal cells, the EGFR ligand is regulated such that optimum cell proliferation occurs as is needed for homeostasis, whereas in malignancy it is overexpressed [36, 48]. Aberrant expression of EGFR is responsible for poor prognosis. Many anticancer drugs target EGFR for therapeutic action. EGFR-targeting drugs are gefitinib and erlotinib for non-small-cell lung cancer, cetuximab for head and neck cancer, erlotinib for pancreatic cancer, lapatinib for breast cancer, and cetuximab and panitumumab for metastatic colorectal cancer [50]. Of these the monoclonal antibodies cetuximab targets the extracellular ligand-binding domain. Gefitinib, erlotinib, lapatinib, and afatinib are small-molecule tyrosine kinase inhibitors which target intracellular domain [14].

The drugs mentioned above not only target and kill cancer cells; they affect the normal cells as well, resulting in side effects. Also, resistance to the drug creeps in which is a major problem of cancer treatment. Ideally, an anticancer drug should elicit its cytotoxicity preferentially to the cancer cells, not harming the normal ones. Plant-derived molecules may solve this problem.

4.4 Benefits of Plant-Derived Medicines

In ancient times Chinese developed their own way of treatment, known as traditional Chinese medicine, where a number of known plants were used. Gastrointestinal problem was treated with cinnamon, astragalus was used to relieve stress, and there are many more traditional medicines. Indian traditional medicine was also well known since ages. Way back plants were used as a whole; later on with the advancement of chemistry, the extracts were isolated and purified, and structures were identified. The first plant-derived medicine was morphine, which is still an important drug. Many other agents were isolated later, like quinine, atropine, caffeine, aspirin, etc., all these are discussed later. By the twentieth century, the active ingredients from plant sources were prescribed with proper dosage. Plant secondary metabolites, which are precursors of drugs, were chemically modified, so as to obtain pharmacologically active forms [23].

Galantamine, a phyto product isolated from *Galanthus woronowii* Losinsk. (Amaryllidaceae) [5], is used to treat Alzheimer's disease. This molecule inhibits acetylcholinesterase (AChE), modulates nicotinic acetylcholine receptor (nAChR), and thereby slows down neurological degeneration [33, 60].

All the plant parts like leaf, flower, stem, bark, root, fruit, and seeds may contain some active ingredients with medicinal value. These parts are rich in alkaloids, flavonoids, phenolics, and tannins [82]. These compounds derived from plant sources have minimal or no toxicity. Many of the phytochemicals possess enormous pharmacological properties. Many adverse health issues can be controlled with the aid of these plant molecules, and this had been practiced in many parts of the world since ages. Some of these are anti-inflammatory; some may be antitumor, or cardiotonic, or anticholinergic.

Cardiotonic agents have favorable effects on the heart. These are acetyldigoxin isolated from foxglove; adoniside from red chamomile; convallatoxin from *Convallaria majalis*, also known as lily-of-the-valley; digitalin, digitoxin, and digoxin from purple foxglove; gitalin from *Digitalis purpurea*; lanatosides A, B, and C from *Digitalis lanata*; ouabain from *Strophanthus gratus*; and scillaren A from Urginea maritima. Many of the phytochemicals show anticancer potential. Betulinic acid from *Betula alba* [61]; camptothecin and topotecan from *Camptotheca acuminata* [15]; colchicine and demecolcine made from *Colchicum autumnale*; etoposide, teniposide, and podophyllotoxin from *Podophyllum peltatum*; irinotecan from *Camptotheca acuminata*; lapachol from *Tabebuia* species; and vinblastine and vincristine from *Catharanthus roseus* are effective drugs due to their antitumor and anticancer activities. They will be discussed elaborately in subsequent sections.

We are constantly exposed to several exogenous factors which cause oxidative stress. Endogenous sources of oxidative stress are also common. These stress factors lead to injury and mutation, influence the intercellular signal transduction and transcription factors, and ultimately pave the way to carcinogenesis. Antioxidants interact with the harmful free radicals and neutralize them; thus they can prevent damage and the onset of carcinogenesis. Many phytochemicals are potential antioxidants which neutralize free radicals removing their power to create damage. Nordihydroguaiaretic acid derived from *Larrea divaricata* (creosote bush) is a good antioxidant. We will see in Chap. 5 that inflammation is a major cause of cancer [75]. Some phytochemicals that act as anti-inflammatory agents are aescin from *Aesculus hippocastanum*, bromelain from pineapple, and borneol, which is obtained from many plants. Some of these agents block the neurotransmitter acetylcholine both in the central and peripheral nervous systems; hence these are also called anticholinergic. Common plants that contain anticholinergic alkaloids are atropine from *Atropa belladonna*, anisodamine and anisodine from *Anisodus tanguticus*, and hyoscyamine and scopolamine from *Hyoscyamus niger*. Muscle relaxants are necessary for coping with muscle spasms, and there are a few of them from plant sources. These are anabesine from *Anabasis aphylla*, cissampeline from *Cissampelos pareira* or velvetleaf, papaverine from *Papaver somniferum* (poppy), and tubocurarine from *Chondrodendron tomentosum*.

Some of the drugs used to suppress or relief cough are known as antitussive; these are bergenin from *Ardisia japonica*; codeine, glaucine, and noscapine from poppy; and rorifone from *Rorippa indica*. Dysentery, an intestinal inflammation, primarily in the colon has some phytoremedy. Aesculetin from *Fraxinus rhynchophylla*, andrographolide and neoandrographolide from *Andrographis paniculata*, berberine from *Berberis vulgaris*, and hemsleyadin from *Hemsleya amabilis* may provide relief from dysentery. Agents that are capable of enhancing the secretion of bile from the liver are coined choleretic. Bile aids in transport of toxic substances out of human body; therefore more is the volume of bile, better is digestion. Plant-derived molecules act in this regard as well.

Curcumin, which is extracted from *Curcuma longa* (turmeric), cynarin from *Cynara scolymus* (artichoke), dandelion root, monoterpenes, a class of hydrocarbons found in the essential oils of many plants, are good choleretic agents. Tranquilizers are agents that can relieve stress. A number of plant-derived molecules act as good tranquilizers. Deserpidine isolated from *Rauvolfia canescens*, kavain from *Piper methysticum*, rescinnamine and reserpine from *Rauwolfia serpentina*, rhomitoxin from the vibrant flower *Rhododendron*, and rotundine from *Stephania sinica* are a few examples. Of these deserpidine, rescinnamine, reserpine, and rhomitoxin control hypertension as well. Rotundine is a good analgesic and sedative. Protoveratrines from *Veratrum album* and tetrandrine from *Stephania tetrandra* also are good antihypertensives. Redness of skin is caused due to dilation of the capillaries and increased blood circulation, and rubefacient agents are possible treatments. Black mustard, a brassica vegetable, contains allyl isothiocyanate, which is a rubefacient. Camphor from *Cinnamomum camphora*, menthol from mentha, and methyl salicylate from *Gaultheria procumbens* are other examples.

Helminth is a parasitic worm. Anthelmintics are agents that expel helminths without causing damage to the host. Arecoline from *Areca catechu*, and quisqualic acid from *Quisqualis indica* are examples of phytochemicals that can cure the parasitic infection. The neurodegenerative disorder Parkinson's disease affects the dopamine producing neurons. The phytochemical L-Dopa isolated from *Mucuna* species may be of help for this state of disease.

Amebiasis is a common problem caused by *Entamoeba histolytica* in tropical and subtropical countries. This is a major cause of morbidity and mortality particularly in developing countries. Metronidazole is commonly used for the treatment of amebiasis, which may show mutagenicity and cerebral toxicity [47]. Other phytoremedies for amebiasis include emetine from *Cephaelis ipecacuanha*, glaucarubin from *Simarouba glauca*, etc. Cinchona trees contain quinine, which is useful against malaria. Quinidine isolated from the same tree shows some antiarrhythmic potential.

Many other health issues have phytoremedies. Theobromine and theophylline are diuretics which may be isolated from *Theobroma cacao*. Sedatives like scopolamine and valepotriates are isolated from the *Datura* species and *Valeriana officinalis*, respectively. Aspirin is an anti-inflammatory pain killer, which was originally derived from the herbs meadowsweet and willow bark. Many other fruits and herbs that we consume also contain aspirin. These are apple, avocado, blueberry, eggfruit, broccoli, cauliflower, cucumber, grape, peach, plum, radish, spinach, and turmeric.

Atropine is used for the treatment of certain types of poisonings due to nerve agents and pesticides. It is normally administered intravenously or intramuscularly. Atropine is also administered prior to anesthesia to decrease mucus secretions, such as saliva. It is used for keeping heartbeat normal during surgery. Atropine has important use for the treatment of ophthalmologic problems like uveitis and early onset of amblyopia (lazy eye). Natural source of atropine is Solanaceae, a family of flowering plants which are annual as well as perennial herbs to vines, shrubs, and trees. Solanaceae family is also known as nightshade family that includes nightshade, Jimson weed, and mandrake. Members of this family contain potent alkaloids, some of which are quite toxic. Common sources of atropine are *Atropa belladonna* and *Datura innoxia*. Members of the *Brugmansia* and *Hyoscyamus* genera are also sources of atropine. However, there are certain side effects of atropine, including nausea, blurred vision, loss of balance, ventricular fibrillation, dizziness, dilated pupils, supraventricular or ventricular tachycardia, photophobia, dry mouth and confusion, deliriant hallucinations, etc.

Spindle fibers divide the genetic material in a cell into two equal halves. Therefore, a spindle poison hampers cell division by disturbing the protein threads connecting the centromere region of the chromosome. Molecules that interfere with spindle formation are therefore effective in case of tumors. Colchicine is a spindle poison used in the treatment of gout [18, 77], Mediterranean fever, Behçet's disease, and pericarditis. Some other spindle poisons are mebendazole, griseofulvin, vinca alkaloids, paclitaxel (Taxol), etc. Colchicine is a toxic alkaloid extracted from plants of the genus *Colchicum*. This alkaloid is contained in the corms of *Colchicum luteum* and the seeds of *Iphigenia*. Colchicine shows toxic effects upon ingestion, inhalation, and absorption in the eyes [2]. Clouding of the cornea and other systemic toxicities may result. Fluid loss through gastrointestinal tract may be life threatening.

Morphine is an age-old pain reliever, isolated from opium. This alkaloid shows its action on the central nervous system (CNS) and smooth muscles [54]. Morphine binds to the opioid receptors in the brain, spine, and other organs, thereby reducing the pain to a great extent and increasing the dopamine level. It is the principal alkaloid in opium, a narcotic and highly addictive, and it induces sleep. Morphine is the most abundant opiate extracted from the dried latex of opium, particularly from the unripe seed pods of the *Papaver somniferum* (poppy). There are a number of routes how morphine can be delivered, e.g., in liquid form, tablets, injectable, and continuous pump infusion. Morphine is popularly used in pain, fibromyalgia, breakthrough pain, kidney stone, and low back pain. It is good in relieving moderate to severe pain, usually after surgery or traumatic injury. Morphine too has several side effects. If taken in a large dose, it may lead to coma. Other side effects are nausea, diarrhea, cramps, anorexia, anxiety, restlessness, irritation, pupil dilation, tremors, tachycardia, etc.

The cardiac glycoside digitoxin is a phytosteroid, better known as a cardiotonic. Digitoxin is structurally similar to that of digoxin, first isolated from *Digitalis lanata*, a foxglove plant. Difference between digoxin and digitoxin is that the former is eliminated via the kidneys, whereas the latter gets eliminated via the liver. The advantage is that digitoxin can be used in patients with compromised kidney function. There are a number of side effects, including anorexia, nausea, vomiting, diarrhea, confused state, visual disturbances, and cardiac arrhythmias. Digitoxin has toxic effects similar to the more commonly used digoxin. Digitoxin toxicity can be treated with antidigoxin antibody fragments [43]. In vitro studies indicate anticancer activity of digitoxin [22, 49]. Anticancer potential of digitoxin has been improved by the discovery of digitoxigenin neoglycosides [44].

Pineapple is popular as a juicy, refreshing, and delicious fruit. Pineapple also has a plethora of health benefits. The active ingredient of pineapple is bromelain, which is present in almost all parts of the plant [3]. It contains a mixture of proteolytic enzymes. Bromelain is sometimes the cause of burning mouth, as the enzymes degrade the proteins of the tissues in the mouth. Bromelain is inflammatory, also used for the treatment of hay fever, ulcerative colitis, pulmonary edema, and debridement. It is also used as a muscle relaxant, stimulator of muscle contractions. It improves the bioavailability of antibiotics and helps in burning fat. It is also a chemopreventive agent. Bromelain is generally safe but may cause stomach irritation and diarrhea in some cases.

Since 1993, almost all paclitaxel produced has been derived from the bark of the Pacific yew [26]. Attempts were made to isolate Taxol from the needles of the tree, but the yield was not enough. This is a promising drug used for cancer treatment. It will be discussed in detail later.

A microorganism of *Plasmodium* group causes malaria, which is a mosquito-borne infection. There are several species of *Plasmodium* that can infect humans, of these *P. falciparum* is the most dangerous; the other forms, namely, *P. vivax*, *P. ovale*, and *P. malariae* also cause malaria but on a milder scale [11]. A naturally occurring alkaloid, quinine is a medication used to treat malaria [51]. Quinine is isolated from the bark of cinchona tree. Malaria is an age-old problem, and way back it was treated with the crude bark extract of this tree [1]. Quinine is very bitter in taste, and it has a number of side effects, including headache, tinnitus (ringing in the ears when there is no external sound), impaired vision, and sweating. Quinine treatment may lead to deafness, low platelet count, and arrhythmia. An individual may become prone to sunburn after quinine therapy.

Another example of phytochemical as a medicine is podophyllotoxin. This is a non-alkaloid toxin lignan, usually extracted from the rhizomes and roots of the *Podophyllum* species [103]. This molecule can stop the replication of both cellular and viral DNA, can halt cell division, and can destabilize microtubules. They can also be used as antiviral, anthelmintic, purgative, and antitumor agent. Antitumor agents etoposide and teniposide are derived from podophyllotoxin [18, 28]. Tissue surrounding the affected site may have irritation like burning, redness, pain, itching, and swelling [45]. Therefore, podophyllotoxin should be applied with care.

Phytomedicines also can control high blood pressure. An alkaloid derived from the roots of *Rauwolfia serpentina* (Indian snakeroot) and *Rauwolfia vomitoria* is reserpine. The Indian name is Sarpagandha, which was used for centuries to treat insanity, fever, and snakebites and also as a tranquilizer [99]. This lipid-soluble alkaloid is capable of penetrating the blood-brain barrier. It inhibits the uptake of norepinephrine into presynaptic storage vesicles by binding and inhibiting catecholamine pump in central and peripheral adrenergic neurons. As a result heart rate decreases and arterial blood pressure is lowered. Reserpine is therefore a drug for treating hypertension, in combination with vasodilators [93]. This agent was employed to control symptoms of dyskinesia (abnormal, uncontrollable, and involuntary movements) in people suffering from Huntington's disease [79]. Reserpine is not free from side effects; it may lead to nasal blockage, nausea, vomiting, anxiety, depression, gastric intolerance, weight gain, gastric ulceration, stomach cramps, and diarrhea. Use of this drug may cause hypotension and bradycardia.

Two other alkaloids from natural sources are vinblastine and vincristine. These are isolated from Madagascar periwinkle, *Catharanthus roseus* (formerly known as *Vinca rosea*). Initially *Vinca rosea* was used in the treatment of diabetes, wasp stings, and eye infections. They have been found to contain more than 70 alkaloids. Some of these alkaloids control blood sugar; some may act as homostatics. They prevent spindle formation and hence regulate cell cycle. Both vinblastine and vincristine bind to the tubulin protein, thereby preventing cell division.

Tubocurarine, a toxic alkaloid, was used as an arrow poison. This alkaloid in conjunction with anesthetics provides skeletal muscle relaxation; hence it is used in surgery or mechanical ventilation. It is not used anymore for clinical anesthesia because safer alternatives, such as cisatracurium and rocuronium, are available. It is also extracted from a plant source, *Chondrodendron tomentosum*, in the bark of the Menispermaceae plant. This toxic plant product found use on arrowheads for hunting animals, because of the inability of tubocurarine to cross mucous membrane and therefore the hunted animal could be consumed without any problem. This alkaloid influences acetylcholinesterase activity and acts as a competitive inhibitor in the nicotinic acetylcholine receptor. The nerve impulse therefore gets blocked [101]. There are certain problems in using this agent. Difficulty in breathing is one problem, even when there is no evidence of impaired oxygen supply. Tubocurarine leads to histamine release, and this effect is associated with bronchospasms, hypotension, and salivary secretions. All these indicate that this agent is not suitable for asthmatics or children [101]. Ganglion-blocking effect by use of tubocurarine [9] may result in hypotension; therefore this agent is good for those affected by myocardial ischemia.

Salicin is an alcoholic β-glucoside. This is generally found in the leaves and bark of willows and poplars. It is an analgesic, antipyretic, and anti-inflammatory agent. Salicin is the origin of aspirin [37]. Salicin is bitter in taste, comparable to that of quinine. Side effects of salicin are nausea, vomiting, rash, dizziness, and shortness of breath. At higher doses salicin may be toxic; it may damage the kidneys and may lead to ulceration of the stomach, diarrhea, bleeding, and digestive discomfort. It is better not to prescribe it to patients with gout, gastritis, asthma, diabetes, hemophilia, stomach ulcers, etc. Also, children, pregnant women, and lactating mothers should not be given salicin.

Milk thistle or *Silybum marianum* is the source of several important phytochemicals. This plant contains many flavoligands and the flavonoid taxifolin. Silymarin is a mixture of flavonolignans including isosilibinin, silibinin, silicristin, and silidianin. This agent is anti-hepatotoxic and can be used in the treatment of liver cirrhosis [73]. Free radicals are generated due to metabolism of toxic substances, causing damage to the cellular membranes and lipoperoxidation. Generation of these free radicals is inhibited by silymarin due to its antioxidant potential. Silymarin increases glutathione levels in the liver, which contributes to the antioxidant defense mechanism. It also increases protein synthesis in hepatocytes. This is used for treatment of liver and gallbladder disease, cirrhosis of the liver, liver toxicity, jaundice, pleurisy, etc. It is also used for diabetes and indigestion. Some side effects of this agent are abdominal pain, nausea, indigestion, diarrhea, rash, allergy, itching, and anorexia.

Another phytochemical that can control rheumatism is harpagoside, isolated from the plant *Harpagophytum procumbens* or devil's claw [64], *Scrophularia ningpoensis*, and *Scrophularia buergeriana* [92]. The active ingredient iridoid glycoside, responsible for the health attributes, has been used for the treatment of fever, diabetes, hypertension, and various blood-related diseases. It is a home remedy for rheumatism. It also shows anti-inflammatory activities [58].

Beta-sitosterol is a type of cholesterol-like substance or phytosterol which is derived from plants and other foods including fish like tuna. Saw palmetto, pumpkin seed, rice bran, soybeans, peanuts, and pecans contain sitosterol. Structurally beta-sitosterol is similar to cholesterol, but it is not as harmful as cholesterol. The LDL or the bad cholesterol is reduced by sitosterol. Symptoms of benign prostatic hyperplasia (BPH) can be greatly relieved by using sitosterol.

Hyoscyamine, a tropane alkaloid, is also a secondary metabolite found in the plant family Solanaceae, and this family includes henbane (*Hyoscyamus niger*), mandrake (*Mandragora officinarum*), tomato (*Solanum lycopersicum*), and nightshade (*Atropa belladonna*). This agent imparts relief to a number of disorders like diverticulitis, peptic ulcers, irritable bowel syndrome, pancreatitis, colic, and interstitial cystitis. It is very much effective in pain control and therefore used to ease neuropathic pain, chronic pain, etc. Hence it is an important part of palliative care. It works better in synergism with opioids. It has some side effects, including dryness of mouth and throat, weight gain, pain in the eye, restlessness, dizziness, arrhythmia, and flushing. Overdose may cause central nervous system symptoms including disorientation, hallucinations, euphoria, and temporary memory loss and finally lead to coma.

Ginkgolide is another phytomedicine, isolated from the root, bark, and leaves of the *Ginkgo biloba* tree. This tree is a native of China. Ginkgolide has anticancer, cardioprotective, neuroprotective, stress-alleviating, and memory-enhancing effects [39, 98]. The neuroprotective action is due to its antiplatelet, antiedemic, antioxidant, pro-hemorrheologic, and microcirculatory action roles [97]. The leaves of *Ginkgo biloba* show no toxic side effects when used in moderation [32]. Ginkgolide finds use in the treatment for cerebrovascular disease and helps to reduce migraine frequency [21, 85, 95].

4.5 Plant-Derived Anticancer Drugs

Treatment of cancer varies depending on several factors, e.g., type of cancer, the stage of cancer, and the overall health of the patient. Often a combinatorial treatment is prescribed, which is known as adjuvant therapy, where surgery is supplemented by chemotherapy or radiotherapy. Chemotherapy employs drugs to kill cancer cells. Generally, adjuvant therapy is given to ensure total destruction of cancer cells, so as to achieve good prognosis. Chemotherapeutic drugs aim at the rapidly dividing cells. Besides killing the cancer cells, these drugs also affect normal cells, causing adverse side effects. These side effects include myelosuppression (decreased production of blood cells), hair loss, cardiotoxicity, neurotoxicity, mucositis (inflammation of digestive tract lining), and immunosuppression. Chemotherapeutic drugs may be administered intravenously, orally, by intramuscular injection, in cycles of alternating treatment with rest periods in between.

The mechanisms by which phytochemicals display their cancer fighting potential are quite complex. Mostly they act via antioxidant mechanisms. They have the power to scavenge free radicals, and at the same time, they are capable of inducing antioxidant defense mechanisms. Many of these phytochemicals activate the Nrf2 pathway and thereby induce antioxidant enzymes and other enzymes that detoxify [40, 68]. Conversion of a procarcinogen to the active carcinogen form is prevented or checked by these phytochemicals [25]. Many of the phytochemicals described before show their anticancer activities in animal models. Nitrosamines are potential carcinogens, but reports reveal that if meat is marinated by rosemary and thyme, the serum levels of heterocyclic amines get reduced appreciably, showing that the phytochemicals are capable of reducing the carcinogenic potential [84].

Phytochemicals have a plethora of advantages. They are nontoxic and comparatively cheap and can be consumed orally, and their mechanisms of action are known for many of them. All these facts are in favor of exploring phytochemicals in development of drugs. Some of the anticancer drugs derived from plant sources, which had passed through clinical trial, are discussed below.

1. Epipodophyllotoxin: These are lignans, extracted from the root of American Mayapple plant or *Podophyllum peltatum*. They are inhibitors of topoisomerase II, which is important for DNA unwinding [88]. The semisynthetic derivatives, etoposide, teniposide, and etoposide phosphate, are used as drugs due to their cytotoxic attributes. These are spindle poisons.
   (i) Etoposide: These anticancer drugs form a ternary complex with DNA and topoisomerase II that restricts the ligation of DNA strands. As a consequence, DNA strands are broken. Brand names are Eposin, Etopophos, VePesid, VP-16, and Toposar.

Typical structure of etoposide is shown below.

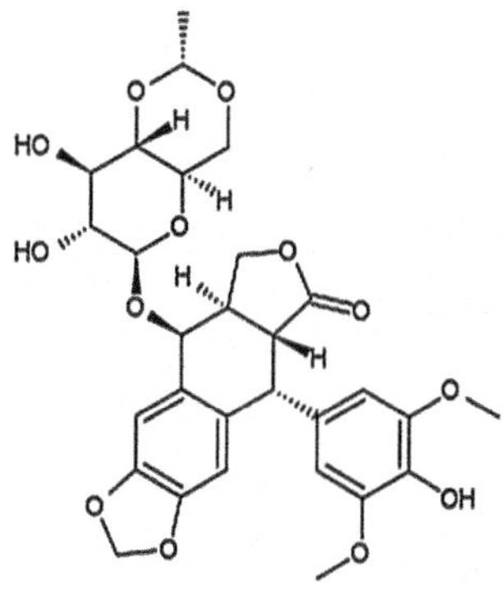

Etoposide is used alone or in conjunction with other medications for the treatment of testicular cancer and lung cancer (e.g., small-cell lung cancer). It is also used in Kaposi's sarcoma, lymphoma, nonlymphocytic leukemia, Ewing's sarcoma, and glioblastoma multiforme. During bone marrow transplant or blood stem cell transplant, this drug is used prior to the commencement of the treatment regimen.

There are a number of side effects, which include lowering of blood pressure. Dosage depends on the body size and medical conditions. Other adverse effects are nausea, vomiting, weakness, diarrhea, anorexia, changes in taste (metallic taste), hair loss, etc. Hair loss gets reversed after completion of the treatment. Mouth sore or sore throat also may occur. Some serious side effects are numbness/tingling of arms/legs, pain in the eye, stomach/abdominal pain, yellowing eyes/skin, sudden changes in vision, dark urine, painful/difficult swallowing, and redness/swelling of veins. Immunity is greatly impaired; therefore ability to fight infections is compromised. Bone marrow suppression may result, resulting in low white blood cell count. Fertility gets lowered in both the sexes.

(ii) Teniposide:

This is another class of podophyllotoxin derivatives which retards the proliferation of cancer cells. Trade names are Vumon, Vee M-26, Veham-Sandoz, and Vehem. This is used in the treatment of Hodgkin's lymphoma, childhood acute lymphoblastic leukemia (ALL), certain brain tumors, and other types of cancer. This drug induces break in DNA and prevents cells from entering the cell cycle; specifically it acts on G2 and S phase of cell cycle, resulting in death of cancer cells.

Structure of teniposide is shown below.

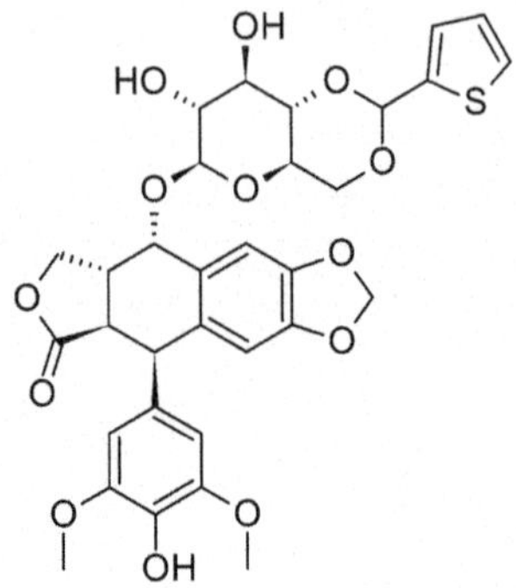

It is a second-line therapy of acute lymphoblastic leukemia (ALL) in combination with other anticancer drugs [29]. It is also used for the treatment of malignant lymphoma, reticulocyte sarcoma, Hodgkin's lymphoma, acute leukemia, primary brain tumors (astrocytoma, glioblastoma, ependymoma), bladder cancer, neuroblastoma, and other solid tumors in children [38]. The side effects include severe bone marrow suppression, gastrointestinal toxicity, stomach pain, diarrhea, anorexia, constipation, drowsiness, headache, dizziness, weakness, hypersensitivity reactions, and alopecia, which can be reversed after the treatment is completed. Mild skin rash is not uncommon. Nausea and vomiting may be severe.

2. Camptothecin:

Camptothecin abbreviated as CPT is an alkaloid, extracted from the bark and stem of *Camptotheca acuminata* (*Camptotheca*). In traditional Chinese medicine, this tree is used to treat cancer [57]. However, solubility of CPT is very poor; hence various synthetic derivatives have been developed. This is a topoisomerase I inhibitor. DNA cleavage is allowed, but ligation is inhibited; therefore DNA strands break. The use of camptothecin systemically is also limited by significant toxicity [83]. The S-phase specificity of this drug class provides some of the rationale for radiosensitization [102]. Derivatives of camptothecin have been synthesized, namely, topotecan, irinotecan, 9-aminocamptothecin, 9-nitrocamptothecin, and belotecan.

(i) Topotecan: This is a derivative of camptothecin, which is soluble in water. The brand names are Hycamptamine, Hycamptin, and Hycamtin. Topotecan is a semisynthetic derivative of camptothecin.

Topotecan's active lactone form intercalates between DNA bases in the cleavage complex of topoisomerase I. This intercalation traps the topoisomerase I in the cleavage complex that binds to DNA, and DNA damage occurs [62].

Structure of topotecan is shown below.

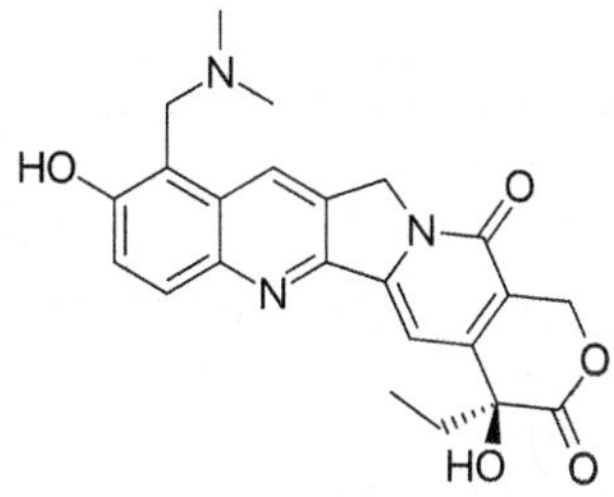

Topotecan is used for the treatment of metastatic cancer of ovary and small-cell lung cancer. This drug is also used in combination with cisplatin to treat cervical cancer, where surgery and radiation therapy are not possible. Side effects include myelosuppression, leukopenia, anemia, neutropenia, thrombocytopenia, nausea, vomiting, diarrhea, stomatitis, constipation, higher susceptibility to infections, and asthenia.

(ii) Irinotecan: This is a chemotherapy drug which shows its action by blocking the enzyme topoisomerase I. As the enzyme topoisomerase I is essential for proliferation, therefore inhibition of the enzyme ceases cancer cell proliferation. This is an alkaloid with a ring structure that is pentacyclic, containing a pyrrole (3, 4 β) quinoline moiety along with an S-configured lactone form and a carboxylate form [7]. Hydrolysis to SN-38 activates irinotecan. SN-38, the antineoplastic drug, is the active metabolite of irinotecan. The inhibition of topoisomerase I by the active metabolite SN-38 inhibits both DNA replication and transcription [97]. To improve the solubility of this drug, tertiary amine cations are employed. This drug is sold in the name Camptosar.

Structure of irinotecan is shown below.

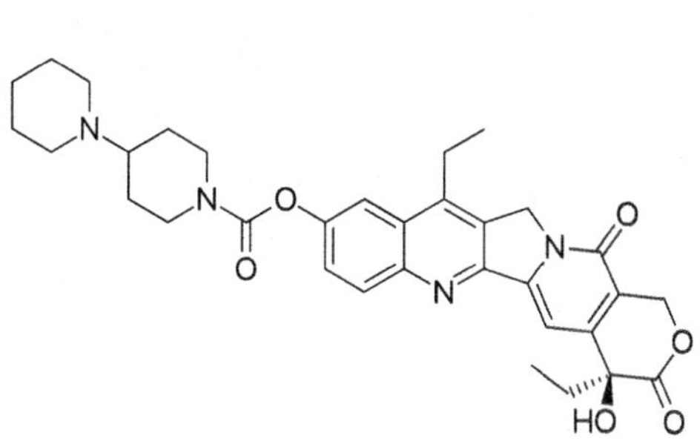

Two major side effects of irinotecan are diarrhea and immunosuppression. Diarrhea could be severe, which causes dehydration and often needs hospitalization. Due to immune suppression, white blood cell count drops and neutropenia results. Other side effects include nausea, vomiting, chest pain, dry cough, wheezing, shortness of breath, redness and numbness of hands and feet, sweating, hot and dry skin, flu-like symptoms, mouth sore, pallor, and hair loss.

3. Taxanes: Taxanes, also known as Taxol, were derived from the bark of Pacific yew tree. These are a class of diterpenes, featuring a taxadiene core. They elicit their action by disruption of microtubule function, which is essential for cell division. They inhibit cell division by stabilizing GDP-bound tubulin and inhibiting the process of cell division by preventing depolymerization. They impart stability to the microtubules and prevent separation of chromosomes during anaphase. Therefore, these groups of compounds are mitotic inhibitors and prevent the progression through the M phase of the cell cycle. These are radiosensitizers. Their solubility in water is poor. Subsequently semisynthetic derivatives have been introduced and proven to be very important anticancer drugs.

Paclitaxel (Taxol) and docetaxel (Taxotere) are widely used as chemotherapy drugs. Cabazitaxel is also FDA approved and employed to treat hormone-refractory prostate cancer.

(i) Paclitaxel: Paclitaxel is derived from extracts of the Pacific yew tree, *Taxus brevifolia* [90]. It disrupts the mitotic spindle dynamics. Its chemical structure is a diterpene having a taxane ring with a four-membered oxetane ring. There is an ester side chain at position C-13 [35].

The structure of paclitaxel is shown below.

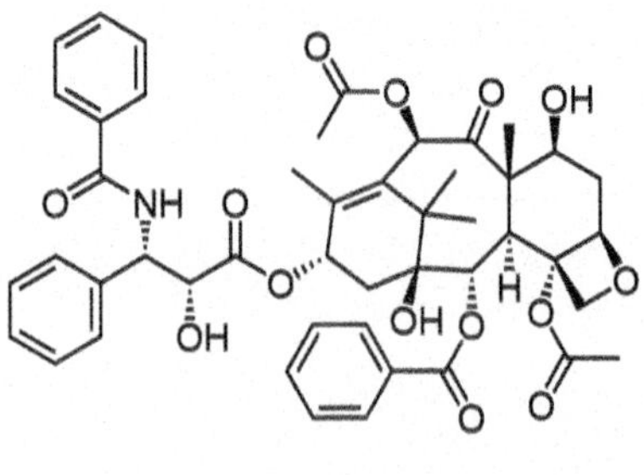

This drug binds to the microtubule polymer, which is unique among chemotherapeutic agents. Paclitaxel prevents cells from progressing from the G2/M phase of the cell cycle as cells are unable to form the normal mitotic apparatus.

Paclitaxel is used in the treatment of ovarian, lung, head and neck, esophageal, breast, prostate, and bladder cancers. It is also used in the treatment of Kaposi's sarcoma.

Side effects of paclitaxel depend on the dose and treatment schedule. Adverse effects of paclitaxel include low blood count, alopecia, myalgia, nail destruction, transient bradycardia and mild elevation of liver enzymes, neuropathy, nausea, vomiting, diarrhea, sore in the mouth, fever, flushing, breathing problem, swelling of feet, low blood pressure, darkening of the skin, and many more. Often resistance to this drug develops.

(ii) Docetaxel: This is another chemotherapeutic drug belonging to the Taxol group, which is extracted from European yew tree *Taxus baccata*. This is more water soluble than paclitaxel. Side effects are similar to that of paclitaxel, but hypersensitivity is less, and chances of neuropathy are also less. Other severe side effects include allergic reactions and chances of future recurrence of cancer. Side effects are common in patients with liver problems.

The structure of docetaxel is shown below.

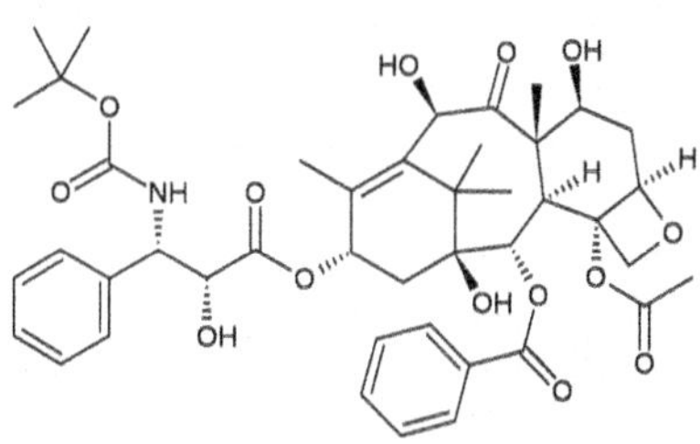

Docetaxel is used for the treatment of head and neck, gastric, breast, lung, ovarian, colorectal, liver, renal, melanoma, and bladder cancers [46]. Docetaxel improves life expectancy and overall quality of life in hormone-refractory prostate cancer [78].

4. Vinca Alkaloids: These are extracted from the periwinkle plant *Catharanthus roseus* (basionym *Vinca rosea*) and other vinca plants. These are a set of antimitotic and anti-microtubule alkaloid agents. They block beta-tubulin polymerization, which is an integral part of cell division. Vinka alkaloids are now synthetically made. Some of the vinka alkaloids are vincristine (VCR), vinorelbine (VRLB), vindesine (VDS), vincaminol, vineridine, and vinburnine. The semisynthetic alkaloid vinorelbine is used to treat non-small-cell lung cancer.
   (i) Vinblastine: Vinblastine, is sold under the brand name Velban. It is a microtubule disruptor, which at a low concentration suppresses microtubule dynamics and at a higher concentration reduces microtubule polymer mass.

Microtubule fragments are produced and enhanced microtubule detachment from spindle poles results in cytotoxicity.

The structure of vinblastine is shown below.

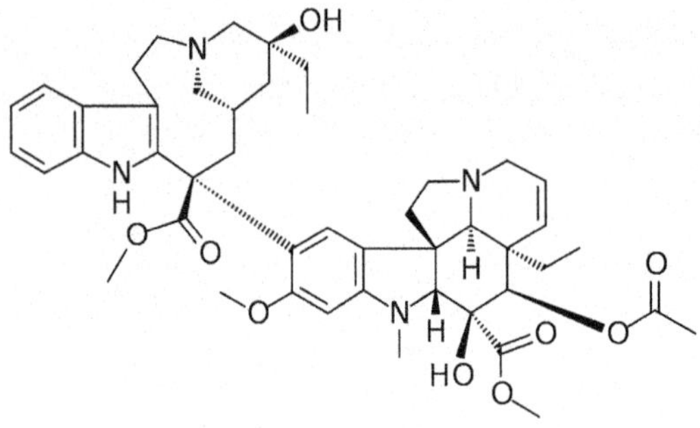

This drug is used to treat Hodgkin's lymphoma, a cancer of a specific type of white blood cells called lymphocytes. This drug is also recommended for non-Hodgkin's lymphoma, mycosis fungoides, testicular cancer, breast cancer, and Kaposi's sarcoma related to acquired immunodeficiency syndrome (AIDS). Vinblastine is used for the treatment of head and neck cancer, cervical cancer, non-small-cell lung cancer, bladder cancer, idiopathic thrombocytopenia purpura, and autoimmune hemolytic anemia. The associated side effects of vinblastine are alopecia, gastrointestinal problems, hypertension, sweating, depression, loss of white blood cells and blood platelets, muscle cramps, vertigo, and headache.

(ii) Vincristine: This drug is available in the market in the name Oncovin. It binds to the tubulin proteins, preventing polymerization of tubulin dimers to form microtubules. As a consequence, chromosomes cannot be separated during metaphase. This results in apoptosis or cell suicide. The vincristine molecule inhibits leukocyte production and maturation. Besides showing its action on cancer cells, vincristine affects all the rapidly dividing cells.

The structure of vincristine is shown below.

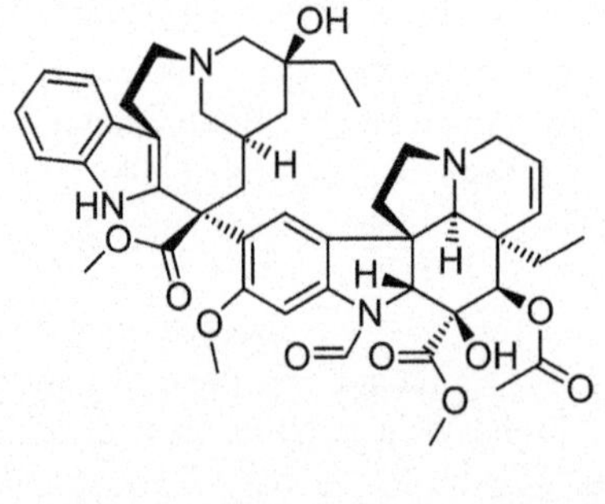

Vincristine is used in non-Hodgkin's lymphoma. This drug is a part of the regimen CHOP (cyclophosphamide, hydroxydaunorubicin, oncovin, prednisone). It is also used to treat neuroblastoma, acute lymphoblastic leukemia, and childhood leukemia; it is used in combination with prednisone. Side effects of vincristine include chemotherapy-induced peripheral neuropathy, hyponatremia, constipation, and alopecia.

(iii) Vinorelbine: This drug also belongs to the vinca alkaloid family. The brand name is Navelbine. It is a chemotherapy drug which is used to treat a number of cancers including non-small-cell lung cancer and breast cancer, which have relapsed. This drug also works on rhabdomyosarcoma, which is an aggressive and malignant form of cancer of skeletal muscle cells that have not fully differentiated. The action of this drug is based on the inhibition of mitosis through interaction with tubulin, thereby inhibiting cell division. It has a few side effects like suppression of the bone marrow, nausea, vomiting, tiredness, numbness, diarrhea, shortness of breath, etc. Other side effects are intense pain, hypersensitivity to cold, poor resistance to infection, bruising or bleeding, anemia, constipation, weakness (asthenia), and phlebitis.

The structure of vinorelbine is shown below.

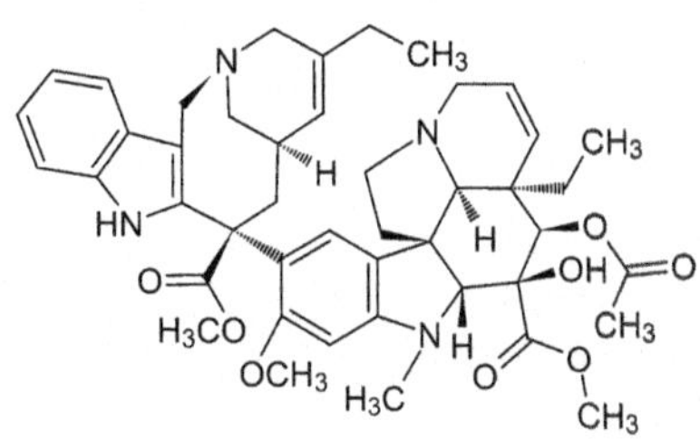

5. Cephalotaxine: It is a benzazepine alkaloid isolated from *Cephalotaxus harringtonii* and *Cephalotaxus fortunei*. Homoharringtonine, also known as omacetaxine mepesuccinate, is a natural product isolated from *Cephalotaxus harringtonii*. Omacetaxine or homoharringtonine is used for treatment of chronic myeloid leukemia, particularly for patients who are resistant to tyrosine kinase inhibitors. A small number of patients having myelodysplastic syndrome and acute myelogenous leukemia have benefited by using this drug. This drug has no effect on solid tumors. The mode of action is by inhibiting protein translation through the prevention of the initial elongation step of protein synthesis and inducing apoptosis.

Structure of cephalotaxine is shown below.

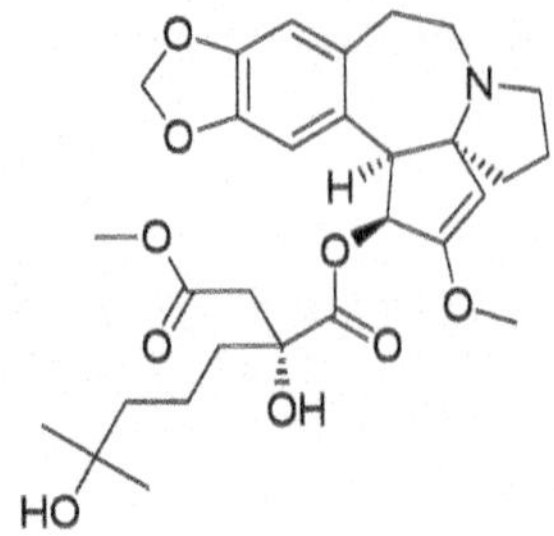

This drug has a number of side effects, which include diarrhea, myelosuppression, nausea, vomiting, fatigue, fever, weakness in muscle, headache, joint pain, cough, alopecia, constipation, edema, rash, back pain, etc.

Therefore, it is to be noted that both taxanes and vinca alkaloids show their action as spindle poison, but the modes of action are different.

4.6 Other Phytochemicals as Anticancer Agents

Chemotherapeutic drugs often show severe side effects. These adverse effects of therapeutics may be curbed by the use of phytochemicals. Phytochemicals that we consume show anticancer properties [91]. Mechanism of action of these plant molecules is yet to be fully understood. Their antioxidant properties render them as possible anticancer agents. They are capable of scavenging free radicals directly or may induce a number of antioxidant enzymes [40, 68]. These plant biomolecules activate Nrf2, which in turn triggers activation of a number of antioxidant enzymes. Inflammation is intricately related to carcinogenesis. Phytochemicals are capable of reducing inflammation. By doing so, they regulate the promotion and progression steps of carcinogenesis. Association between oxidative stress and activation of NF-κB with inflammation is known. Pro-inflammatory genes, which are also oncogenes, are regulated by these factors. A large number of plant-derived molecules have been found to inhibit NF-κB.

Thiol containing phytochemicals prevent the conversion of pro-carcinogens to their active forms [40]. Curcumin has been found to inhibit the carcinogenesis caused by triclocarban in in vitro model. Kaempferol enhances antioxidant enzyme activities in humans [41]. In an in vivo model, indole-3-carbinol had been found to decrease the chance of lung carcinogenesis [52]. Similar effect of indole-3-carbinol was observed in a clinical study in the urine marker of DNA damage among the smokers [31]. Onions are rich in quercetin. Increased intake of quercetin has been found to diminish the level of oxidative metabolites [34]. Quercetin also aids in improvement of mitochondrial dysfunctions due to exposure to 3-nitropropionic acid [63, 72]. Heterocyclic amines are carcinogenic to humans. Rosemary and thyme, during marination of meat, reduce this carcinogenicity [84].

It has become evident that phytochemicals have a great impact on carcinogenesis via amelioration of several steps in signaling, associated with growth, invasion, and metastasis [94]. The fruit pomegranate contains an abundance of the polyphenol ellagic acid. It has been found to inhibit cell growth and induce programmed cell death in human prostate cancer cells which are sensitive to androgen [76]. Extract from pomegranate inhibits metastasis in an in vitro study involving hormone (estrogen)-sensitive and hormone-resistant breast cancer cell lines. Several markers involved in cell adhesion and cell migration are affected by this molecule in cancer cells but not in normal cells [87]. This active molecule has been found to inhibit a chemokine which attracts breast cancer cells to the bone, thereby preventing migration and metastasis [70]. Curcumin retards growth of cancer cell growth by cell cycle inhibition, besides it induces apoptosis and inhibits invasion and migration [20, 65, 67, 100]. Stem cells that result in breast carcinoma are negatively regulated by curcumin; and it is worth mentioning that curcumin acts preferentially on the cancer stem cells, without causing any harm to the normal breast stem cells [89]. Curcumin results in diminished expression of Bcl-2 by modulating miRNA expression in breast cancer cells [7]. Stabilization of tumor suppressor genes is facilitated by curcumin in colorectal cancer cell lines [71]. EGCG from green tea inhibits DNA synthesis, angiogenesis, and dedifferentiation and thereby inhibits cancer cell proliferation [22, 81, 104]. Ornithine decarboxylase is an enzyme that guides cells to proliferate fast and evade apoptosis. EGCG also blocks this enzyme and acts as an anticancer agent [5, 86].

Epigenetics plays a crucial role in carcinogenesis. Resveratrol is a good regulator of epigenetics and therefore controls proliferation of cell, cell survival, and apoptosis in prostate cancer. It does so by modulation of gene expression through deacetylation of the FOXO transcription factor [25]. Caffeic acid and phenethyl ester decrease cell motility and therefore inhibit metastasis via negative modulation of NF-κB signaling [31]. Epithelial-to-mesenchymal transition (EMT) plays a pivotal role in metastasis. Luteolin has been found to inhibit EMT, tumor growth, and metastasis. This is how this compound can impede cancer initiation and development [31].

Some of the phytochemicals can impact carcinogenesis through a hormonal mechanism. Isoflavones and lignans which are found in soy products are phytoestrogenic. Legumes and some cruciferous vegetables are also phytoestrogenic compounds. They bind to the estrogen receptor, though the binding is weak. By such binding they do not stimulate proliferation of cells but can block the binding of endogenously produced harmful estrogens [59, 69]. A study named "Shanghai Breast Cancer Survival Study" was conducted, where it was observed that women consuming high amount of isoflavone- and flavanone-rich foods had less risk of death [80]. Some phytoestrogenic molecules have been reported to affect 5-alpha reductase, thereby lowering endogenous testosterone levels. Probably due to this reason, risk of prostate cancer is low in a population where intake of beans and pulses is higher [55].

Some other phytochemicals have shown promise as anticancer compounds in in vitro studies and animal models. Biswas et al. [8] have mentioned a comprehensive list of the plants that have such beneficial phytochemicals. These plants include *Acacia pennatula, Aglaia sylvestre, Allium sativum, Amorphophallus campanulatus, Andrographis paniculata, Annona atemoya, Baliospermum montanum, Barleria prionitis, Basella rubra, Bleekeria vitensis, Calotropis gigantea, Colubrina macrocarpa, Combretum caffrum, Commiphora opobalsamum, Curcuma domestica, Datura metel, Ficus benghalensis, Flacourtia ramontchi, Hemiangium excelsum, Hygrophila spinosa, Juniperus indica, Madhuca indica, Moringa oleifera, Nigella sativa, Oroxylum indicum, Pandanus odoratissimum, Paris polyphylla, Penicillium notatum, Pistacia lentiscus, Phyllanthus niruri, Picrorhiza kurroa, Piper longum, Podophyllum hexandrum, Prosopis cineraria, Pterospermum acerifolium, Raphanus sativus, Rubia cordifolia, Salvia officinalis, Semecarpus anacardium, Tabebuia impetiginosa, Teucrium polium, Tinospora cordifolia, Vitis vinifera*, and *Ziziphus mauritiana* [8].

Diospyrin, isolated from *Diospyros montana*, induces programmed cell death in different cancer cells via activation of caspases [12]. Herbal drugs are known for their therapeutic efficacy. Some examples are *Tiliacora racemosa* and oil of *Semecarpus anacardium* nut. They have been tested in various cancer cells for apoptosis-inducing properties [13]. Some other drugs derived from plants, e.g., colchicinamide, curcumol, curdione, d-tetrandrine, gossypol, 10-hydroxycamptothecin, indirubin, lycobetaine, and monocrotaline, may have some medicinal attributes, but in order to establish these compounds in treatment regimen, extensive research is needed [16].

Many plant-derived molecules show their potential on different cancers. These phytochemicals along with their sources and types of cancer models where they are effective and some of their probable molecular targets are tabulated in Appendix II.

4.7 Conclusion

Phytochemicals have been found to be effective for their anticancer properties in various models. These results, some in vivo and some in vitro, may not be translated in human population without proper clinical trials. But, preliminary results give an indication that these biomolecules may be ideal candidates for anticancer drug development. We have discussed the role phytochemicals play in treating many diseases including cancer. We have seen the chemical structures and classification of phytochemicals in Chap. 3. These two chapters together present function of many phytochemicals in treating many diseases, including cancer. We have presented the classification of chemotherapy drugs and in particular discussed in details the plant-derived chemotherapy drugs. Though our focus is treatment of cancer, it is important to understand how widespread is the use of phytochemicals for treating other diseases. Hence, we have reviewed the roles phytochemicals play as medicines in a broad range of diseases.

References

1. J. Achan, A.O. Talisuna, A. Erhart, A. Yeka, J.K. Tibenderana, F.N. Baliraine, P.J. Rosenthal, U. D'Alessandro, Quinine, an old anti-malarial drug in a modern world: Role in the treatment of malaria. Malar. J. **10**(1), 1–12 (2011)
2. Y. Alster, D. Varssano, A. Loewenstein, M. Lazar, Delay of corneal wound healing in patients treated with colchicine. Ophthalmology **104**(1), 118–119 (1997)
3. Z.I. Arshad, A. Amid, F. Yusof, I. Jaswir, K. Ahmad, S.P. Loke, Bromelain: An overview of industrial application and purification strategies. Appl. Microbiol. Biotechnol. **98**(17), 7283–7297 (2014)
4. U. Bachrach, Y.C. Wang, Cancer therapy and prevention by green tea: Role of ornithine decarboxylase. Amino Acids **22**(1), 1–13 (2002)
5. M.J. Balunas, A.D. Kinghorn, Drug discovery from medicinal plants. Life Sci. **78**, 431–441 (2005)
6. U. Banik, S. Parasuraman, A.K. Adhikary, N.H. Othman, Curcumin: The spicy modulator of breast carcinogenesis. J. Exp. Clin. Cancer Res. **36**, 98 (2017)
7. S. Basu, M. Zeng, T. Yin, S. Gao, M. Hu, Development and validation of an UPLC-MS/MS method for the quantification of irinotecan, SN-38 and SN-38 glucuronide in plasma, urine, feces, liver and kidney: Application to a pharmacokinetic study of irinotecan in rats. J. Chromatogr. B Anal. Technol. Biomed. Life Sci. **1015–1016**, 34–41 (2016)
8. J. Biswas, M. Roy, A. Mukherjee, Anticancer drug development based on phytochemicals. J. Drug Discov. Deve. Deliv. **2**(1), 1012–1017 (2015)
9. W.C. Bowman, S.N. Webb, Neuromuscular blocking and ganglion blocking activities of some acetylcholine antagonists in the cat. J. Pharm. Pharmacol. **24**(10), 762–772 (1972)
10. Cancer Treatment & Survivorship Facts & Figures 2016-2017. (American Cancer Society, Atlanta, 2016)
11. H. Caraballo, Emergency department management of mosquito-borne illness: Malaria, dengue, and west nile virus. Emerg. Med. Pract. **16**(5), 1–23 (2014). Archived from the original on 2016-08-01. "Malaria Fact sheet N°94". WHO. March 2014. Archived from the original on 3 September 2014. Retrieved 28 August 2014
12. S. Chakraborty, M. Roy, B. Hazra, R.K. Bhattacharya, Induction of apoptosis in human cancer cell lines by diospyrin, a plant derived bisnaphthoquinonoid and its synthetic derivatives. Cancer Lett. **188**, 85–93 (2002)
13. S. Chakraborty, M. Roy, A.K. Taraphdar, R.K. Bhattacharya, Cytotoxic effect root extract of Tiliacora racemosa and oil of Semecarpus anacardium nut in human tumor cells. Phytother. Res. **18**(8), 595–600 (2004)
14. K. Chanprapaph, V. Vachiramon, P. Rattanakaemakorn, Epidermal growth factor receptor inhibitors: A review of cutaneous adverse events and management. Dermatol. Res. Pract. **2014**, 734249 (2014)
15. G.M. Cragg, D.J. Newman, A tale of two tumor targets: topoisomerase I and tubulin. The Wall and Wani contribution to cancer chemotherapy. J. Nat. Prod. **67**(2), 232–244 (2004)
16. G. Cravotto, L. Boffa, L. Genzini, D. Garella, Phytotherapeutics: An evaluation of the potential of 1000 plants. J. Clin. Pharm. Ther. **35**(1), 11–48 (2010)
17. N. Dalbeth, T.J. Lauterio, H.R. Wolfe, Mechanism of action of colchicine in the treatment of gout. Clin. Ther. **36**(10), 1465–1479 (2014)
18. Y. Damayanthi, J.W. Lown, Podophyllotoxins: Current status and recent developments. Curr. Med. Chem. **5**(3), 205–252 (1998)
19. Y. DENG, E. VERRON, R. ROHANIZADEH, Molecular mechanisms of anti-metastatic activity of curcumin. Anticancer Res. **36**(11), 5639–5647 (2016)
20. P.M. Dewick, *Medicinal Natural Products: A Biosynthetic Approach*, 3rd edn. (Wiley, Chichester, 2012), pp. 230–232
21. Q.P. Dou, Molecular mechanisms of green tea polyphenols. Nutr. Cancer **61**(6), 827–835 (2009)

22. H.A. Elbaz, T.A. Stueckle, W. Tse, Y. Rojanasakul, C.Z. Dinu, Digitoxin and its analogs as novel cancer therapeutics. Exp. Hematol. Oncol. **1**(1), 4 (2012)
23. M. Frarag, M. Salih Mohammed, I. Foud, W. Ahmed, M. Malik, The role of natural products in drug discovery and development. World J. Pharm. Res. **4**(3) (2015)
24. G.F.S. Fernandes, G.D.B. Silva, A.R. Pavan, D.E. Chiba, C.M. Chin, J.L.D. Santos, Epigenetic regulatory mechanisms induced by resveratrol. Nutrients **9**(11), 1201 (2017)
25. A.V. Gasper, A. Al-Janobi, J.A. Smith, J.R. Bacon, P. Fortun, C. Atherton, M.A. Taylor, C.J. Hawkey, D.A. Barrett, R.F. Mithen, Glutathione S-transferase M1 polymorphism and metabolism of sulforaphane from standard and high-glucosinolate broccoli. Am. J. Clin. Nutr. **82**, 1283–1291 (2005)
26. H. Gersmann, J. Aldred, Medicinal tree used in chemotherapy drug faces extinction. The Guardian. Archived from the original on 16 February 2017 (2011)
27. D.S. Goodsell, The molecular perspective: Methotrexate. Oncologist **4**(4), 340–341 (1999)
28. M. Gordaliza, P.A. García, J.M. del Corral, M.A. Castro, M.A. Gómez-Zurita, Podophyllotoxin: Distribution, sources, applications and new cytotoxic derivatives. Toxicon **44**(4), 441–459 (2004)
29. J.L. Grem, D.F. Hoth, B. Leyland-Jones, S.A. King, R.S. Ungerleider, R.E. Wittes, Teniposide in the treatment of leukemia: A case study of conflicting priorities in the development of drugs for fatal diseases. J. Clin. Oncol. **6**(2), 351–379 (1988)
30. S.C. Gupta, J.H. Kim, S. Prasad, B.B. Aggarwal, Regulation of survival, proliferation, invasion, angiogenesis, and metastasis of tumor cells through modulation of inflammatory pathways by nutraceuticals. Cancer Metastasis Rev. **29**(3), 405–434 (2010)
31. S.S. Hecht, S.G. Carmella, P.M. Kenney, S.H. Low, K. Arakawa, M.C. Yu, Effects of cruciferous vegetable consumption on urinary metabolites of the tobacco-specific lung carcinogen 4-(methylnitrosamino)-1-(3-pyridyl)-1-butanone in singapore Chinese. Cancer Epidemiol. Biomark. Prev. **13**(6), 997–1004 (2004)
32. T. Heinonen, W. Gaus, Cross matching observations on toxicological and clinical data for the assessment of tolerability and safety of Ginkgos biloba leaf extract. Toxicology **327C**(1), 95–115 (2015)
33. M. Heinrich, H.L. Teoh, Galanthamine from snowdrop—the development of a modern drug against Alzheimer's disease from local Caucasian knowledge. J. Ethnopharmacol. **92**(2–3), 147–162 (2004)
34. T.M. Henagan, W.T. Cefalu, D.M. Ribnicky, R.C. Noland, K. Dunville, W.W. Campbell, L.K. Stewart, L.A. Forney, T.W. Gettys, J.S. Chang, C.D. Morrison, In vivo effects of dietary quercetin and quercetin-rich red onion extract on skeletal muscle mitochondria, metabolism, and insulin sensitivity. Genes Nutr. **10**(1), 2 (2015)
35. S.B. Horwitz, Taxol (paclitaxel): Mechanisms of action. Ann. Oncol. **5**(Suppl 6), S3–S6 (1994)
36. N.E. Hynes, H.A. Lane, ERBB receptors and cancer: The complexity of targeted inhibitors. Nat. Rev. Cancer **5**(5), 341–354 (2005)
37. M. Islam, C.M.D. Rafael, Anticancer prospection of salicin, a historical origin of aspirin. Int. J. Med. **5**, 264–265 (2017)
38. W. Jasek (ed.), *Austria-Codex (in German)*, 62nd edn. (Österreichischer Apothekerverlag, Vienna, 2007), pp. 8855–8856
39. G. Jin, X.R. He, L.P. Chen, The protective effect of ginko bilboa leaves injection on the brain dopamine in the rat model of cerebral ischemia/reperfusion injury. Afr. Health Sci. **14**(3), 725–728 (2014)
40. I. Johnson, Phytochemicals and cancer. Proc. Nutr. Soc. **66**, 207–215 (2007)
41. H.Y. Kim, O.H. Kim, M.K. Sung, Effects of phenol-depleted and phenol-rich diets on blood markers of oxidative stress, and urinary excretion of quercetin and kaempferol in healthy volunteers. J. Am. Coll. Nutr. **22**(3), 217–223 (2003)
42. B. Kumar, S. Singh, I. Skvortsova, V. Kumar, Promising targets in anti-cancer drug development: Recent updates. Curr. Med. Chem. **24**(42), 4729–4752 (2017)

43. V. Kurowski, H. Iven, H. Djonlagic, Treatment of a patient with severe digitoxin intoxication by Fab fragments of anti-digitalis antibodies. Intensive Care Med. **18**(7), 439–442 (1992)
44. J.M. Langenhan, N.R. Peters, I.A. Guzei, F.M. Hoffmann, J.S. Thorson, Enhancing the anticancer properties of cardiac glycosides by neoglycorandomization. Proc. Natl. Acad. Sci. U. S. A. **102**(35), 12305–12310 (2005)
45. E. Longstaff, G. von Krogh, Condyloma eradication: self-therapy with 0.15–0.5% podophyllotoxin versus 20–25% podophyllin preparations–an integrated safety assessment. Regul. Toxicol. Pharmacol. **33**(2), 117–137 (2001)
46. K.A. Lyseng-Williamson, C. Fenton, Docetaxel: A review of its use in metastatic breast cancer. Drugs **65**(17), 2513–2531 (2005)
47. M. Martínez-Castillo, J. Pacheco-Yepez, N. Flores-Huerta, P. Guzmán-Téllez, R.A. Jarillo-Luna, L.M. CárdenaJs-Jaramillo, R. Campos-Rodríguez, M. Shibayama, Flavonoids as a natural Treatment against Entamoeba histolytica. Front. Cell. Infect. Microbiol. **8**, Article 209 (2018)
48. J. Mendelsohn, J. Baselga, The EGF receptor family as targets for cancer therapy. Oncogene **19**(56), 6550–6565 (2000)
49. L. Menger, E. Vacchelli, O. Kepp, A. Eggermont, E. Tartour, L. Zitvogel, G. Kroemer, L. Galluzzi, Trial watch: Cardiac glycosides and cancer therapy. Oncoimmunology **2**(2), e23082 (2013)
50. H. Modjtahedi, S. Essapen, Epidermal growth factor receptor inhibitors in cancer treatment: Advances, challenges and opportunities. Anti-Cancer Drugs **20**(10), 851–855 (2009)
51. F. Mojab, Antimalarial natural products: a review. Avicenna J. Phytomed. **2**(2), 52–62 (2012)
52. M.A. Morse, S.D. LaGreca, S.G. Amin, F.-L. Chung, Effects of Indole-3-carbinol on lung tumorigenesis and DNA methylation induced by 4-(Methylnitrosamino)-l-(3-pyridyl)-l-butanone (NNK) and on the metabolism and disposition of NNK in A/J Mice. Cancer Res. **50**, 2613–2617 (1990)
53. N. Murawski, M. Pfreundschuh, New drugs for aggressive B-cell and T-cell lymphomas. Lancet Oncol. **11**(11), 1074–1085 (2010)
54. P.B. Murphy, M.J. Barrett, Morphine, [Updated 2019 Feb 28]. In: StatPearls [Internet]. Treasure Island (FL): StatPearls Publishing; 2019 January. Available from: https://www.ncbi.nlm.nih.gov/books/NBK526115/
55. T.M. Nicholson, W.A. Ricke, Androgens and estrogens in benign prostatic hyperplasia: Past, present and future. Differentiation **82**(4–5), 184–199 (2011)
56. N. Normanno, A. de Luca, C. Bianco, L. Strizzi, M. Mancino, M.R. Maiello, A. Carotenuto, G. De Feo, F. Caponigro, D.S. Salomon, Epidermal growth factor receptor (EGFR) signaling in cancer. Gene **366**(1), 2–16 (2006)
57. J. O'Leary, F.M. Muggia, Camptothecins: A review of the development and schedules of administration. Eur. J. Cancer **34**(10), 1500–1508 (1998)
58. K.S. Park, A systematic review on anti-inflammatory activity of Harpagoside. J. Biochem. Mol. Biol. Res. **2**(3), 166–169 (2016)
59. H.B. Patisaul, W. Jefferson, The pros and cons of phytoestrogens. Front. Neuroendocrinol. **31**(4), 400–419 (2010)
60. T. Pirttila, G. Wilcock, L. Truyen, C.V. Damaraju, Long-term efficacy and safety of galantamine in patients with mild-to-moderate Alzheimer's disease: Multicenter trial. Eur. J. Neurol. **11**(11), 734–741 (2004)
61. E. Pisha, H. Chai, I.S. Lee, T.E. Chagwedera, N.R. Farnsworth, G.A. Cordell, C.W. Beecher, H.H. Fong, A.D. Kinghorn, D.M. Brown, M.C. Wani, M.E. Wall, T.J. Hieken, T.K. Das Gupta, J.M. Pezzuto, Discovery of betulinic acid as a selective inhibitor of human melanoma that functions by induction of apoptosis. Nat. Med. **1**(10), 1046–1051 (1995)
62. Y. Pommier, Topoisomerase I inhibitors: Camptothecins and beyond. Nat. Rev. Cancer **6**(10), 789–802 (2006)
63. X. Qian, T. Melkamu, P. Upadhyaya, F. Kassie, Indole-3-carbinol inhibited tobacco smoke carcinogen-induced lung adenocarcinoma in A/J mice when administered during the post-initiation or progression phase of lung tumorigenesis. Cancer Lett. **311**(1), 57–65 (2011)

64. K. Raedschelders, D.M. Ansley, D.D. Chen, The cellular and molecular origin of reactive oxygen species generation during myocardial ischemia and reperfusion. Pharmacol. Ther. **133**, 230–255 (2012)
65. A.H. Rahmani, M.A. Al Zohairy, S.M. Aly, M.A. Khan, Curcumin: A potential candidate in prevention of cancer via modulation of molecular pathways. Biomed. Res. Int. **2014**, 761608 (2014)
66. P.T.R. Rajagopalan, Z. Zhang, L. McCourt, M. Dwyer, S.J. Benkovic, G.G. Hammes, Interaction of dihydrofolate reductase with methotrexate: Ensemble and single-molecule kinetics. Proc. Natl. Acad. Sci. U. S. A. **99**(21), 13481–13486 (2002)
67. J. Ravindran, S. Prasad, B.B. Aggarwal, Curcumin and Cancer cells: How many ways can curry kill tumor cells selectively? AAPS J. **11**(3), 495–510 (2009)
68. D.J. Reuland, S. Khademi, C.J. Castle, D.C. Irwin, J.M. McCord, B.F. Miller, K.L. Hamilton, Upregulation of phase II enzymes through phytochemical activation of Nrf2 protects cardiomyocytes against oxidant stress. Free Radic. Biol. Med. **56**, 102–111 (2013)
69. I.M.C.M. Rietjens, J. Louisse, K. Beekmann, The potential health effects of dietary phytoestrogens. Br. J. Pharmacol. **174**(11), 1263–1280 (2017)
70. A. Rocha, L. Wang, M. Penichet, M. Martins-Green, Pomegranate juice and specific components inhibit cell and molecular processes critical for metastasis of breast cancer. Breast Cancer Res. Treat. **136**(3), 647–658 (2012)
71. G. Sa, T. Das, Anti cancer effects of curcumin: Cycle of life and death. Cell Div **3**, 14 (2008)
72. R. Sandhir, Mehrotra, Quercetin supplementation is effective in improving mitochondrial dysfunctions induced by 3-nitropropionic acid: Implications in Huntington's disease, Biochimica et Biophysica Acta (BBA) - Mol. Basis Dis. **1832**(3), 421–430 (2013)
73. R. Saller, R. Brignoli, J. Melzer, R. Meier, An updated systematic review with meta-analysis for the clinical evidence of silymarin. Forsch. Komplementmed. **15**(1), 9–20 (2008)
74. F.P. Santos, H. Kantarjian, A. Quintás-Cardama, J. Cortes, Evolution of therapies for chronic myelogenous leukemia. Cancer J. **17**(6), 465–476 (2011)
75. R.-N. Seth, Cancer mechanisms, why cancer and inflammation? YALE J. Biol. Med. **79**, 123–130 (2006)
76. P. Sharma, S.F. McClees, F. Afaq, Pomegranate for prevention and treatment of cancer: An update. Molecules **22**(1), 177 (2017)
77. P.G. Shekelle, S.J. Newberry, J.D. FitzGerald, A. Motala, C.E. O'Hanlon, A. Tariq, A. Okunogbe, R. Shanman, Management of Gout: A systematic review in support of an American College of Physicians Clinical Practice Guideline. Ann. Intern. Med. **166**(1), 37–51 (2017)
78. M. Shelley, C. Harrison, B. Coles, J. Staffurth, T.J. Wilt, M.D. Mason, Chemotherapy for hormone-refractory prostate cancer. Cochrane Database Syst. Rev. **4**, CD005247 (2006)
79. S. Howard, Illustrated pharmacology memory cards: Pharmnemonics. Minireview, p. 13. ISBN 1-59541-101-1. (2008)
80. X.O. Shu, Y. Zheng, H. Cai, K. Gu, Z. Chen, W. Zheng, W. Lu, Soy food intake and breast Cancer survival. JAMA **302**(22), 2437–2443 (2009)
81. B.N. Singh, S. Shankar, R.K. Srivastava, Green tea catechin, epigallocatechin-3-gallate (EGCG): Mechanisms, perspectives and clinical applications. Biochem. Pharmacol. **82**(12), 1807–1821 (2011)
82. S. Singh, B. Sharma, S.S. Kanwar, A. Kumar, Lead phytochemicals for anticancer drug development. Front. Plant Sci. **7**, 1667 (2016)
83. W.J. Slichenmyer, E.K. Rowinsky, R.C. Donehower, S.H. Kaufmann, The current status of camptothecin analogues as antitumor agents. J. Natl. Cancer Inst. **85**(4), 271–291 (1993)
84. J.S. Smith, 2007. and The Food Safety Consortium. Brush on the marinade, hold off the cancerous compounds. Science Daily 2007; June 28
85. K. Stromgaard, K. Nakanishi, Chemistry and biology of terpene Trilactones from Ginkgo Biloba. Angew. Chem. Int. Ed. **43**(13), 1640–1658 (2004)
86. S.-Y. Sun, N. Hail, R. Lotan, Apoptosis as a novel target for Cancer chemoprevention. J. Natl. Cancer Inst. **96**(9), 662–672 (2004)

87. D.N. Syed, J.C. Chamcheu, V.M. Adhami, H. Mukhtar, Pomegranate extracts and Cancer prevention: Molecular and cellular activities. Anti Cancer Agents Med. Chem. **13**(8), 1149–1161 (2013)
88. C.H. Takimoto, E. Calvo, Principles of Oncologic Pharmacotherapy, in *Cancer Management: A Multidisciplinary Approach*, ed. by R. Pazdur, L. D. Wagman, K. A. Camphausen, W. J. Hoskins, Cmp United Business Media, USA (2009)
89. W.F. Taylor, E. Jabbarzadeh, The use of natural products to target cancer stem cells. Am. J. Cancer Res. **7**(7), 1588–1605 (2017)
90. R. Thirumaran, G.C. Prendergast, P. Gilman, Chapter 7: Cytotoxic chemotherapy in clinical treatment of cancer, in *Cancer Immunotherapy: Immune Suppression and Tumor Growth*, ed. by G. C. Prendergast, E. M. Jaffee, Academic Press, USA (Academic/Elsevier, Amsterdam/Boston, 2007), pp. 101–116
91. R. Thomas, E. Butler, F. Macchi, M. Williams, Phytochemicals in cancer prevention and management? British J. Med Pract. **8**(2), a815 (2015)
92. J. Tian, X. Ye, Y. Shang, Y. Deng, K. He, X. Li, Preparative isolation and purification of harpagoside and angroside C from the root of Scrophularia ningpoensis Hemsley by high-speed counter-current chromatography. J. Sep. Sci. **35**, 2659–2664 (2012)
93. C. Tsioufis, C. Thomopoulos, Combination drug treatment in hypertension. Pharmacol. Res. **125**(Pt B), 266–271 (2017)
94. S. Upadhyay, M. Dixit, Role of polyphenols and other phytochemicals on molecular signaling. Oxidative Med. Cell. Longev. **2015**, 504253 (2015)
95. S. Usai, L. Grazzi, G. Bussone, Gingkolide B as migraine preventive treatment in young age: Results at 1-year follow-up. Neurol. Sci. **32**(Suppl 1), I97–I99 (2011)
96. M.E. Van Meter, E.S. Kim, Bevacizumab: Current updates in treatment. Curr. Opin. Oncol. **22**, 586–591 (2010)
97. W. Voigt, S. Matsui, M.B. Yin, W.C. Burhans, H. Minderman, Y.M. Rustum, Topoisomerase-I inhibitor SN-38 can induce DNA damage and chromosomal aberrations independent from DNA synthesis. Anticancer Res. **18**(5A), 3499–3505 (1998)
98. W.P. Wang, N. Liu, Q. Kang, Simultaneous determination by UPLC-MS/MS of seven bioactive compounds in rat plasma after oral administration of Ginkgos biloba tablets: Application to a pharmacokinetic study. J. Zhejiang Univ. Sci. B. **15**(11), 929–939 (2014)
99. T. Weber, *Rauwolfia*: Gandhi's Favourite Tranquiliser? South Asia J. South Asian Stud. **41**(3), 567–578 (2018)
100. R. Wilken, M.S. Veena, M.B. Wang, E.S. Srivatsan, Curcumin: A review of anti-cancer properties and therapeutic activity in head and neck squamous cell carcinoma. Mol. Cancer **10**, 12 (2011)
101. I.U. Willcockson, A. Hong, R.P. Whisenant, J.B. Edwards, H. Wang, H.K. Sarkar, S.E. Pedersen, Orientation of d-tubocurarine in the muscle nicotinic acetylcholine receptor-binding site. J. Biol. Chem. **277**(44), 42249–42258 (2002)
102. C.D. Willey, J.A. Bonner, Interaction of chemotherapy and radiation, in *Clinical Radiation Oncology*, 3rd edn., Elsevier, Philadelphia (2012)
103. H. Xu, M. Lv, X. Tian, A review on hemisynthesis, biosynthesis, biological activities, mode of action, and structure-activity relationship of podophyllotoxins: 2003-2007. Curr. Med. Chem. **16**(3), 327–349 (2009)
104. C.S. Yang, H. Wang, J.X. Chen, J. Zhang, Effects of tea Catechins on Cancer signaling pathways. Enzyme **36**, 195–221 (2014)

Molecular Mechanisms of Phytochemical Actions in Cancer

5

Abstract

Extensive research has shown that a broad range of phytochemicals are important in chemoprevention and chemosensitization. Phytochemicals not only work on genetic pathways; they are responsible for modulating epigenetic control mechanisms like histone acetylation and methylation, DNA methylation, and miRNA posttranslational silencing. We start the chapter with a review of these epigenetic control mechanisms. Inflammation and oxidative stress are two most important causes of carcinogenesis, and we review the molecular mechanisms involved in these two processes, in particular the Keap1-Nrf2 pathway. We then discuss the roles of phytochemicals in mitigating oxidative stress and inflammation. Chemoprevention is important for reducing the burden of cancer worldwide; we review the roles of phytochemicals in chemoprevention next. Finally we discuss how phytochemicals can sensitize cancer cells for enhancing the effects of chemotherapeutic drugs.

5.1 Introduction

The aim of this chapter is to understand the molecular mechanisms through which phytochemicals participate in cancer prevention and therapy. We discuss the molecular mechanisms through which phytochemicals act on key pathways. There are two main causes of cancer: the first is mutations that can permanently alter the DNA sequence. The proteins translated from mutated genes change structurally and functionally, affecting important genetic pathways. The second cause is the malfunctioning of epigenetic control mechanisms. We have seen in Chap. 2 that many of the important pathways act through epigenetic controls like histone acetylation and DNA methylation. Hence it is not surprising that many of these pathways do not function in most cancers as they should in healthy tissues. Phytochemicals mostly work to prevent such aberrations in epigenetic control

M. Roy, A. Datta, *Cancer Genetics and Therapeutics*,
https://doi.org/10.1007/978-981-13-9471-3_5

mechanisms, help in chemoprevention, and also help in chemosensitization, i.e., prevent the resistance against anticancer chemotherapeutic agents in cancer cells. We will examine the molecular mechanisms of phytochemicals in all of these roles in details. A key pathway that mediates antioxidant and anti-inflammatory responses to prevent cancer is the Keap1-Nrf2 signaling pathway. Phytochemicals can act as chemopreventive agents by enhancing the effects of this pathway. We will review this pathway next.

The chapter starts with examining three key epigenetic control mechanisms, histone acetylation, DNA methylation, and miRNA silencing, as phytochemicals have effects on all three of these mechanisms. Next we discuss the important Keap1-Nrf2 pathway. Following this we discuss the roles of phytochemicals in mitigating inflammation and oxidative stress, in particular by enhancing the effects of the Keap1-Nrf2 pathway. We then consider the molecular mechanisms of phytochemicals in chemoprevention and chemosensitization.

5.2 Epigenetic Control

The central dogma of biology is the following: genes are transcribed to mRNAs, and mRNAs are translated to proteins. Mutations in genes change the structure and function of the corresponding proteins, and they malfunction, disrupting the normal functions of the proteins.

The functions of proteins can also be disrupted if the rate of production of proteins is reduced, resulting in nonavailability or partial availability of the protein for a particular metabolic activity. It is also possible that an increase in the rate of production of a protein may increase some metabolic activities. Transcriptional and posttranscriptional control mechanisms affect the rate of production of proteins. We discuss three such epigenetic mechanisms, histone acetylation/deacetylation and DNA methylation which are transcriptional control mechanisms that can increase or decrease the rate of production of proteins, whereas miRNA silencing is a posttranscriptional control mechanism in which the translation rate of mRNA to protein is reduced through silencing of mRNA by miRNA. We follow mainly the classic textbook by Alberts et al. [1] for discussing these three epigenetic control mechanisms. Aberrations in epigenetic control are possible to control and reverse, unlike mutations in genes. Hence many drugs or therapeutic agents target epigenetic control mechanisms.

5.2.1 Histone Modification

Eukaryotic DNA is tightly packed inside the nucleus of a cell. As a result the access of transcription factors (proteins that initiate transcription to mRNA) to specific genes is tightly controlled. This packing is due to two classes of proteins that bind to the DNA to form eukaryotic chromosomes, histones, and chromosomal proteins. The complex of these two classes of proteins with DNA is called chromatin.

Chromatins are organized from nucleosome core particles which are made of eight histone proteins, and two consecutive nucleosome core particles are connected by linker DNA. We refer the reader to the book by Alberts et al. [1] for more details of this packaging of DNA in the nucleus. For our purpose we need to understand histone acetylation, the key mechanism that allows the unpacking of the DNA so that transcription factors have easier access to specific genes. Histone deacetylation on the other hand makes the packing tighter preventing access to transcription factors. Similarly, methylation of histones also can increase or decrease the transcription of genes.

DNA and histones are tightly bound in the nucleosome. The concentration of histone is so high that the weight of histones matches the weight of DNA [1]. There are two parts of a histone, a core fold and an N-terminal tail. These tails undergo different modifications including acetylation, methylation, and phosphorylation, changing the chromatin structure. In particular these modifications destabilize the chromatin structure and allow better access of transcription factors to genes.

The N-terminal end of the histones contains the NH_2 and COOH tails. Histone acetylation occurs along the NH_2 tails of histones. Acetyl groups are added by histone acetyltransferases (HATs). The positive charge of the lysine amino acid residue in the tail is removed as a result of the acetylation. This makes it difficult for the histone to neutralize the charge of the DNA wound around it, and the DNA becomes loose, allowing better access for transcription factors. Hence HATs act as transcriptional activators. The opposite process is deacetylation, when the acetyl group is removed from the lysine residue by histone deacetylase (HDAC), resulting in tighter packing of histone and DNA. HDACs act as transcriptional repressors.

Histone methylation has a finer control in the sense that it can increase or decrease transcription of genes depending on the number of amino acids in histones that are methylated and also the number of methyl groups that are attached. Methylation is done by histone methyltransferases (HMT) and demethylation by histone demethylases (HDM).

The proteins involved in histone phosphorylation are histone phosphatases. Histone phosphorylation has similar effects as histone acetylation and histone methylation, namely, relaxing the DNA-histone packing so that transcription factors and other proteins get access to the DNA. One of the most important examples of histone phosphorylation is cellular responses to DNA damage. Chromatin domains around DNA breakage sites are separated due to histone phosphorylation. The mechanism of histone modification is illustrated in Fig. 5.1.

5.2.2 DNA Methylation

DNA methylation is the process through which methyl groups are added to DNA molecules by a class of enzymes called DNA methyltransferases, in particular the two bases cytosine and adenine can be methylated. DNA methylation is a very important epigenetic control mechanism, as we have seen in Chap. 2. The main effect of DNA methylation is the prevention of gene transcription. Transcription

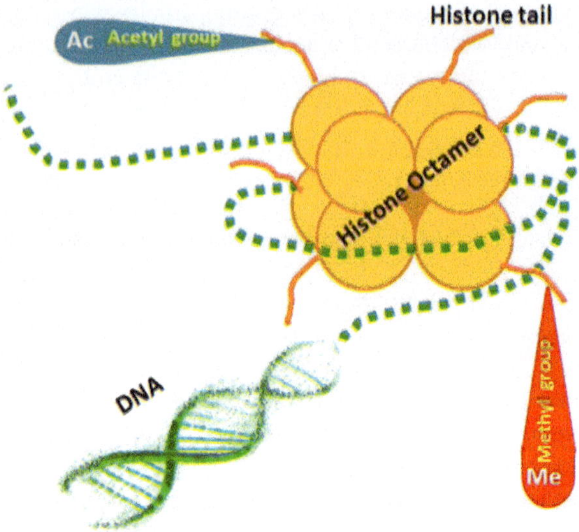

Fig. 5.1 DNA is wrapped around the histone octamer. Both acetylation and methylation occur at the histone tails

factors are a class of proteins that have DNA-binding domains. Transcription factors bind to specific DNA sequences at the promoter regions near the transcription start sites of a gene. They help in recruiting RNA polymerases that are responsible for transcription. Methylation of specific bases in the promoter regions prevents the binding of transcription factors and hence prevents the transcription of the corresponding gene.

CpG islands are repeated sequences of cytosine and guanine nucleotides from the 5' to the 3' end of DNA, usually several hundred base pairs long. As cytosine can be methylated, CpG islands are targets for DNA methylation. Though CpG islands can occur anywhere in the genome, their presence in the promoter regions of genes makes them an important target for epigenetic control through DNA methylation. About 70% of the promoter regions of genes in the human genome contain CpG islands near transcriptional start sites. Most such CpG islands are hypomethylated (methylated at less than normal levels); in fact hypomethylation is a genome-wide normal process. However many CpG islands associated with tumor repressor genes are hypermethylated in many cancers, whereas CpG islands are hypomethylated in many oncogenes. As a result the transcription of tumor repressor genes is reduced, and transcription of oncogenes is enhanced. There are several drugs that target the DNA methyltransferases to prevent hypermethylation of CpG islands. DNA methylation is illustrated in Fig. 5.2.

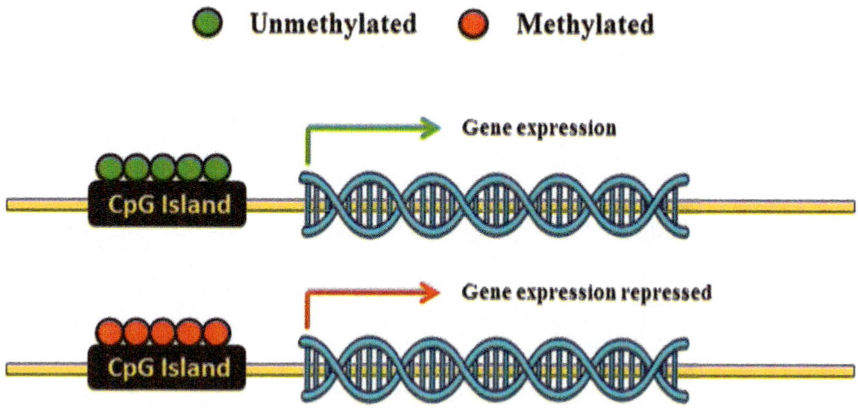

Fig. 5.2 Methylation at the CpG islands represses gene expression

5.2.3 miRNAs

miRNAs (microRNA) are small noncoding RNAs (they do not code for any protein) that play very important roles in epigenetic control. miRNAs are usually 18–25 nucleotides long, and their role is in posttranscriptional silencing of mRNAs. miRNAs bind to the 3' untranslated regions of mRNAs to prevent the translation of mRNAs into proteins. Many miRNA families are highly conserved across eukaryotes indicating that this epigenetic control is of ancient origin. miRNA expression is highly correlated with DNA methylation and histone acetylation [2]. For example, approximately half of the known miRNA genes are targets for DNA CpG methylation. Also the transcriptions of many miRNA genes are regulated through histone acetylation. On the other hand, miRNAs modulate the translations of DNA methyltransferases and HDACs. Different miRNAs can act as either oncogenes or tumor suppressor genes; miRNAs play major roles in epithelial-mesenchymal transition (EMT) that is a root cause of metastasis. miRNAs are novel biomarkers as well as therapeutic targets in cancer initiation and metastasis due to their role in EMT. The action of miRNA in suppressing mRNA translation is illustrated in Fig. 5.3.

5.3 Keap1-Nrf2 Pathway

The Kelch-like ECH-associated protein (Keap1) is a very important participant in the Keap1-Nrf2 pathway. This is a complex pathway triggered due to oxidative stress. Other important participants in this pathway are antioxidant response elements (AREs) which are transcriptional regulatory elements in untranslated regions of many genes. We base our review of this pathway on the papers by Raghunath et al. [3] and Nguyen et al. [4].

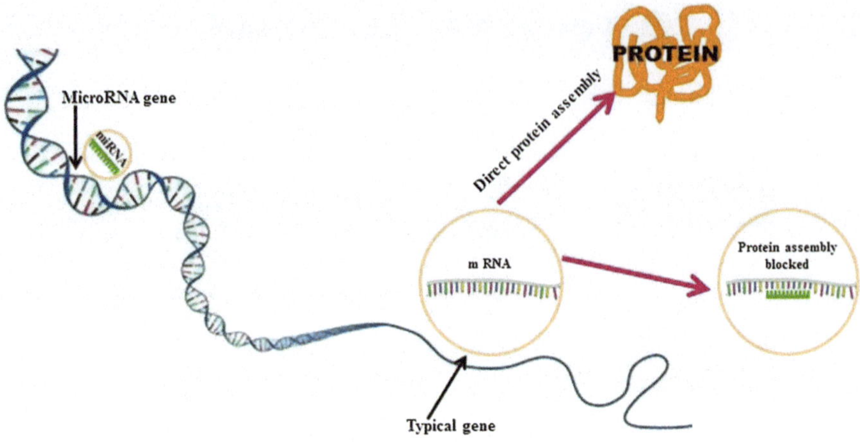

Fig. 5.3 miRNA binds at the 3'UTR of mRNA and inhibits mRNA translation

Nuclear factor-like 2, or Nrf2, is a transcription factor that regulates the activation of proteins that act against oxidative damage. We first discuss the mechanism of detoxification of toxic chemicals and xenobiotics (foreign chemicals) by the proteins that prevent oxidative damage and then discuss the specifics of the Keap1-Nrf2 pathway.

The detoxification in the cell happens in a regulated manner in three phases [3]. Cytochrome P450 superfamily of enzymes are responsible for metabolizing the xenobiotics in the first phase. Conjugating enzymes including glutathione S-transferases are responsible for conjugating reactive electrophile species with glutathione (GSH) in the second phase. Finally transporter proteins eliminate the GSH conjugates in the final phase. A common characteristic of all the proteins involved in these three phases is that many of them have AREs as transcriptional regulatory elements and the transcription factor Nrf2 acts on the AREs to increase the transcription of these proteins.

Nrf2 is encoded by the NFE2L2 gene. Nrf2 stays in the cytoplasm when there are no xenobiotics or oxidative stress in the cell. The average lifetime of Nrf2 is very short under unstressed conditions. Nrf2 is kept in the cytoplasm by a group of proteins; most important among them are Keap1 and Cullin 3. These two proteins degrade Nrf2 through ubiquitination. Oxidative stress in the cell disrupts the Keap1-Cullin 3 complex, and Nrf2 is not ubiquitinated anymore, resulting in a higher concentration of Nrf2 in the cytoplasm. It is then transported to the nucleus and binds to the ARE in the promoter regions of many antioxidative genes. This results in increased transcription of these genes and the start of the three-phase mechanism for the removal of xenobiotics. The Keap1-Nrf2 pathway is illustrated in Fig. 5.4.

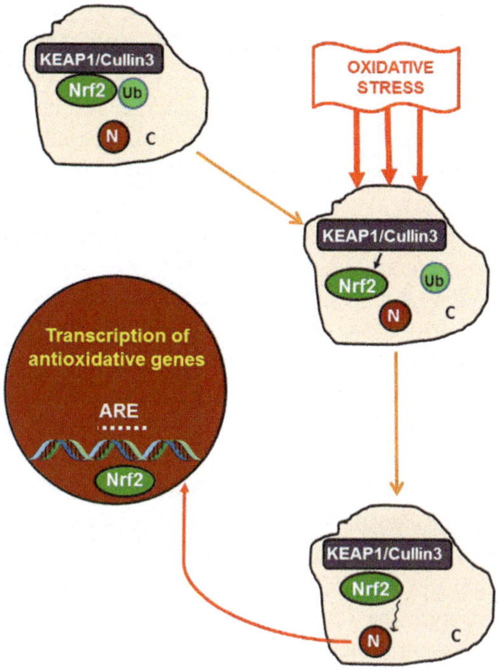

Fig. 5.4 Keap1-Nrf2 pathway: Under normal circumstances, Nrf2 is degraded by the Keap1/Cullin3 ubiquitination complex. Nrf2 is released from this complex under oxidative stress and translocated to the nucleas, where it binds to ARE

5.4 Oxidative Stress

Reactive oxygen species (ROS) and antioxidant molecules are in an equilibrium in normal cells, as ROS are usually neutralized at the rate they are produced. ROS are produced in normal cells as a result of normal metabolism, and it is not harmful in low concentrations. However higher concentrations of ROS can damage large molecules like lipids, proteins, and DNA. In particular a higher concentration of ROS may damage DNA at a rate that the cellular DNA repair mechanisms cannot cope with. There is now clear evidence of a connection between oxidative stress and carcinogenesis. Oxidative stress is significantly increased due to the presence of environmental carcinogens like alcohol and cigarette smoke. Another major cause of oxidative stress is inflammation like burns and ultra violet radiation. ROS can cause DNA damage through chromosomal translocations, point mutations, insertions, and deletions. ROS can accelerate other deleterious effects through multiple pathways and can cause initiation, progression, and metastasis in cancer [5].

ROS can affect the normal metabolism in cells through epigenetic intervention, like histone acetylation and DNA methylation, and can cause carcinogenesis. The homeobox gene Cdx1 is expressed in the intestine, and it is known to transcribe a tumor suppressor protein [6]. It has been shown that increased oxidative stress due to H_2O_2 intervention can increase the activity of DNA methyltransferase 1 (DNMT1) and histone deacetylase 1 (HDAC1) that in turn inhibit the expression of Cdx1 in colorectal cancer [7]. Mahalingaiah et al. [8] reported the malignant transformation of human renal tubular epithelial cells (HK-2) due to chronic oxidative stress. They reported aberrant expression profiles of epigenetic regulatory genes for both DNA methylation (DNMT1, DNMT3a) and histone modification genes (HDAC1, HMT1, and HAT1) that resulted in increased survival and malignant transformation of HK-2. Ito et al. [9] have shown a connection between oxidative stress and a reproductive disorder endometriosis, through aberrant DNA methylation and histone modification. All three epigenetic control mechanisms, DNA methylation, histone modification, and miRNA posttranscriptional control, have been implicated in aging due to oxidative stress [10]. The effect of histone modification and DNA methylation due to oxidative stress has been observed in human breast cancer cells. MCF-7 breast cancer cells transformed into estrogen-independent aggressive forms from estrogen-dependent nonaggressive breast cancer cells upon incubation with H_2O_2 [11].

The most potent biomarker for oxidative stress is the hydroxyl radical or OH-. However there are many other known biomarkers [2, 12] of cancer. 8-Hydroxy-2'-deoxyguanosine (8-OHdG) is a by-product of DNA oxidation. Oxidized bases in the DNA are normally removed by enzymes that are part of the base excision repair pathway [13]. However an excessive amount of oxidized bases is a cause of cancer. Nitrotyrosine (ONOO-) is a product of tyrosine (one of the nonessential amino acids) nitration caused by reactive nitrogen species. Glutathione S-transferase-pi (GST-pi), superoxide ion radical (O_2/H_2O_2), and inducible nitric oxide synthase (iNOS) are some of the other major markers of oxidative stress.

5.5 Inflammation

Inflammation is critical for tumor initiation and progression, and the physiological responses to inflammation are very complex [14]. Normal inflammation is usually self-limiting, in the sense that the agents involved in healing processes do not accumulate beyond a certain concentration. However the agents that are beneficial in normal circumstances can negatively affect the healing process, and that results in pathogenesis and neoplastic progression.

Inflammation causes both DNA mutation and epigenetic changes resulting in tumor initiation and progression. DNA mutations occur in normal cells, but DNA repair machinery usually can recover from these mutations to a large extent. However, an inflammatory environment prevents recovery from DNA mutations, and as a result an excessive rate of mutations may trigger carcinogenesis. Cytokines (extracellular immunological signaling peptides) are secreted by immune cells and mediate the inflammatory process by reducing tissue damage and improving healing. However cytokines are known to be associated with inflammation and cancer initiation and involved in cell proliferation, migration, and angiogenesis. Some of the most important cytokines in this respect are tumor necrosis factor alpha (TNF-α), transforming growth factor beta 1 (TGF-β1), interleukin-1β, and IL-6 [15].

Nitric oxide (NO) is another molecule that plays an important role in inflammation pathogenesis. NO is a signaling molecule that is produced in the endothelial cells by the nitric oxide synthase (NOS) family of enzymes, in particular by inducible NOS (iNOS). NO has an anti-inflammatory role in normal conditions; however an excessive concentration of NO, or production of NO in abnormal conditions, has a pro-inflammatory effect [16].

There is increasing evidence that the Nrf2 pathway is involved in persistent inflammation [17]. Hyperoxia (excessive oxygen supply) is a common treatment in many pulmonary diseases; however the effects of oxidant exposure for patients recovering from oxidant-mediated lung injury are not well understood. Reddy et al. [17] found that Nrf2-deficient mice die even under elevated but sublethal hyperoxia, thereby establishing a link between inflammation and the Nrf2 pathway.

Chronic inflammation also causes epigenetic alterations due to DNA and histone modifications. Such aberrations in epigenetic control affect the expression of a host of genes associated with inflammation [2]. DNA hypo- and hypermethylations are associated with several inflammation-related diseases including cystic fibrosis and obesity. Similarly histone acetylation also has been implicated in inflammation-related diseases. For example, the transcriptional coactivator histone acetyltransferase (HAT) has been found to have aberrant expressions in inflammatory diseases [2].

5.6 Roles of Dietary Phytochemicals in Oxidative Stress

We have seen classifications of phytochemicals in Chap. 3. Steinmetz and Potter [18] identified many classes of phytochemicals from plant diets like fruits, nuts, and vegetables. These include flavonoids, isoflavonoids, lignin, phenolic acid, carotenoids, phytosterols, phytostanols, tocotrienols, and organosulfur compounds. We will discuss the molecular mechanisms of phytochemicals in chemoprevention and chemosensitization later. First we will discuss the roles phytochemicals play in mitigating oxidative stress and inflammation.

5.6.1 Phytochemicals and Oxidative Stress

Reactive oxygen species (ROS) is either produced endogenously due to normal metabolism or enter cells from external sources. ROS is necessary in low concentrations for performing many important metabolic functions in cells. These functions include homeostatic signaling for normal cell survival and proliferation, mainly through modifications of phosphates and kinases [19]. Nicotinamide adenine dinucleotide phosphate (NADP+) and its reduced form NADPH are required to maintain the balance of ROS in normal cells. This balance is destroyed either due to external environmental sources of ROS or due to overproduction of ROS due to abnormal cellular activities, and as a result ROS may cause damage to DNA and other important biomolecules like lipids and proteins. There are some innate defense mechanisms for reducing oxidative stress in the cellular environment [20]. These include enzymatic scavengers of ROS including superoxide dismutases (SOD), glutathione peroxide, and catalase, as well as nonenzymatic agents like glutathione, various vitamins (mainly A, C, and E), and flavonoids. However such innate mechanisms are incapable of reducing ROS in high concentrations, and as a result many different enzymes and pathways are activated in high ROS concentrations.

High oxidative stress can activate many major signaling pathways depending on the cell types and duration and magnitudes of the stress [20]. These pathways include ATM (ataxia-telangiectasia mutated), ERK (extracellular signal-regulated kinase), HSF1 (heat shock transcription factor 1), JAK (Janus protein kinase), PKC (protein kinase C), PLC-γ1 (phospholipase C-γ1), STAT (signal transducers and activators of transcription), NFκB (nuclear factor κB), and JNK (c-Jun N-terminal kinase) [20].

MAPKs (mitogen-activated protein kinase) can be divided into three subfamilies, ERK, JNK, and p38. We have reviewed the MAPK/ERK signaling pathway in Chap. 2. Here we discuss the JNK and p38 pathways and their interrelationships with the ERK pathway. Our review is based on the paper by Martindale and Holbrook [20]. The ERK, JNK, and p38 pathways are activated by the MAPK kinase in a cascading fashion. The main mechanism is the phosphorylation of a range of transcription factors by MAPK. The phosphorylation occurs in a chain, a MAPK kinase (MAPKK) phosphorylates MAPK, and MAPKK is in turn phosphorylated by a MAPK kinase kinase (MAPKKK). The JNK and p38 pathways are related to stress and quite often activated together. These two pathways are called stress-activated protein kinase (SAPK). We briefly review the SAPK pathway as we have not discussed this pathway earlier.

The JNK pathway has been investigated for over two decades; however, the mechanisms of its activation are still controversial due to the many MAPKs involved in this pathway [21]. In fact the role of the JNK pathway has been identified as mediating apoptosis during excessive oxidative stress, whereas some other studies have identified pro-survival functions of JNK. It is possible that the activation of the JNK pathway during apoptosis is a failed attempt at activating pro-survival responses that were overwhelmed by other pro-apoptotic responses [20, 21]. The findings for the p38 pathway are similar; different studies have found this pathway to be both pro-apoptotic and pro-survival [20].

5.6.1.1 Phenolic Phytochemicals and Oxidative Stress

Phenolic phytochemicals have been studied for their antioxidant activities for a long time. Their antioxidant property comes from the phenolic hydroxyl groups that provide hydrogen atoms for capturing ROS. Both H_2O_2 and its product OH are responsible for cancer cell proliferation through signal transduction. Phenolic phytochemicals inhibit the effects of H_2O_2 on several pathways including MAPK, NF-κB, and AP-1.

The effect of the green tea phenolic phytochemical, epigallocatechin gallate (EGCG) has been studied in cultured human epidermal keratinocytes when the cells are given oxidative stress through ultraviolet radiation (UV-B) [22]. The ultraviolet radiation increased the concentration of intracellular H_2O_2 concentration and increased the phosphorylation of MAPK, JNK, and p38. It was found that EGCG inhibited the activation of MAPK by scavenging H_2O_2.

One of the most important phenolic compounds is resveratrol, found in grapes. Manna et al. [23] induced the tumor necrosis factor (TNF) on different cell lines including Jurkat, H4 glioma, HeLa, and U-937. TNF is a trigger for NF-κB activation; however they showed that resveratrol prevented the activation of NF-κB. TNF also increased the concentration of ROS and activated MAPKK and JNK in their study, but resveratrol inhibited these effects as well. The efficacy of resveratrol pre-treatment in reducing effects due to oxidative stress has been reported in several cell lines. Yu et al. [24] induced oxidative stress in HeLa cell lines using phorbol myristate acetate (PMA) which is an activator of protein kinase C (PKC) and NF-κB and ultraviolet radiation (UV-C). Transcription of a reporter gene construct containing AP-1 response elements was enhanced when HeLa cells were exposed either to PMA or to UV-C. However, the transcriptional activation of this reporter gene construct decreased when the cells were pre-treated with resveratrol. They reported similar inhibition of MAPK and JNK through pre-treatment of HeLa cells by resveratrol. It was inferred that most probably resveratrol inhibited both protein tyrosine kinase and protein kinase C since inhibitors of these enzymes inhibited MAPK and AP-1.

The inhibition of cancer cell proliferation by phenolic compounds has been reported in several studies. Inhibition of cancer cell proliferation is due to inducing apoptosis as well as through arresting cell cycle. The mechanism of inhibition is through scavenging ROS, in particular H_2O_2 by phenolic compounds, as ROS are necessary for cancer cells to divide and proliferate uncontrollably. Apigenin is a phenolic phytochemical of the flavone class. It has been shown that apigenin inhibited growth and induced apoptosis in human anaplastic thyroid carcinoma cells [25]. Apigenin has also multiple other roles in suppressing various other cancers through autophagy, preventing cell migration, and triggering immune response [26, 27]. Resveratrol also plays similar roles in cell cycle arrest. For example, Ahmad et al. [28] reported G_1-phase cell cycle arrest in human epidermoid carcinoma A431 cells by resveratrol. Treatment by resveratrol suppressed free radical formation by 12-O-tetradecanoylphorbol-13-acetate (TPA) in human leukemia HL-60 cells [29]. Cyclin and cyclin-dependent kinases (CDK) are important for cell cycle progression. Ahmad et al. [28] also reported that the expression of CDK inhibitor p21 was increased due to resveratrol. She et al. [30] reported that resveratrol acts through the JNK pathway, phosphorylating and activating p53 to induce apoptosis. p53 is an important transcription factor that activates the Waf-1 gene. Waf-1 codes for p21 that induces apoptosis. Other phenolic compounds like quercetin and genistein have shown to induce apoptosis in pancreatic carcinoma cells [31].

We have till now discussed the ability of phenolic phytochemicals in scavenging ROS, in other words their antioxidant properties. However quite paradoxically, phenolic phytochemicals also have strong prooxidant properties, in the sense that they can increase the concentrations of ROS and put cells under extreme stress due to elevated concentrations of ROS under certain experimental conditions. As a result apoptosis can occur due to extreme oxidative stress induced by phenolic compounds. Most probable cause of apoptosis in cells due to phenolic compounds is oxidative damage to DNA. Single-strand breakage of Jurkat T-lymphocyte cells was reported using EGCG [32] and curcumin [33]. Though EGCG caused elevated concentration of H_2O_2 in the cell culture media, curcumin most probably increased the concentration of H_2O_2 in the intracellular environment. Similar dual antioxidant and prooxidant roles are known for phenolic compounds derived from tea [34]. It is a natural question to ask how the prooxidant effects of phenolic compounds affect normal cells. It is still an active research area; however, it can be said that most probably cancer cells are more severely affected due to the prooxidant effects of phenolic compounds, as they are already under strong oxidative stress. This is why cancer drugs like cisplatin kill cancer cells more effectively in comparison with normal cells.

5.6.1.2 Isothiocyanates and Oxidative Stress

Some of the isothiocyanates (ITC) have antioxidant properties, and some have prooxidant properties. Several ITCs, including phenylmethyl ITC, benzyl ITC, and phenylethyl ITC, have mainly prooxidant properties [35]. They inhibit the proliferation of cancer cells by inducing oxidative stress by increasing the levels of ROS in a manner similar to phenolic phytochemicals.

Kassie et al. [36] reported a dose-dependent increase in concentration of thiobarbituric acid-reactive substances when HepG2 hepatoma cells were incubated with benzyl isothiocyanate. As thiobarbituric acid-reactive substances are a measure of lipid peroxidation due to ROS, this showed a direct connection between benzyl isothiocyanate and ROS concentration. High concentration of benzyl isothiocyanate caused DNA damage due to elevated levels of ROS. Kassie et al. [36] also reported the reduction of such DNA damage by antioxidants like ascorbate and β-carotene. Direct increase of ROS concentration due to benzyl isothiocyanate has also been reported by Nakamura et al. [37]. ROS levels were about 50-fold higher when epithelial cells of rat liver were treated with benzyl isothiocyanate.

That elevated ROS levels associated with isothiocyanates are directly responsible for apoptosis has been demonstrated in several studies. An isothiocyanate called sulforaphane induced cell cycle arrest and apoptosis in human colon cancer cells HT-29, even though the levels of p53 protein did not increase [38]. However this behavior seems to be dependent on cell lines. Fimognari et al. [39] reported that sulforaphane increased p53 expression in Jurkat T-lymphocyte cells. There is also evidence that both phenylethyl isothiocyanate and phenylmethyl isothiocyanate increase apoptotic signaling most probably through JNK activation [40].

Sulforaphane is one of the most studied isothiocyanates for its extensive antioxidant properties [41, 42]. Though the exact molecular mechanisms are still being researched, sulforaphane has been found to be effective in many human cancers in reducing cell viability. These cancers include cervical cancer, pancreatic cancer, hepatocellular carcinoma, and ovarian cancer. The most probable reason for this could be the increased antioxidant effects of sulforaphane, similar to the phenolic phytochemicals [41].

5.7 Roles of Phytochemicals in Inflammation

Inhibition of various inflammatory mediators like cytokines and NO is important for controlling inflammatory diseases including cancer. Phytochemicals are very important for controlling these inflammatory mediators. The role of phytochemicals in controlling certain cancers has been confirmed through in vitro and in vivo studies. Phytochemicals mainly mitigate the effects of pro-inflammatory cytokines through kinases, such as protein kinase C and mitogen-activated protein kinase. NF-κB is one of the most important transcription factors that control the transcription of many cytokines [43]. Phytochemicals inhibit NF-κB by altering its capacity to bind to DNA. This in turn reduces the transcription of cytokines.

RAW264.7 is a murine macrophage cell line that is often used for testing the human de novo response of natural products. Lipopolysaccharides (LPS) are large molecules that are found on the outer membranes of gram-negative bacteria and induce the secretion of pro-inflammatory agents like cytokines and NO in most cell lines. Flavanones like hesperidin, neohesperidin, narirutin, and naringin as well as polymethoxylated flavones are the main phytochemicals present in citrus peel extracts [44]. Etoh et al. [45] have reported the reduction in release of TNF-α and NO from RAW 264.7 cells when the cells were stimulated by LPS. Moreover citrus peel extracts suppressed the expressions of NF-κB and iNOS in the same cell lines. Blackberry is rich in phytochemicals like flavonoids, anthocyanins, cyanidin, and ellagic acid. It has been reported that blackberry extract reduced the release of NO in a dose-dependent way in LPS-induced RAW 264.7 cell lines [46]. Blueberries are a rich source of phytochemicals such as resveratrol, flavonoids, anthocyanidins, anthocyanins, and ellagic acid. Lau et al. [47] reported that when phenolic extracts from blueberries were applied to LPS-activated BV2 microglia cells, the production of TNF-α, NO, and IL-1β was inhibited. Moreover the protein and mRNA levels of COX-2 and iNOS were also significantly reduced in the same study.

There are many similar studies of inhibitory effects of phytochemicals on pro-inflammatory agents. Though we have only examined a few of these studies, the reader can get a detailed view of these inhibitory effects from the excellent recent review article by Zhu et al. [48].

5.8 Molecular Mechanism of Phytochemicals in Chemoprevention

Though there are significant genetic risks for many cancers, it has now been proven that lifestyle plays an important role in etiology of cancer. The role of lifestyle risk factors like smoking, alcohol consumption, and obesity has been linked to occurrence of several cancers. It has been suspected for a long time that dietary intakes or healthy food habits also play important roles in cancer prevention. Though there were enough anecdotal evidence for this, recent large-scale scientific studies have proven this indisputably, for example, one such recent study has been reported by Grosso et al. [49]. We have noted earlier that reactive oxygen species (ROS) are natural occurrences in cells; however increased concentrations of ROS can tilt the balance towards uncontrolled proliferation of cells. We have noted earlier that the Keap1-Nrf2 is a very important pathway for keeping ROS concentration under control in cells. It has been observed that phytochemicals present in many fruits and vegetables act directly on the Nrf2 pathway, thus establishing a link between healthy diet and chemoprevention, i.e., suppression of onset of tumors. Similarly phytochemicals play important roles in metastasis. The aim of this section is to examine this link through understanding of how different dietary phytochemicals intervene in the Nrf2 pathway for chemoprevention, as well as the roles of phytochemicals in preventing metastasis.

5.8.1 Intervention of Phytochemicals in the Keap1-Nrf2 Pathway

We discuss the roles of dietary phytochemicals in suppression of the onset of tumors by preventing cell proliferation. We examine the roles of polyphenols like curcumin, genistein, and epigallocatechin and also the role of isothiocyanates and triterpenoids.

Curcumin is the main component of the popular Indian spice turmeric. Curcumin is a lipophilic polyphenol, diferuloylmethane, with known antioxidative properties. Curcumin is a popular ingredient in many traditional Indian and Chinese medicines. There is mounting evidence that curcumin is a promising polyphenol for preventing cancer, diabetes, neurological disorders, and kidney disorders [50]. We discuss the effects of curcumin on the Nrf2 pathway based on the review article by Li et al. [2] and Shu et al. [51]. The effect of curcumin on the Nrf2 pathway has been studied in [52, 53]. Das and Vinayak [52] studied Nrf2 activation in the mouse model. They induced Dalton's lymphoma in the livers of mice. Nrf2 pathway and phase II antioxidant enzymes were suppressed when lymphoma was progressing in mice. After that they administered curcumin for nine consecutive days through intraperitoneal injections. The Nrf2 pathway was activated, and the expression levels of several phase II enzymes like GST, GR, and NQO1 were restored after administering curcumin. This result shows that curcumin could be an effective preventive agent for lymphoma. Chen et al. [53] worked on MCF-7 human breast cancer cells. Curcumin had dual effects; it increased the expression of Nrf2 as well as decreased the expression of flap endonuclease 1 (Fen1). An endonuclease is an enzyme that cleaves the phosphodiester bonds of DNA, and Fen1 is a specific endonuclease that is known to be active in breast cancer. They also showed that application of curcumin increases the translocation of Nrf2 from the cytoplasm to the nucleus, thus facilitating the activation of the Nrf2 pathway.

Epigallocatechin gallate (EGCG) is the major active ingredient in green tea, and genistein is an isoflavonoid derived from soybean. Both of these polyphenols have important effects on the Nrf2 pathway. Polychlorinated biphenyl (PCB) is known for its environmental toxicity. Han et al. [54] reported that EGCG increased the transcription of Nrf2 and other genes in the Keap1-Nrf2 signaling pathway, in endothelial cells affected by PCB 126-induced inflammation. EGCG also reduced the binding of the transcription factor NF-κB to DNA. EGCG also has preventive effects on epithelial-mesenchymal transition (EMT), one of the key factors for cancer metastasis. Kanlaya et al. [55] demonstrated that EGCG has a protective effect on oxalate-induced EMT in tubular cells. First, MDCK renal tubular cells were treated with 0.5 mM sodium oxalate for 24 h. This reduced the expression levels of epithelial markers, as well as increased the expressions of vimentin and fibronectin, indicative of EMT. However when the same cells were treated by 25 μM EGCG for an hour before the treatment with sodium oxalate, these changes were prevented by EGCG inducing the expressions of antioxidant enzymes related to Nrf2. Genistein has antioxidant potential in several tissues and cell lines. Zhang et al. [7] studied the effect of genistein on human endothelial cells EA.hy926 when the cells were treated with hydrogen peroxide. When the cells were pre-treated with genistein, the expression levels of Nrf2, HO-1, and PPARγ were increased. Heme oxygenase 1 (HO-1) is an enzyme of the heat shock protein family and known to reduce oxidative stress due to H_2O_2 [56]. Peroxisome proliferator-activated receptor-γ (PPARγ) is a nuclear receptor that has been implicated in many diseases including cancer. Nrf2, HO-1, and PPAR-γ protected the EA.hy926 cells from H_2O_2 damage [7]. Similarly, genistein was shown to induce higher expressions of Nrf2 and phase II detoxification genes in Caco-2 human colon cancer cells [57]. Caco-2 cells were first treated with genistein and then with H_2O_2; the genistein pre-treatment significantly reduced H_2O_2-induced cell death and increased the expression of HO-1.

Cruciferous vegetables are rich in phytochemicals including isothiocyanates (ITCs), and consumption of these vegetables may reduce the risks of contracting many human cancers [58]. The main isothiocyanates in these vegetables are allyl isothiocyanate (AITC), benzyl isothiocyanate (BITC), phenethyl isothiocyanate (PEITC), and sulforaphane (SFN). AITC is present in mustard, cabbage, and horseradish; BITC and PEITC are present in water cress and garden cress; and SFN is present in brassicas like cauliflower, broccoli, and Brussels sprouts [2]. Many in vitro studies have confirmed that ITCs have strong pharmacological activities against many kinds of cancers. Also some in vivo studies have indicated that the bioavailability of ITCs is quite good [59].

Wattenberg [60, 61] experimented with the anticarcinogenic potential of BITC in some of the early works. BITC inhibited mammary tumor formation due to the administration of a known carcinogen 7,12-dimethylbenz[a]anthracene (DMBA) to Sprague-Dawley rats. BITC was administered 4 hours before the administration of DMBA. Addition of BITC to a diet containing DMBA also prevented neoplasm in the stomach in ICR/HA mice [60, 61]. There is mounting evidence that the chemopreventive effects of ITCs are through the Nrf2/ARE pathway. In particular, ITCs induce the expression of phase II enzymes like UGT, GST, and NQO-1; as a result phase II enzymes participate in xenobiotic elimination. We have previously discussed the connection between phase II enzymes and Nrf2. The promoter regions of phase II enzymes share the common sequence ARE, and the increased transcription of phase II enzymes is related to Nrf2 binding to ARE. This connection between Nrf2 and ARE has been demonstrated in several studies. Mice that are deficient in Nrf2 are susceptible to carcinogenesis and respond poorly to preventive treatments [62, 63]. Hong et al. [64] demonstrated a direct link between the release of Nrf2 in the cytoplasm and the Keap1-Nrf2 ubiquitination complex due to sulforaphane (SFN). Hong et al. [64] showed that sulforaphane reacts with the thiol group of Keap1 releasing Nrf2 from the Keap1-Nrf2 complex. Nrf2 is translocated to the nucleus with increased concentration due to this mechanism. The mechanism due to PEITC seems to be different, as PEITC does not seem to react with the Keap1-Nrf2 complex [65]. Xu et al. [65] proposed a MAPK-dependent mechanism, since JNK1 and ERK attenuated PEITC-induced ARE activity. They also demonstrated that Nrf2 was released from the Keap1-Nrf2 complex when Nrf2 was phosphorylated.

5.8.2 Intervention of Phytochemicals in the Apoptosis Pathways

We have reviewed the intrinsic and extrinsic apoptosis pathways in Chap. 1, and several phytochemicals have chemopreventive effects through triggering the apoptosis pathways [51]. The Bcl-2 family of genes play very intricate roles in the intrinsic apoptotic pathway; in particular the complex balance of expression levels of the pro-apoptotic and antiapoptotic members of this family is important for the intrinsic pathway. Phytochemicals intervene in the apoptosis pathways through various mechanisms. Some of the examples are increasing the expressions of p53 and Bax [66], through depolarization of mitochondrial transmembrane potential, generation of H_2O_2, and release of cytochrome c [67] and through dissipating mitochondrial membrane potential, upregulation of pro-apoptotic Bax expression, and downregulation of antiapoptotic Bcl-2 expression [68].

The review article by Taraphdar et al. [69] cites many early works on phytochemicals inducing apoptosis, including phenolic compounds, flavonoids, tea polyphenols, and alkaloids. We mention some of the examples here, referring the reader to their paper [69] for further details. *Selaginella tamariscina* is a plant used extensively in Korean and Chinese traditional medicine. It contains several beneficial phytochemicals including phenol, sterol, alkaloid, aliphatic acid, and terpenoid. Lee et al. [70] studied the effects of the extracts from this plant on human ovarian cancer cell line A-2780 and human leukemia cell line U-937. The extracts increased the expression of p53 gene and also arrested the cells at the G1 phase of the cell cycle, most probable causes for inducing apoptosis. Ren and Tang [71] showed that the extract from the plant *Solanum muricatum* inhibits tumor growth by inducing apoptosis through cytotoxic activities on a broad range of human cancer cells, including ovarian (OVCAR), liver (QSY7721,SK-HEP1), colon (HT-29), lung (NCI-H-209), stomach (MKN-45), prostrate (PC-3, DUI-45), and breast (MDA-MB-435). Similarly, Inoue et al. [72] showed that gallic acid induces apoptosis in various transformed cell lines including HL-60 (human promyelocytic leukemia), HeLa (human epithelial carcinoma), KB (human epidermoid carcinoma), PLC/PRF/5 (human hepatoma), P-388 D1 (mouse lymphoid neoplasm), and dRLh-84 (rat hepatoma). Apoptosis was induced most probably through reactive oxygen species generation including H_2O_2.

Roy et al. [73] investigated apoptosis on different cancer cell lines due to several natural phenolic compounds including curcumin, yakuchinone B, resveratrol, and capsaicin. They used four different human cancer cell lines in their study, chronic myelogenous leukemia (K-562), breast adenocarcinoma (MCF-7), myeloblastic leukemia (HL-60), and cervical epithelial carcinoma (HeLa). The phenolic compounds were cytotoxic and inhibited growth; however apoptosis was induced in different cells by different phenolic compounds studied through dose-dependent applications of the phenolic compounds. The mechanism of apoptosis was through chromatin condensation and nuclear fragmentation. Curcumin was the most cytotoxic towards K-562, HL-60, and MCF-7, but HeLa cells were unaffected. Yakuchinone B was overall less active than curcumin but had cytotoxic effects on all four cell lines. Resveratrol was cytotoxic only in leukemic cells, and capsaicin was marginally cytotoxic on all cells. However none of the phenolic compounds were cytotoxic on lymphocytes isolated from normal human blood. Chakraborty et al. [74] studied the effect of diospyrin, a bisnaphthoquinonoid natural product, and its three synthetic derivatives on four human cancer cell lines: chronic myelogenous leukemia (K-562), breast adenocarcinoma (MCF-7), acute myeloblastic leukemia (HL-60), and cervical epithelial carcinoma (HeLa). D7, one of the synthetic derivatives of diospyrin, exhibited the highest cytotoxicity and induced apoptosis in all the cell lines, which was evident from chromatin condensation and nuclear fragmentation. Similarly, theaflavin, a phytochemical present in black tea, caused apoptosis in human leukemia cells [75, 76] and human leukemia K562 cells [77].

Mukherjee et al. [78] showed that curcumin induced apoptosis in HL-60, the human leukemia cells, through the inhibition of telomerase activity. Mukherjee et al. [79] showed in a study that histone deacetylase was inhibited by curcumin resulting in upregulation of p21 and apoptosis in human breast cancer cells. As we have discussed before, histone deacetylases (HDAC) can inhibit gene expression by preventing the access of transcription factors to specific genes. Roy et al. [80] showed that PEITC can inhibit HDACs in breast cancer cells. Sarkar et al. [81] showed that sulforaphane induced apoptosis in breast cancer cells by targeting heat shock proteins. Sarkar et al. [82] further demonstrated that PEITC also induces cell cycle arrest and apoptosis by targeting heat shock proteins.

5.9 Molecular Mechanisms of Phytochemicals in Chemosensitization

The aim of chemosensitization is to enhance the activity or efficacy of chemotherapeutic drugs by using secondary drugs, without increasing the toxic side effects. We discuss in this section the use of phytochemicals as chemosensitizing agents, used with cancer drugs. A tumor has a complex environment comprising malignant cells that have uncontrollable growth and proliferation, as well as normal cells that are required for normal functioning of tissues.

The aim of chemotherapeutic drugs is to induce apoptosis in the malignant cells while keeping the normal cells unharmed. No chemotherapeutic drug can achieve this in practice, as the pathways they target for inducing apoptosis are similar both in malignant and healthy cells, and hence healthy cells are also affected. Moreover this problem is exacerbated when chemotherapeutic drugs are administered in higher dosages for eliminating malignant cells. Most chemotherapeutic drugs induce apoptosis through genotoxic stress, and as the malignant cells become more and more immune to these stresses, the dosage needs to be increased, affecting healthy cells. Such high dosage of chemotherapeutic drugs has many short- and long-term side effects due to the death of normal cells. These side effects include lack of digestive track lining due to loss of intestinal epithelial cells and hair fall due to the loss of hair keratinocyte cells. Chemosensitizing agents are important so that chemotherapeutic drugs can be administered at relatively low doses for inducing apoptosis in malignant cells and also preventing these side effects. We first discuss some of the challenges in chemosensitization and then discuss the different phytochemicals that have gone through clinical trials as chemosensitization agents.

5.9.1 Challenges in Chemosensitization

We need to briefly review the apoptosis pathways from Chap. 1 to understand the intricacies that the chemotherapeutic drugs have to overcome in order to induce apoptosis in malignant cells. Both the intrinsic and extrinsic pathways can confer resistance to chemotherapy through upregulation of antiapoptotic genes involved in these pathways. We give a few representative examples of chemoresistance and refer the reader to the review articles [83, 84].

5.9.1.1 Chemoresistance in the Intrinsic Apoptosis Pathway

At the heart of the intrinsic apoptotic pathway is the release of first group of proteins including cytochrome c and Smac/DIABLO through mitochondrial pores. Cytochrome c binds with procaspase-9, and the cascading activation of the caspase family of proteins starts. The intrinsic apoptosis pathway is controlled by the Bcl-2 family of proteins that are the main agents for opening the mitochondrial pores. There are about 25 genes in the Bcl-2 family; some are antiapoptotic and some pro-apoptotic. The pro-apoptotic proteins open the mitochondrial pores, and the antiapoptotic proteins prevent this. These proteins form complexes in normal cellular environments; however the pro-apoptotic proteins are released from these complexes under stress signals. Chemoresistance results when the proteins participating in the intrinsic pathway are aberrantly expressed in cells. Upregulation of the antiapoptotic proteins and downregulation of pro-apoptotic proteins result in chemoresistance.

Some studies have indeed shown the connection between upregulation of the antiapoptotic Bcl-2 proteins and the tumor malignancy and poor response to chemotherapy and radiation therapy [85–88].

5.9.1.2 Chemoresistance in the Extrinsic Apoptosis Pathway

The extrinsic apoptosis pathway is triggered by external signals or ligands received through receptors on the cellular membrane. Two of the most common ligand/receptor pairs are FasL/FasR (CD95L/CD95) and TNFα/TNFR1. Once the ligand is bound to the receptor, cytoplasmic adapter proteins with matching death domains bind to the receptor. FADD is the adapter protein for FasL/FasR and TRADD; RIP and FADD are the adapter proteins for TNFα/TNFR1. A death-inducing signaling complex (DISC) is formed next by binding with procaspase-8, and the autocatalytic activation of the procaspases begins, and next the execution pathway is triggered. It is possible to disrupt the extrinsic apoptosis pathway at two points, preventing the extracellular ligands from binding to the receptors or preventing the adapter proteins from binding to the ligand/receptor complex. Both of these mechanisms confer chemoresistance as we discuss below.

The extrinsic pathway is disrupted by soluble receptors that act like decoys instead of FasL or CD95 and bind to CD95L or FasR, thus disrupting the pathway. Two such decoy-soluble receptors are known, soluble CD95 (sCD95) and decoy receptor 3 (DcR3). sCD95 is expressed in many tumors, and sera of cancer patients have elevated levels of sCD95. In particular, higher levels of sCD95 have been associated with poor prognosis of melanoma patients [89, 90]. Cellular FLICE (FADD-like IL-1β-converting enzyme)-inhibitory protein (c-FLIP) is a very important antiapoptotic protein. c-FLIP interferes with the extrinsic pathway at the level of death receptors. c-Flip prevents the formation of DISC by binding to FADD and/or caspase-8 or -10 [91].

5.9.1.3 Downregulation of Pro-apoptotic Genes

Tumors also can acquire resistance to apoptosis by mutating or downregulating pro-apoptotic genes, thus disrupting the apoptosis pathways. The pro-apoptotic Bcl-2 family gene BAX is mutated in several cancers [92, 93]. It has been observed that there is a direct correlation between poor response to chemotherapy and reduced expression of BAX [94]. Many other pro-apoptotic genes are downregulated in several cancers; we refer to the review article by Igney and Krammer [83], as well as the references in their paper for details, and mention a few of these downregulated genes here.

The APAF1 protein is a very important part of the apoptosome that forms after the release of cytochrome c from the mitochondria. APAF1 is often transcriptionally inactivated through methylation in melanoma, and several APAF1-negative melanomas respond to chemotherapy poorly. Caspase-8 is often inactivated through methylation in neuroblastomas, and caspase-8-deficient neuroblastoma cells are resistant to chemotherapy. The death receptor CD95 is downregulated in many tumor cells, including neoplastic colon epithelium, hepatocellular carcinoma, and melanomas. The downregulation of CD95 causes immune evasion and chemoresistance.

5.9.2 Phytochemicals as Chemosensitizers

A large-scale literature review by Vinod et al. [84] has shown that many phytochemicals have been identified as potential chemosensitizers. These phytochemicals are mainly phenolic compounds, phenols, phenolic acids, polyphenols, and polyphenolic flavonoids. The main phytochemicals that inhibit the antiapoptotic pathways are curcumin, quercetin, resveratrol, EGCG, genistein, etc. Though in vitro analyses have shown their chemosensitization potentials, and some of these compounds have gone through clinical trials, their in vivo effectiveness is hampered due to poor bioavailability. We discuss issues related to bioavailability in details in Chap. 8; here we briefly review the chemosensitization potentials of phytochemicals following the recent review article by Vinod et al. [84]. We refer to their article for more details.

5.9.2.1 Curcumin

Curcumin has been found to have chemosensitization effects in association with two drugs, 5FU and gemcitabine. 5FU (fluorouracil) is a chemotherapy drug used in several different cancers including stomach, esophageal, breast, bowel, skin, and pancreatic cancer. Gemcitabine is a medication used to treat breast, pancreatic, ovarian, non-small cell lung, and bladder cancer. There is strong evidence that one of the chemosensitization effects of curcumin is through the NF-κB pathway.

Gemcitabine increases the expression levels of NF-κB in several pancreatic carcinoma cell lines [95], conferring chemoresistance. Curcumin inhibits this over expression of NF-κB resulting in inhibition of proliferation and angiogenesis [96]. Curcumin also enhances the cytotoxicity of gemcitabine through downregulation of COX-2 and ERK-1/2 levels [97]. Curcumin enhances the efficacy of mitotic chemotherapeutic agents, for example, in vitro and in vivo studies have confirmed that antitumor effects of the drug paclitaxel are enhanced by curcumin [84]. Paclitaxel upregulates NF-κB, Akt, and the antiapoptotic members of the Bcl-2 family; curcumin on the other hand downregulates these genes. Curcumin has many other chemosensitization effects through other pathways [84].

5.9.2.2 Quercetin

The flavonoid quercetin is present in many foods, including red onion, green and black tea, red wine, and leafy vegetables. It has been confirmed as a potential chemosensitization agent in many studies, for anticancer drugs including antitumor antibiotics like doxorubicin, TRAIL, and alkylating agents like DTIC and cisplatin. The overexpression of HIF1-α due to hypoxia has been implicated in many cancers. HIF1-α is an important transcription factor that is involved in the transcriptions of scores of genes, some among those genes are involved in angiogenesis, and hence its upregulation can reduce the effects of chemotherapeutic drugs. It has been shown that quercetin downregulates H1F1-α and hence makes cancer cells sensitive to doxorubicin [98].

Quercetin also helps in activating or enhancing immune responses against tumors. One of the major side effects of chemotherapy is immune dysfunction. It has been shown that quercetin induced tumor-specific T-cell responses, promoting lymphocyte proliferation [84]. Another major chemosensitizing effect of quercetin is the sensitization of cancer cells to TRAIL-mediated apoptosis through acting upon the components of death-inducing signaling complex (DISC) of the apoptosis pathway [99].

5.9.2.3 Resveratrol

TRAIL is the main drug that is potentiated by resveratrol. The chemosensitization action of resveratrol seems to be the downregulation of antiapoptotic members of the Bcl-2 family of genes, NF-κB pathway, and the inhibitor of apoptosis (IAP) family of proteins [100]. Resveratrol seems to induce the distribution of death receptors on lipid rafts in colon cancer cells. This sensitizes the colon cancer cells to death receptors like TRAIL and TNF. It has also been reported that resveratrol upregulates the pro-apoptotic members like Bax, Bak, Puma, Noxa, Bim, and Trail-R1/R2, as well as downregulates antiapoptotic members like Bcl-2 and Bcl-x_L, of the Bcl-2 family of genes [101].

5.9.2.4 EGCG

EGCG (epigallocatechin gallate) is a key polyphenol in green tea. The main chemosensitization action of EGCG is overcoming TRAIL resistance in many kinds of cancer cells. EGCG modulates the key proteins involved in intrinsic and extrinsic apoptosis pathways, e.g., TRAIL-R1, FADD, c-FLIP, Bcl-2, and Bcl-x_L. These modulations sensitize LNCaP cells that are TRAIL-resistant, and the cells respond to apoptosis mediated by TRAIL. The combination of TRAIL and EGCG also significantly reduced the proliferation and migration of LNCaP cells [102]. EGCG mitigates many side effects of the drug IFN-α (interferon-α) through chemosensitization effects. IFN-α is used in therapy for high-risk melanoma patients. Both in vitro and in vivo studies have shown that EGCG in combination with IFN-α increases FasL-mediated apoptosis through the external apoptosis pathway in melanoma cells. The same study also reported the upregulation of FasL and downregulation of NF-κB [103]. EGCG also downregulates EGFR, mTOR, Akt, and NF-κB in breast cancer cells and acts a chemosensitizer for the drug 4-hydroxy tamoxifen [104].

EGCG acts as a chemosensitizer for several other drugs used in chemotherapy, including paclitaxel, alkylating agents like DTIC, 5-FU, cisplatin gemcitabine, and sulindac, through various mechanisms like downregulation of antiapoptotic members of Bcl-2 family of genes, increasing mitochondrial membrane depolarization (thus facilitating the release of cytochrome c), and inhibition of EGFR [84].

5.9.2.5 Genistein

Chemosensitizing effects of genistein have been reported in combination with several drugs used in chemotherapy. The main molecular mechanisms involved seem to be the downregulation of NF-κB and Akt, as well as the modulation of TRAIL, or its downstream genes. Genistein has been reported to be a promising chemosensitizer along with antimetabolites like 5-FU and gemcitabine in osteosarcoma [105]. Genistein also sensitizes pancreatic cancer cells through the same mechanism when gemcitabine and erlotinib are used in combination [106]. Trail receptor 2 or Trail-R2 (also known as death receptor 5, or DR5) is a cell surface receptor that mediates apoptosis by binding to TRAIL. Genistein upregulates TRAIL-R2 and sensitizes gastric adenocarcinoma cells to TRAIL-induced apoptosis [107].

It has been reported that genistein also downregulates Akt for sensitizing cancer cells to the apoptotic effects induced by TRAIL [108]. Genistein also has chemosensitization effects in combination with other chemotherapeutic drugs through downregulation of NF-κB, e.g., with the mitotic inhibitor docetaxel in prostate cancer, with docetaxel and cisplatin in pancreatic cancer, and with CHOP (cyclophosphamide, doxorubicin, and prednisone) in large cell lymphoma [84]. Genistein has chemosensitizing effects by downregulating Akt in acute myeloid leukemia in conjunction with the drug cytosine arabinoside (Ara-C) and with perifosine in prostate cancer [84].

5.10 Conclusion

We have discussed in this chapter the molecular mechanisms that are important for understanding the effects of phytochemicals in chemoprevention and chemosensitization. Phytochemicals not only are able to modulate expressions of genes, but these modulations affect epigenetic control mechanisms like histone modification (through acetylation and methylation), DNA methylation, and miRNA silencing. We have reviewed these epigenetic control mechanisms in details. We have also reviewed the roles of oxidative stress and inflammation in carcinogenesis and how phytochemicals can prevent these mechanisms by intervening in various genetic pathways including the Keap1-Nrf2 pathway. The roles of phytochemicals are now firmly established in preventing the initiation and progression of carcinogenesis. We have discussed the chemopreventive effects of phytochemicals in details, with a focus on their actions on different genes and pathways. We have finally discussed the molecular mechanisms of different dietary phytochemicals in chemosensitization, i.e., augmenting the effects of chemotherapeutic drugs by administering phytochemicals in conjunction with these drugs.

References

1. B. Alberts, A.D. Johnson, J. Lewis, D. Morgan, M. Raff, K. Roberts, P. Walter, *Molecular Biology of the Cell*, 6th edn. (Garland Science, New York, 2015)
2. W. Li, G. Yue, C. Zhang, R. Wu, A.Y. Yang, J. Gasper, A.-N.T. Kong, Dietary phytochemicals and cancer chemoprevention: a perspective on oxidative stress, inflammation, and epigenetics. Chem. Res. Toxicol. **29**, 2071–2095 (2016)
3. A. Raghunath, K. Sundarraj, R. Nagarajan, F. Arfuso, J. Bian, A.P. Kumar, G. Sethi, E. Perumal, Antioxidant response elements: discovery, classes, regulation and potential applications. Redox Biol. **17**, 297–314 (2018)
4. T. Nguyen, P. Nioi, C.B. Pickett, The Nrf2-antioxidant response element signaling pathway and its activation by oxidative stress. J. Biol. Chem. **284**(20), 13291–13295 (2009)
5. G.-Y. Liou, P. Storz, Reactive oxygen species in cancer. Free Radic. Res. **44**(5) (2010). https://doi.org/10.3109/10715761003667554
6. A. Hryniuk, S. Grainger, J.G. Savory, D. Lohnes, Cdx1 and Cdx2 function as tumor suppressors. J. Biol. Chem. **289**(48), 33343–33354 (2014)
7. R. Zhang, K. Kang, K.C. Kim, S.Y. Na, W.Y. Chang, G.Y. Kim, H.S. Kim, J.W. Hyun, Oxydative stress causes epigenetic alteration of CDX1 expression in colorectal cancer cells. Gene **524**, 214–219 (2013)
8. P.K.S. Mahalingaiah, L. Ponnusamy, K.P. Singh, Oxidative stress-induced epigenetic changes associated with malignant transformation of human kidney epithelial cells. Oncotarget **8**(7), 11127–11143 (2017)
9. F. Ito, Y. Yamada, A. Shigemitsu, M. Akinishi, H. Kaniwa, R. Miyake, S. Yamanaka, H. Kobayashi, Role of oxidative stress in epigenetic modification in endometriosis. Reprod. Sci. **24**(11), 1493–1502 (2017)
10. A. Guillaumet-Atkins, Y. Yañez, M..D. Peris-Diaz, I. Calabria, C. Palanca-Ballester, J. Sandoval, Epigenetics and oxidative stress in aging. Oxid. Med. Cell. Longev. **2017**, 9175806 (2017)
11. P.K. Mahalingaiah, L. Ponnusamy, K.P. Singh, Chronic oxidative stress causes estrogen-independent aggressive phenotype, and epigenetic inactivation of estrogen receptor alpha in MCF-7 breast cancer cells. Breast Cancer Res. Treat. **153**, 41–56 (2015)
12. P. Katakwar, R. Metgud, S. Naik, R. Mittal, Oxidative stress markers in oral cancer: a review. J. Cancer Res. Ther. **12**, 438–446 (2016)
13. M.S. Cooke, M.D. Evans, M. Dizdaroglu, J. Lune, Oxidative DNA damage: mechanisms, mutation, and disease. FASEB J. **17**(10), 1195–214 (2003)
14. L.M. Coussens, Z. Werb, Inflammation and cancer. Nature **19**, 860–867 (2002)
15. P. Thejass, G. Kuttan, Inhibition of endothelial cell differentiation and proinflammatory cytokine production during angiogenesis by allyl isothiocyanate and phenyl isothiocyanate. Integr. Cancer Ther. **6**, 389–399 (2007)
16. J.N. Sharma, A. Al-Omran, S.S. Parvathy, Role of nitric oxide in inflammatory diseases. Inflammopharmacology **15**(6), 252–259 (2007)
17. N.M. Reddy, S.R. Kleeberger, T.W. Kensler, M. Yamamoto, P.M. Hassoun, S.P. Reddy, Disruption of Nrf2 impairs the resolution of hyperoxia-induced acute lung injury and inflammation in mice. J. Immunol. **182**, 7264–7271 (2009)
18. K.A. Steinmetz, J.D. Potter, Vegetables, fruit, and cancer. I. Epidemiology. Cancer Causes Control **2**(5), 325–357 (1991)
19. R.A. Cairns, I.S. Harris, T.W. Mak, Regulation of cancer cell metabolism. Nat. Rev. Cancer **11**(2), 85–95 (2011)
20. J.L. Martindale, N.J. Holbrook, Cellular response to oxidative stress: signalling for suicide and survival. J. Cell. Physiol. **192**(1), 1–15 (2002)
21. A. Zeke, M. Misheva, A. Reményi, M.A. Bogoyevitch, JNK signaling: regulation and functions based on complex protein-protein partnerships. Microbiol. Mol. Biol. Rev. **80**(3), 793–835 (2016)

22. S.K. Katiyar, F. Afaq, K. Azizuddin, H. Mukhtar, Inhibition of UVB-inducedoxidative stress-mediated phosphorylation of mitogen-activated protein kinase signaling pathways in cultured human epidermal keratinocytes by green tea polyphenol epigallocatechin-3-gallate. Toxicol. Appl. Pharmacol. **176**, 110–117 (2001)
23. S.K. Manna, A. Mukhopadhyay, B.B. Aggarwal, Resveratrol suppresses TNF-induced activation of nuclear transcription factors NF-κB, activator protein-1, and apoptosis: potential role of reactive oxygen intermediates and lipid peroxidation. J. Immunol. **164**, 6509–6519 (2000)
24. R. Yu, V. Hebbar, D.W. Kim, S. Mandlekar, J.M. Pezzuto, A.N. Kong, Resveratrol inhibits phorbol ester and UV-induced activator protein 1 activation by interfering with mitogen-activated protein kinase pathways. Mol. Pharmacol. **60**, 217–224 (2001)
25. F. Yin, A.E. Giuliano, A.J. van Herle, Signal pathways involved in apigenin inhibition of growth and induction of apoptosis of human anaplastic thyroid cancer cells (ARO). Anticancer Res. **19**, 4297–4303 (1999)
26. X. Yan, M. Qi, P. Li, Y. Zhan, H. Shao, Apigenin in cancer therapy: anti-cancer effects and mechanisms of action. Cell Biosci. **7**, 50 (2017)
27. B. Sung, H.Y. Chung, N.D. Kim, Role of apigenin in cancer prevention via the induction of apoptosis and autophagy. J. Cancer Prev. **21**(4), 216–226 (2016)
28. N. Ahmad, V.M. Adhami, F. Afaq, D.K. Feyes, H. Mukhtar, Resveratrol causes WAF-1/p21-mediated G 1-phase arrest of cell cycle and induction of apoptosis in human epidermoid carcinoma A431 cells. Clin. Cancer Res. **7**, 1466–1473 (2001)
29. J.-H. Ko, G. Sethi, J.-Y. Um, M.K. Shanmugam, F. Arfuso, A.P. Kumar, A. Bishayee, K.S. Ahn, The role of resveratrol in cancer therapy. Int. J. Mol. Sci. **18**(12), 2589 (2017)
30. Q.B. She, C. Huang, Y. Zhang, Z. Dong, Involvement of c-jun NH_2-terminal kinases in resveratrol-induced activation of p53 and apoptosis. Mol. Carcinogenesis **33**, 244–250 (2002)
31. M. Mouria, A.S. Gukovskaya, Y. Jung, P. Buechler, O.J. Hines, H.A. Reber, S.J. Pandol, Food-derived polyphenols inhibit pancreatic cancer growth through mitochondrial cytochrome C release and apoptosis. Int. J. Cancer **98**, 761–769 (2002)
32. M.K. Johnson, G. Loo, Effects of epigallocatechin gallate and quercetin on oxidative DNA damage to cellular DNA. Mutat. Res. **459**, 211–218 (2000)
33. M.R. Kelly, J. Xu, K.E. Alexander, G. Loo, Disparate effects of similar phenolic phytochemicals as inhibitors of oxidative damage to cellular DNA. Mutat. Res. **485**, 309–318 (2001)
34. X. Mao, C. Gu, D. Chen, B. Yu, J. He, Oxidative stress-induced diseases and tea polyphenols. Oncotarget **8**(46), 81649–81661 (2017)
35. G. Loo, Redox-sensitive mechanisms of phytochemical-mediated inhibition of cancer cell proliferation. J. Nutr. Biochem. **14**(2), 64–73 (2003)
36. F. Kassie, B. Pool-Zobel, W. Parzefall, S. Knasmuller, Genotoxic effects of benzyl isothiocyanate, a natural chemopreventive agent. Mutagenesis **14**, 595–603 (1999)
37. Y. Nakamura, H. Ohigashi, S. Masuda, A. Murakami, Y. Morimitsu, Y. Kawamoto, T. Osawa, M. Imagawa, K. Uchida, Redox regulation of glutathione S-transferase induction by benzyl isothiocyanate: correlation of enzyme induction with the formation of reactive oxygen intermediates. Cancer Res. **60**, 219–225 (2000)
38. L. Gamet-Payrastre, P. Li, S. Lumeau, G. Cassar, M.-A. Dupont, S. Chevolleau, N. Gasc, J. Tulliez, F. Terce, Sulforaphane, a naturally occurring isothiocyanate, induces cell cycle arrest and apoptosis in HT29 human colon cancer cells. Cancer Res. **60**, 1426–1433 (2000)
39. C. Fimognari, M. Nusse, R. Cesari, R. Iori, G. Cantelli-Forti, P. Hrelia, Growth inhibition, cell cycle arrest and apoptosis in human T-cell leukemia by the isothiocyanate sulforaphane. Carcinogenesis **23**, 581–586 (2002)
40. Y.-R. Chen, W. Wang, A.-N.T. Kong, T.-H. Tan, Molecular mechanisms of c-Jun N-terminal kinase-mediated apoptosis induced by anticarcinogenic isothiocyanates. J. Biol. Chem. **273**(3), 1769–1775 (1998)
41. S.M. de Figueiredo, S.A. Filho, J.A. Nogueira-Machado, R.B. Caligiorne, The anti-oxidant properties of isothiocyanates: a review. Recent Pat. Endocr. Metab. Immune Drug Discov. **7**(3), 213–225 (2013)

42. L. Valgimigli, R. Iori, Antioxidant and pro-oxidant capacities of ITCs. Environ. Mol. Mutagen. **50**, 222–237 (2009)
43. T. Liu, L. Zhang, D. Joo, S.-C. Sun NF-κB signaling in inflammation. Signal Transduct. Target. Ther. **2**, 17023 (2017)
44. N. M'hiri, I. Ioannou, M. Ghoul, N.M. Boudhrioua, Phytochemical characteristics of citrus peel and effect of conventional and nonconventional processing on phenolic compounds: a review. Food Rev. Int. **33**(6), 587–619 (2017)
45. T. Etoh, Y.P. Kim, M. Hayashi, M. Suzawa, S. Li, C. Ho, K. Komiyama, Inhibitory effect of a formulated extract from multiple citrus peels on LPS-induced inflammation in RAW 246.7 macrophages. Funct. Foods Health Dis. **3**, 242–253 (2013)
46. A. Fazio, P. Plastina, J. Meijerink, R.F. Witkamp, B. Gabriele, Comparative analyses of seeds of wild fruits of Rubus and Sambucus species from Southern Italy: fatty acid composition of the oil, total phenolic content, antioxidant and anti-inflammatory properties of the methanolic extracts. Food Chem. **140**, 817–824 (2013)
47. F.C. Lau, D.F. Bielinski, J.A. Joseph, Inhibitory effects of blueberry extract on the production of inflammatory mediators in lipopolysaccharide-activated BV2 microglia. J. Neurosci. Res. **85**, 1010–1017 (2007)
48. F. Zhu, B. Du, B. Xu, Anti-inflammatory effects of phytochemicals from fruits, vegetables and food legumes: a review. Crit. Rev. Food Sci. Nutr. **58**(8), 1260–1270 (2018)
49. G. Grosso, F. Bella, J. Godos, S. Sciacca, D. Del Rio, S. Ray, F. Galvano, E.L. Giovannucci, Possible role of diet in cancer: systematic review and multiple meta-analyses of dietary patterns, lifestyle factors, and cancer risk. Nutr. Rev. **75**(6), 405–419 (2017)
50. S.S. Boyanapalli, A.-N. Kong, "Curcumin, the king of spices": epigenetic regulatory mechanisms in the prevention of cancer, neurological and inflammatory diseases. Curr. Pharmacol. Rep. **1**, 129–139 (2015)
51. L. Shu, K.-L. Cheung, T.O. Khor, C. Chen, A.-N. Kong, Phytochemicals: cancer chemoprevention and suppression of tumor onset and metastasis. Cancer Metastasis Rev. **29**, 483–502 (2010)
52. L. Das, M. Vinayak, Long term effect of curcumin in restoration of tumour suppressor p53 and phase-II antioxidant enzymes via activation of Nrf2 signaling and modulation of inflammation in prevention of cancer. PLoS One **10**, e0124000 (2015)
53. B. Chen, Y. Zhang, Y. Wang, J. Rao, X. Jiang, Z. Xu, Curcumin inhibits proliferation of breast cancer cells through Nrf2-mediated downregulation of Fen1 expression. J. Steroid Biochem. Mol. Biol. **143**, 11–18 (2014)
54. S.G. Han, S.S. Han, M. Toborek, B. Hennig, EGCG protects endothelial cells against PCB 126-induced inflammation through inhibition of AhR and induction of Nrf2-regulated genes. Toxicol. Appl. Pharmacol. **261**, 181–188 (2012)
55. R. Kanlaya, S. Khamchun, C. Kapincharanon, B. Thongboonkerd, Protective effect of epigallocatechin-3-gallate (EGCG) via Nrf2 pathway against oxalate-induced epithelial mesenchymal transition (EMT) of renal tubular cells. Sci. Rep. **6**, 30233 (2016)
56. Z. Jian, K. Li, L. Liu, Y. Zhang, Z. Zhou, C. Li, T. Gao, Heme oxygenase-1 protects human melanocytes from H2O2-induced oxidative stress via the Nrf2-ARE pathway. J. Invest. Dermatol. **131**(7), 1420–1427 (2011)
57. X. Zhai, M. Lin, F. Zhang, Y. Hu, X. Xu, Y. Li, K. Liu, X. Ma, X. Tian, J. Yao, Dietary flavonoid genistein induces Nrf2 and phase II detoxification gene expression via ERKs and PKC pathways and protects against oxidative stress in Caco-2 cells. Mol. Nutr. Food Res. **57**, 249–259 (2013)
58. J.V. Higdon, B. Delage, D.E. Williams, R.H. Dashwood, Cruciferous vegetables and human cancer risk: epidemiologic evidence and mechanistic basis. Pharmacol. Res. **55**(3), 224–236 (2007)
59. X. Wu, Q.-h. Zhou, K. Xu, Are isothiocyanates potential anti-cancer drugs? Acta Pharmacol. Sin. **30**(5), 501–512 (2009)
60. L.W. Wattenberg, Inhibition of carcinogenic effects of polycyclic hydrocarbons by benzyl isothiocyanate and related compounds. J. Natl. Cancer Inst. **58**, 395–398 (1977)

61. L.W. Wattenberg, Inhibition effects of benzyl isothiocyanate administered shortly before diethylnitrosamine or benzo[a]pyrene on pulmonary and forestomach neoplasia in A/J mice. Carcinogenesis **8**, 1971–1973 (1987)
62. M. Ramos-Gomez, M.K. Kwak, P.M. Dolan, K. Itoh, M. Yamamoto, P. Talalay, T.W. Kensler, Sensitivity to carcinogenesis is increased and chemoprotective efficacy of enzyme inducers is lost in nrf2 transcription factor-deficient mice. Proc. Natl. Acad. Sci. **98**, 3410–3415 (2001)
63. T.O. Khor, M.T. Huang, A. Prawan, Y. Liu, X. Hao, S. Yu, W.K. Cheung, J.Y. Chan, B. Reddy, C.S. Yang, A.N. Kong, Increased susceptibility of Nrf2 knockout mice to colitis-associated colorectal cancer. Cancer Prev. Res. **1**, 187–191 (2008)
64. F. Hong, M.L. Freeman, D.C. Liebler, Identification of sensor cysteines in human Keap1 modified by the cancer chemopreventive agent sulforaphane. Chem. Res. Toxicol. **18**, 1917–1926 (2005)
65. C. Xu, X. Yuan, Z. Pan, G. Shen, J.H. Kim, S. Yu, T.O. Khor, W. Li, J. Ma, A.N. Kong, Mechanism of action of isothiocyanates: the induction of ARE-regulated genes is associated with activation of ERK and JNK and the phosphorylation and nuclear translocation of Nrf2. Mol. Cancer Ther. **5**, 1918–1926 (2006)
66. W. Tang, J.W. Liu, W.M. Zhao, D.Z. Wei, J.J. Zhong, Ganoderic acid T from Ganoderma lucidum mycelia induces mitochondria mediated apoptosis in lung cancer cells. Life Sci. **80**(3), 205–211 (2006)
67. J.Y. Cheung, R.C. Ong, Y.K. Suen, V. Ooi, H.N. Wong, T.C. Mak, K.P. Fung, B. Yu, S.K. Kong, Polyphyllin D is a potent apoptosis inducer in drug-resistant HepG2 cells. Cancer Lett. **217**(2), 203–211 (2005)
68. M.S. Lee, J.C. Yuet-Wa, S.K. Kong, B. Yu, V.O. Eng-Choon, H.W. Nai-Ching, T.M. Chung-Wai, K.P. Fung, Effects of polyphyllin D, a steroidal saponin in *Paris polyphylla*, in growth inhibition of human breast cancer cells and in xenograft. Cancer Biol. Ther. **4**(11), 1248–1254 (2005)
69. A.K. Taraphdar, M. Roy, R.K. Bhattacharya, Natural products as inducers of apoptosis: implication for cancer therapy and prevention. Curr. Sci. **80**(11), 1387–1396 (2001)
70. I.-S. Lee, A. Nishikawa, F. Furukawa, K. Kasahara, S.-U. Kim, Effects of Selaginella tamariscina on in vitro tumor cell growth, p53 expression, G1 arrest and in vivo gastric cell proliferation. Cancer Lett. **144**, 93–99 (1999)
71. W. Ren, D.G. Tang, Extract of Solanum muricatum (Pepino/CSG) inhibits tumor growth by inducing apoptosis. Anticancer Res. **19**(1A), 403–408 (1999)
72. M. Inoue, R. Suzuki, T. Koide, N. Sakaguchi, Y. Ogihara, Y. Yabu, Antioxidant, gallic acid, induces apoptosis in HL-60RG cells. Biochem. Biophys. Res. Commun. **204**(2), 898–904 (1994)
73. M. Roy, S. Chakraborty, M. Siddiqi, R.K. Bhattacharya, Induction of apoptosis in tumor cells by natural phenolic compounds. Asian Pac. J. Cancer Prev. **3**(1), 61–67 (2002)
74. S. Chakrabarty, M. Roy, B. Hazra, R.K. Bhattacharya, Induction of apoptosis in human cancer cell lines by diospyrin, a plant derived bisnaphthoquinonoid and its synthetic derivatives. Cancer Lett. **188**, 85–93 (2002)
75. T. Kundu, S. Dey, M. Roy, M. Siddiqi, B.K. Bhattacharya, Induction of apoptosis in human leukemia cells by black tea and its polyphenol theaflavin. Cancer Lett. **230**(1), 111–121 (2005)
76. T. Kundu, R.K. Bhattacharya, M. Siddiqi, M. Roy, Correlation of apoptosis with comet formation induced by tea polyphenols in human leukaemia cells. J. Environ. Pathol. Toxicol. Oncol. **24**(2), 89–102 (2005)
77. S. Chakraborty, T. Kundu, R.K. Bhattacharya, M. Siddiqi, M. Roy, Tea induced apoptosis in human leukemia cell K562 as assessed by comet formation. Asian Pac. J. Cancer Prev. **7**(2), 201–207 (2006)
78. S. Mukherjee, U. Ghosh, N.P. Bhattacharya, R.K. Bhattacharya, S. Dey, M. Roy, Curcumin induced apoptosis in human leukemia cell HL-60 is associated with inhibition of telomerase activity. Mol. Cell. Biochem. **297**, 31–39 (2007)

79. S. Mukherjee, R. Sarkar, J. Biswas, M. Roy, Curcumin inhibits histone deacetylase leading to cell cycle arrest and apoptosis via upregulation of p21 in breast cancer cell lines. Int. J. Green Nanotechnol. **4**, 183–197 (2012)
80. M. Roy, S. Mukherjee, J. Biswas, Inhibition of an epigenetic modulator, histone deacetylase by PEITC in breast cancer - a detailed mechanistic approach. Int. J. Ther. Appl. **5**, 1–13 (2012)
81. R. Sarkar, S. Mukherjee, J. Biswas, M. Roy, Sulphoraphane, a naturally occurring isothiocyanate induces apoptosis in breast cancer cells by targeting heat shock proteins. Biochem. Biophys. Res. Commun. **427**, 80–85 (2012)
82. R. Sarkar, S. Mukherjee, M. Roy, Targeting heat shock proteins (HSPs) by phenethyl isothiocyanate results in cell cycle arrest and apoptosis of human breast cancer cells. Nutr. Cancer **65**(3), 1–14 (2013)
83. F.H. Igney, P.H. Krammer, Death and anti-death: tumor resistance to apoptosis. Nat. Rev. Cancer **2**, 277–288 (2002)
84. B.S. Vinod, T.T. Maliekal, R.J. Anto, Phytochemicals as chemosensitizers: from molecular mechanism to clinical significance. Antioxid. Redox Signal. **18**(11), 1307–1348 (2013)
85. L. Campos, J.P. Rouault, O. Sabido, P. Oriol, N. Roubi, C. Vasselon, E. Archimbaud, J.P. Magaud, D. Guyotat, High expression of BCL-2 protein in acute myeloid leukemia cells is associated with poor response to chemotherapy. Blood **81**, 3091–3096 (1993)
86. M. Weller, U. Malipiero, A. Aguzzi, J.C. Reed, A. Fontana, Protooncogene bcl-2 gene transfer abrogates Fas/APO-1 antibody-mediated apoptosis of human malignant glioma cells and confers resistance to chemotherapeutic drugs and therapeutic irradiation. J. Clin. Invest. **95**, 2633–2643 (1995)
87. O. Hermine, C. Haioun, E. Lepage, M.F. d'Agay, J. Briere, C. Lavignac, G. Fillet, G. Salles, J.P. Marolleau, J. Diebold, F. Reyas, P. Gaulard, Prognostic significance of BCL-2 protein expression in aggressive non-Hodgkin's lymphoma. Groupe d'Etude des Lymphomes de l'Adulte (GELA). Blood **87**, 265–272 (1996)
88. E. Coustan-Smith, A. Kitanaka, C.H. Pui, L. McNinch, W.E. Evans, S.C. Raimondi, F.G. Behm, M. Aricò, D. Campana, Clinical relevance of BCL-2 overexpression in childhood acute lymphoblastic leukemia. Blood **87**, 1140–1146 (1996)
89. S. Ugurel, G. Rappl, W. Tilgen, U. Reinhold, Increased soluble CD95 (sFas/CD95) serum level correlates with poor prognosis in melanoma patients. Clin. Cancer Res. **7**, 1282–1286 (2001)
90. C.D. Gerharz, U. Ramp, M. Déjosez, C. Mahotka, B. Czarnotta, U. Bretschneider, I. Lorenz, M. Müller, P.H. Krammer, H.E. Gabbert, Resistance to CD95 (APO-1/Fas)-mediated apoptosis in human renal cell carcinomas: an important factor for evasion from negative growth control. Lab. Invest. **79**, 1521–1534 (1999)
91. A.R. Safa, Roles of c-FLIP in apoptosis, necroptosis, and autophagy. J. Carcinog. Mutagen. Suppl. 6, 003 (2013). https://doi.org/10.4172/2157-2518.S6-003
92. N. Rampino, H. Yamamoto, Y. Ionov, Y. Li, H. Sawai, J.C. Reed, M. Perucho, Somatic frameshift mutations in the BAX gene in colon cancers of the microsatellite mutator phenotype. Science **275**, 967–969 (1997)
93. J.P. Meijerink, E.J. Mensink, K. Wang, T.W. Sedlak, A.W. Slöetjes, T. de Witte, G. Waksman, S.J. Korsmeyer, Hematopoietic malignancies demonstrate loss-of-function mutations of BAX. Blood **91**, 2991–2997 (1998)
94. S. Krajewski, C. Blomqvist, K. Franssila, M. Krajewska, V.M. Wasenius, E. Niskanen, S. Nordling, J.C. Reed, Reduced expression of proapoptotic gene BAX is associated with poor response rates to combination chemotherapy and shorter survival in women with metastatic breast adenocarcinoma. Cancer Res. **55**, 4471–4478 (1995)
95. A. Arlt, A. Gehrz, S. Müerköster, J. Vorndamm, M.L. Kruse, U.R. Fölsch, H. Schäfer, Role of NF-κB and Akt/PI3K in the resistance of pancreatic carcinoma cell lines against gemcitabine-induced cell death. Oncogene **22**(21), 3243–3251 (2003)

96. A.B. Kunnumakkara, S. Guha, S. Krishnan, P. Digaradjane, J. Gelovani, B.B. Aggarwal, Curcumin potentiates anti-tumor activity of gemcitabine in an orthotopic model of pancreatic cancer through suppression of proliferation, angiogenesis, and inhibition of nuclear factor-κB-regulated gene products. Cancer Res. **67**, 3853–3861 (2007)
97. S. Lev-Ari, A. Vexler, A. Starr, M. Ashkenazy-Voghera, J. Greif, D. Aderka, R. Ben-Yosef, Curcumin augments gemcitabine cytotoxic effect on pancreatic adenocarcinoma cell lines. Cancer Invest. **25**, 411–418 (2007)
98. G. Du, H. Lin, M. Wang, S. Zhang, X. Wu, L. Lu, L. Ji, L. Tu, Quercetin greatly improved therapeutic index of doxorubicin against 4T1 breast cancer by its opposing effects on H1F-1alpha in tumor and normal cells. Cancer Chemother. Pharmacol. **65**, 277–287 (2010)
99. F.H. Psahoulia, K.G. Drosopoulos, L. Doubravska, L. Andera, A. Pintzas, Quercetin enhances TRAIL-mediated apoptosis in colon cancer cells by inducing the accumulation of death receptors in lipid rafts. Mol. Cancer Ther. **6**, 2591–2599 (2007)
100. S.C. Gupta, R. Kannappan, S. Reuter, J.H. Kim, B.B. Aggarwal, Chemosensitization of tumors by resveratrol. Ann. N. Y. Acad. Sci. **1215**, 150–160 (2011)
101. S. Shankar, Q. Chen, I. Siddiqui, K. Sarva, R.K. Srivastava, Sensitization of TRAIL-resistant LNCaP cells by resveratrol (3, 4', r tri-hydroxystilbene): molecular mechanisms and therapeutic potential. J. Mol. Signal. **2**, 7 (2007)
102. I.A. Siddiqui, A. Malik, V.M. Adhami, M. Asim, B.B. Hafeez, S. Sarfaraz, H. Mukhtar, Green tea polyphenol EGCG sensitizes human prostrate carcinoma LNCaP cells to TRAIL-mediated apoptosis and synergistically inhibits biomarkers associated with angiogenesis and metastasis. Oncogene **27**, 2055–2063 (2008)
103. M. Nihal, H. Ahsan, I.A. Siddiqui, H. Mukhtar, N. Ahmad, G.S. Wood, Epigallocatechin-3-gallate (EGCG) sensitizes melanoma cells to interferon induced growth inhibition in a mouse model of human melanoma. Cell Cycle **8**, 2057–2063 (2009)
104. M.J. Scandlyn, E.C. Stuart, T.J. Somers-Edgar, A.R. Menzies, R.J. Rosengren, A new role for tamoxifen in oestrogen receptor-negative breast cancer when it is combined with epigallocatechin gallate. Br. J. Cancer **99**, 1056–1063 (2008)
105. B. Zhang, Z.L. Shi, B. Liu, X.B. Yan, J. Feng, H.M. Tao, Enhanced anticancer effect of gemcitabine by genistein in osteosarcoma: the role of Akt and nuclear factor-κB. Anticancer Drugs **21**, 288–296 (2010)
106. B.F. El-Rayes, S. Ali, I.F. Ali, P.A. Philip, J. Abbruzzese, F.H. Sarkar, Potentiation of the effect of erlotinib by genistein in pancreatic cancer: the role of Akt and nuclear factor-κB. Cancer Res. **66**, 10553–10559 (2006)
107. C.Y. Jin, C. Park, J. Cheong, B.T. Choi, T.H. Lee, J.D. Lee, W.H. Lee, G.Y. Kim, C.H. Ryu, Y. H. Choi, Genistein sensitizes TRAIL-resistant human gastric adenocarcinoma AGS cells through activation of caspase-3. Cancer Lett. **257**, 56–64 (2007)
108. S.Y. Park, D.W. Seol, Regulation of Akt by EGF-R inhibitors, a possible mechanism of EGF-R inhibitor-enhanced TRAIL-induced apoptosis. Biochem. Biophys. Res. Commun. **295**, 515–518 (2002)

Improvement of Cancer Therapy Using Phytochemicals

6

Abstract

Better prognosis of cancer patients undergoing chemotherapy depends on the success of the drugs used in chemotherapy. However, cancer cells develop drug resistance, and overcoming drug resistance is one of the challenges in chemotherapy. Moreover, the prognosis is often poor if metastasis, the most dangerous phase of cancer sets in. We discuss the challenges in better prognosis in this chapter and how phytochemicals help in better prognosis in cancer in all phases of the disease. We extensively review the current literature that shows that phytochemicals are effective in both preventing drug resistance and preventing the onset of metastasis, by acting on the key genes and pathways involved in these processes.

6.1 Introduction

The prospect of recovery from a disease is coined as prognosis. It is a prediction regarding the course of a disease, i.e., chance of recovery or survival. Prognosis is a statistical estimation based on past experience. Cancer fighters and their relatives are often anxious about the future of the treatment regimen. But, a definite prognosis for a particular patient is hard to come by. The prognosis depends on several important deciding factors, e.g., type of cancer, its subtypes, location, stage of the disease, grade of the disease, molecular profile, and if the disease has spread to other parts of the body. Apart from these, age of the patient, general health condition, and how an individual responds to the treatment contribute to prognosis.

Survival rate of a particular type of cancer is a statistical measure how many patients with a particular type of cancer and at the same stage survive. This is a gross measure that is expressed as a percentage. It is imprecise to some extent as patients may die due to some other illness, e.g., a cancer patient may die due to heart attack. The 5-year survival rate is often used as a prognostic measure for cancer patients,

M. Roy, A. Datta, *Cancer Genetics and Therapeutics*,
https://doi.org/10.1007/978-981-13-9471-3_6

and this is calculated from the point of diagnosis. Survival rates may be absolute or relative. Absolute survival rate states the percentage of patients who remain alive for 5 years after diagnosis. Generally relative survival rate is used, which is defined as the number of patients who are alive for 5 years after diagnosis of the disease divided by the number of healthy people of the same gender and age, who are alive for 5 years. The third term that is used as a measure of prognosis is disease-free survival or relapse-free survival rate. It is the period a patient survives without any symptoms of cancer, after completion of treatment.

Some other terms used in prognosis include median survival rate, which is the time after which half of the people with a certain type of cancer are alive and the rest died. This is often used for describing prognosis of aggressive cancers. In order to describe the response to a particular treatment, progression-free survival is used. This is a measure to find how long the treatment strategy works. During this period cancer does not relapse again.

6.2 Challenges in Prognosis

There are many factors that affect prognosis; these are:

(i) Type and site of cancer: Prognosis for different cancers is different. Some may have a good prognosis; others may not; their fatality rate may be high. Cervical cancer, early-stage breast cancer, testicular cancer, prostate cancer, melanoma, etc. are some of the cancers that have good prognosis. On the other hand, pancreatic cancer and lung cancer have poor prognosis.

(ii) Stage of cancer: It is important to determine the stage of a cancer to decide upon a treatment strategy. Knowledge about the stage of cancer helps doctors to plan treatment, so that the prognosis is good.

(iii) Metastasis: Metastasis is one of the major causes of mortality and poor prognosis. So, it is important to determine whether the cancer has metastasized.

(iv) Grade of a cancer: It is a description of a tumor based on the extent of abnormality of cancer cells. Grade may give a hint whether the cancer can spread to other parts of the body. When the organization of tumor tissues is close to that of a normal tissue, it is termed well differentiated. On the contrary, a poorly differentiated cancer is one with very little normal tissue structures and more abnormal-looking cells. A high-grade cancer usually carries a poor prognosis.

(v) Age and general health of the patient when the disease is diagnosed: Because of the age factor, compromised treatment is given to elderly patients. This may be a reason for poor treatment outcome in elderly patients [10]. Overall health condition is also a deciding factor that affects prognosis.

(vi) Response to treatment: This can be defined in several ways: complete response, partial response, stable disease, and disease progression. Complete response indicates disappearance of tumor, and automatically the tumor markers that are high in cancer condition come back to the normal value. Partial response, as evident from its name, indicates that the disease has been controlled, but has not been eradicated. Tumor markers in this case may go down. Stable disease means that the cancer has not grown further nor regressed, and as a consequence, the markers remain almost unaltered. Disease progression indicates that the cancer has grown and the markers go up.

Quite often resistance to chemotherapy leads to poor outcome of the treatment. Initially the growth might have responded to chemotherapy, but after a certain time, it starts growing again. Intervention with a new drug might help to overcome the situation.

There may be several explanations for development of resistance to chemotherapy. Any drug cannot kill all the cells, and the cells that survive may undergo mutations and become resistant to the drug. Population of drug-resistant cells increases and they may overshoot the number of drug-sensitive cells. Gene amplification may be another big problem. Multiple copies of a particular gene in cancer cells pose serious problems. Amplification of genes leads to overproduction of proteins and the chemotherapeutic drug may become ineffective. Sometimes, the drug may be pumped out of the cancer cell, making it unavailable. Permeability glycoprotein (P-gp) is also known as multidrug resistance protein 1 (MDR1) or ATP-binding cassette subfamily B member 1 (ABCB1). This protein is also known as cluster of differentiation 243 (CD243). It is an important protein of the cell membrane that pumps many foreign substances out of the cells. Cellular uptake of drugs and their metabolites are determined by transporter actions, which greatly impact on the distribution, toxicology, and effectiveness of drugs. A protein is responsible for transportation of the drug across the cell membrane; therefore, if that protein ceases to work, cancer cells remain devoid of drugs. Chemotherapeutic drugs kill cancer cells, and to achieve that, they target DNA and damage them. If the DNA repair system efficiently repairs the damage, then the cell survives.

Cancer cells sometimes may acquire some mechanisms that make drugs ineffective. This may lead to development of drug resistance, a phenomenon when the cancer cells become tolerant to the drugs. The detailed mechanism of development of drug resistance will be discussed in subsequent section. Therefore, in order to improve efficacy of chemotherapy, control of drug resistance and metastasis need to be addressed.

6.3 Development of Drug Resistance

Drug resistance is a major problem for any treatment regimen. This problem was first detected when some bacteria became resistant to certain antibiotics, causing serious threat to human health. Later on, development of resistance to therapeutic drugs has been noticed in various disease conditions, reducing the effectiveness of the treatment protocol. In case of cancer, drug resistance to therapeutics poses serious problems and results in poor prognosis. When the resistance develops to more than one drug, it is called multidrug resistance. Resistance may appear because of mutations developed, which alter the structure of the cell, biochemical pathways, and the target sites of the drug. Ability of the drug at its best is hindered leading to drug resistance.

There are different mechanisms that can promote drug resistance directly or indirectly. The mechanisms that can act independently or in combination through various signal transduction pathways are drug efflux, epithelial-mesenchymal transition (EMT), drug inactivation, drug target alteration, epigenetics, DNA damage repair, and cell death inhibition. Resistance to drugs may be due to a number of mechanisms. These include compromised drug uptake, enhancement in drug efflux, evasion of drug-induced apoptosis, activation of detoxification enzymes, and DNA repair machinery [18]. There are certain factors like drug administration and its distribution, metabolism, and excretion which are responsible for best outcome of therapy. The plasma membrane is also a big hindrance why a drug cannot act at its fullest efficacy. Therefore, transportation of the drug within the cell is very important. Reduced expression of solute carriers is therefore a big problem. Solute carriers are a family of proteins comprising uptake transporters. Genes of this family encode transporters including passive transporters, ion-coupled transporters, and exchangers [27]. These transporters help in the uptake of nucleotide analogs to the cell, which are used as cancer therapeutics [126]. Genetic polymorphism of these transport proteins may vary from person to person; hence potentiality of the drugs may not be same for all individuals receiving therapy.

Development of drug resistance may also be caused by ATP-dependent drug efflux pumps. Lipid bilayer also plays a role in drug uptake. Lipid metabolism may be altered in drug-resistant cells; as a consequence, lipid bilayer properties may be different, and hence drug uptake may be compromised [20]. Even the availability in the cell does not ensure the fullest potential of the drug. The phase I (or oxidative metabolism) and phase II enzymes (or the enzymes responsible for conjugation) within the cell also play a part in this regard. Phase I enzyme action is mediated by cytochrome P450 enzymes (CYPs) and epoxide hydrolases.

Often cancer cells become dependent on oncogenes for their malfunctioning and therefore targeting these oncogenes is the underlying principle of targeted therapy [100]. One obstacle for this therapy is the development of resistance due to mutation of the target protein [19]. This type of mutations is probably caused by the therapeutic drugs [34]. Moreover, targeting of one protein may not be efficacious, because the cells may follow another survival pathway for their existence [13, 14]. In such cases it is always preferable to target all the pathways involved [64].

Development of resistance to one drug often confers resistance to other drugs, a phenomenon called cross-resistance [111]. Loss of a transporter can lead to resistance to that particular drug and at the same time may result in resistance to other drugs that use the same transporters. Cells in a tumor are heterogeneous in nature, with different genetic makeups [63], and during the course of tumorigenesis, genetic alterations may result, giving them growth advantage. A population of drug-resistant cells that adapt themselves to this environment becomes dominant and starts growing, making the tumor resistant to therapeutics [21]. Mutation in a drug-binding site or changes in drug transporters also render growth advantage.

6.3.1 Models of Drug Resistance

In order to understand the origin of resistant cell population in a tumor, two models may be of help: the first is the environment-mediated drug resistance (EMDR), and the second is the stem cell model (CSC).

In environment-mediated drug resistance (EMDR), tumor cells are protected from the initial effect of therapy. Subsequently complex and permanently acquired resistance may be developed in response to the therapy. This type of resistance is due to an adaptive response, between cancer cells and the tumor microenvironment. At the beginning of treatment, this interaction may be targeted to prevent onset of resistance to therapeutics [65].

Cancer stem cells (CSCs) are a subset of cells within the tumor with the potential for self-renewal, differentiation, and tumorigenicity. CSCs deviate from normal stem cells through dysregulation of self-renewal pathways [125]. They are situated in a microenvironment called niche, which is made of fibroblasts and endothelial, mesenchymal, and immune cells, which also play pivotal roles in drug resistance. They are responsible for chemoresistance and eventually lead to failure in therapy. Cells in the microenvironment help CSCs to sustain by triggering signaling pathways. Extracellular matrix also provides shelter to CSCs and protects them from therapeutic drugs, thus conferring a growth advantage [83].

6.4 Metastasis

Metastasis is a process where a tumor is spread to a distant part of the body from its original site. When cancer cells break away from the primary tumor site, they may be carried via blood stream or the lymphatic system. They may settle at a new site and start growing. Cancer cells may spread to the lymph nodes, liver, bones, lungs, or brain. They may also metastasize to the abdominal cavity or the pleural space. Metastasis is one of the causes of cancer morbidity and mortality, hence resulting in poor prognosis.

Epithelial cells line the boundary of many organs. They tend to adhere and provide protection from the environment. Loosely packed mesenchymal cells, having no polarity, on the other hand may remain isolated and can migrate to other parts. Tissues are composed of cells, as well as extracellular space containing carbohydrates and proteins. This extracellular space is called the extracellular matrix (ECM), where the cells have a tendency to stay together. For metastasis, the first requisite is to shatter these bonds between the cells, so that they can move to a distant organ of the body.

Cell-cell linkage is facilitated by immunoglobulins and cadherins, while these cells get linked to the ECM with the help of integrins. Several proteins are involved in cell-cell adhesions, and E-cadherin is the most important among these proteins. E-cadherin is responsible for contact inhibition of normal cells. Loss of E-cadherin function is synonymous with loss of contact inhibition, which means cells can grow in piles forming a tumor. E-cadherin function is missing in migrating cells; in its place N-cadherin is found in abundance. N-cadherin aids the cancer cells to pass through the blood vessels, thus helping them to migrate from their site of origin. Tumor cells may migrate as individual cells, or as a stream of loosely attached cells, or as a collection of adherent cells [7]. These migrating cells are somewhat different in morphology. Regular shape of a normal cell is changed to a rugged look. These cells get detached from the ECM, which eventually is degraded by some proteins, particularly matrix metalloproteases. The metastatic cell is characterized by increased motility and evasion to apoptosis. EMT plays a pivotal role in cell migration and occurs through degradation of ECM, cessation of E-cadherin, loss of cell adhesion junctions, and cell polarity. These events cumulatively promote migration of invasive mesenchymal cells. In the new location, the reverse transition is needed, i.e., mesenchymal to epithelial (MET) transition for formation and growth of a new tumor at a distant site.

Transforming growth factor beta (TGF-β) plays a dual role; it may act as a tumor suppressor by checking cell cycle; conversely it may promote malignant growth by activation of EMT, thereby supporting invasion and metastasis. EMT is supported by a plethora of transcription factors; of these, Snail, Slug, Twist, and ZEB1/2 are worth mentioning. TGF-β binds to its receptor, thereby phosphorylating Smad2 and Smad3 and activating them. These proteins in turn bind to Smad4, to form a complex that translocates to the nucleus. In the nucleus it activates Snail/Slug, ZEB1/2, and Twist. These transcription factors suppress the epithelial markers and concomitantly trigger the mesenchymal proteins.

A number of steps are involved for a growth to metastasize. First cancer cells get detached from the primary site and then spread through the circulatory and lymphatic systems, evading immune attacks. They finally settle at distant capillary beds, invade, and start growing in new organs [97]. In order to understand the process of metastasis, one should focus on the origin of metastatic cells due to (i) epithelial to mesenchymal transition (EMT), (ii) stem cell origin, (iii) macrophage facilitation of metastasis, and (iv) myeloid cell origin.

Epithelial cancers with poor differentiation are aggressive and are prone to spread to distal parts than differentiated cancers [40]. It is postulated that epithelial stem cells or differentiated cells in due course are transformed to a tumor with metastatic features [97]. Many types of cancers originate in epithelial tissues. Normal epithelial cells can proliferate giving rise to adenoma. Genetic and epigenetic alterations later on lead to carcinoma in situ. In this stage the tumor still has an intact basement membrane, which gets damaged later. Abnormalities in cell-cell or cell-matrix interactions might occur during progression of tumors. The neoplastic cells are mostly of mesenchymal nature, lacking cell-cell adhesion, and the loss of adhesion helps them to spread to distant organs. Invasive tumor cells are transported through the lymphatic system or blood vessels. A new tumor may develop at the new site.

The progress of an initiated tumor cell toward metastasis is very complex. The different stages are local invasion, intravasation (invasion of cancer cells to lymphatic or blood vessels through basal membranes), survival in the circulation, extravasation (migration of cancer cells from blood or lymphatic vessels to the tissues), and colonization. Metastasis, the deadliest part of cancer, starts with a small number of cells that leave the primary site and colonize a distant part [55]. The tumor initiating or cancer stem cells are a subset of tumor cells having characteristics similar to normal stem cells. They possess the property of self-renewal and differentiation [86]. These cells are the basis of tumor formation, and they drive invasion, metastasis, and recurrence of the disease. There are a number of signaling pathways, which let the CSCs to generate tumor cells. Some of these pathways are Notch, Hedgehog, nuclear factor kappa-light-chain-enhancer of activated B cells (NF-κB), Wnt, and TGF-β [11]. CSCs are also cause of resistance to therapy, both chemoresistance and radioresistance. The CSCs can be identified and isolated by cell surface markers. For a best outcome of treatment, the CSCs need to be eliminated, because they are the root cause of poor prognosis. Origin of the metastatic cancer cells is the stem cells, precisely the undifferentiated and semi-differentiated cells. Genetic and biochemical similarities are present in CSCs and tumor cells; both of these utilize energy from anaerobic sources for their sustenance [97].

6.5 Macrophages

Macrophages are a type of white blood cell, engulfing cellular debris, cancer cells, etc. They also digest foreign particles which are not usually present in healthy cells. This is known as phagocytosis, where pathogens are destroyed. Macrophages boost up the immune system and have an anti-inflammatory potential. Macrophages therefore protect us against infections. Malignant tumors contain a number of macrophages, known as tumor associated macrophages (TAM), that can enhance tumor inflammation and angiogenesis and initiate metastasis. These TAMs are involved in all steps of metastasis. Neoplastic and nonneoplastic macrophages are often indistinguishable in the tumor microenvironment. This is due to the fact that the expression of genes and morphology and function of these cells are quite similar [33, 62].

In order for absorption of nutrients, growth, and proliferation, cells need energy. This is obtained from breakdown of nutrients by a process called respiration. The energy is stored as adenosine triphosphate (ATP). It is thought that metastatic cancers arise from respiratory insufficiency. Chronic inflammation in the tumor microenvironment can disrupt mitochondrial respiration [15, 96]. Aerobic respiration occurs in three stages: *glycolysis*, where glucose gets divided into two molecules of pyruvate; the *Krebs cycle*, where these pyruvate molecules enter the matrix of a mitochondria; and *electron transport*, where the energy from nicotinamide adenine dinucleotide (NADH) and flavin adenine dinucleotide (FADH2), which result from the Krebs cycle, is transferred to ATP. Due to insufficient respiration, aerobic glycolysis, also called Warburg effect arises. This effect is named after Otto Warburg, whose hypothesis explains the root cause of cancer. His hypothesis states that tumorigenesis is caused by insufficient cellular respiration caused by insult to mitochondria [117].

Whenever an immune response is needed, glucose uptake of the immune cells gets enhanced. Under these circumstances, they tend to drift toward aerobic glycolysis rather than metabolizing glucose via normal respiration. This warrants rapid cell division. This metabolic switch is utilized by cancer cells in order to maintain uncontrolled growth. Tumors are devoid of oxygen, and as a consequence of this, stress hypoxia-inducible factor (HIF) gets induced. HIF in turn activates several genes that support aerobic glycolysis and suppress genes that aid in normal respiration. HIF therefore plays a pivotal role in switching the metabolism of cancer cells from normal respiration to aerobic glycolysis. Besides, other factors also have a lead role.

Phosphatidylinositol 3-kinases (PI3Ks), a family of enzymes that control cell growth, proliferation, differentiation, motility, survival, and intracellular trafficking, are intricately associated with carcinogenesis. PI3Ks get activated by outside signals and start functioning on cellular growth, promote angiogenesis, and activate aerobic glycolysis. They inhibit programmed cell death and all these aid in supporting cancer cells for their survival. PI3K activates AKT and in turn mammalian target of rapamycin (mTOR). HIF gets activated by mTOR, and finally all these proteins help in switching metabolism of cancer cell from normal respiration to aerobic glycolysis. PI3K is regulated by another molecule phosphatase and tensin homolog (PTEN). There is a crosstalk between the signaling molecules, and therefore these pathways are interlinked and regulate one another. The key molecule in carcinogenesis is p53 which detects damage to the genetic material. p53 curbs aerobic respiration and thereby stimulates normal respiration. Unfortunately, p53 is mutated in most of the cancers, shattering the balance between normal respiration and aerobic glycolysis. This chain of events orchestrates carcinogenesis.

Cancer cells secrete lactic acid, which induces TAM, and formation of new blood vessels is facilitated. HIF pathway has a lead role in this regard. HIF is accumulated in the cells during hypoxia and gets transported to the nucleus where transcription of a number of genes occurs. One of the target genes is vascular endothelial growth factor (VEGF), which promotes angiogenesis. Once angiogenesis is on, tumor cells look for the endothelial cells and also pericytes, which give support to the nascent blood vessels. Endothelial cells hire pericytes through a cell signaling pathway. Endothelial cells secrete platelet-derived growth factor (PDGF), whose receptors reside on the cell surface of pericytes. Pericytes then secrete more and more VEGF, facilitating the growth of new blood vessels. Therefore, VEGF plays a lead role in angiogenesis. Oncoproteins also aid in activation of VEGF. Inhibition of angiogenesis is very important, but cancer cells fail to produce these inhibitors. Some of the inhibitors of angiogenesis are under the regulation of p53, but as mentioned before p53 is often mutated or lost in cancer cells. This leads to more and more angiogenesis. Thrombospondin 1 is an angiogenesis inhibitor. Other angiogenesis inhibitors that are used for cancer treatment are Avastin or bevacizumab, which slow down the growth of new blood cells [8, 113].

Phytochemicals may be exploited for the improvement of prognosis by their multi-targeting potential, viz., their antiproliferative, pro-apoptotic, anti-metastatic, anti-angiogenic, and chemo-enhancing attributes. These may contribute to overcome the obstacles in cancer therapy and hence may improve prognosis.

6.6 Overcoming Drug Resistance Using Phytochemicals

Drug resistance and metastasis need to be fixed to achieve better prognosis in cancer. Safe and alternative strategies to resolve the issues are the need of the hour. Plant-derived active ingredients are nontoxic and hence safe. They are also economical. Therefore, they have gained huge attention.

Diagnostic tools, therapeutic measures, and awareness have progressed a lot; even then prognosis is a major challenge for breast cancer. The main reasons for poor outcome are development of drug resistance, associated side effects, metastasis, and relapse of tumor growth. These complications need urgent attention and remedy. Plant-derived molecules, being nontoxic and safe, may be a respite from associated side effects. Phytochemicals also modulate a number of genes and proteins, modulate signal transduction pathways, and therefore may have a modulatory action on cell proliferation, invasion, angiogenesis, and metastasis. They may target breast cancer stem cells and often overcome mechanisms responsible for drug resistance. They may sensitize cancer cells and may potentiate the efficacy of anticancer drugs. In this way, these biomolecules may act as adjuvants with chemotherapeutic drugs, improving the therapeutic efficacy.

Camellia sinensis, popularly known as tea, is widely consumed all over the world. Consumption of this beverage reduces the risk of various cancers like skin, oral, esophageal, stomach, liver, pancreas, breast, leukemia, etc. Catechins, the active compounds of tea, are mainly responsible for this reduction of risk. Catechins like epigallocatechin-3-gallate (EGCG) inhibit the expression of P-glycoprotein in multidrug-resistant human epidermal carcinoma cell line KB-C2 cells. It interacts with the substrate binding sites of P-glycoprotein, thereby reducing its action [45]. EGCG, by inhibiting cancer stem cells, may help to tackle drug resistance in cancer [16]. Resistance to 5-fluorouracil (5-FU) during therapy of gastric cancer (GC) limits its efficacy. EGCG has been reported to revert 5-FU resistance in in vitro and in vivo study models by inactivating TFAP2A/VEGF pathway and reducing levels of MDR-1 and P-gp [107]. Resistance to multiple drugs in cancer therapy results due to overexpression of ABC transporters. Chemosensitizers inhibit the activity of these transporters leading to greater intracellular concentration of drugs. In vitro studies show that curcumin reduced P-gp activity in Caco-2 and CEM cells resistant to doxorubicin. EGCG negatively affected P-gp expression in drug-resistant Caco-2 cells. Thus, these phytochemicals, when administered to cells in conjunction with doxorubicin, can revert drug resistance [54]. Enhanced expression of P-gp and multidrug resistance proteins (MRP1) leads to efflux of cisplatin from cells, culminating into drug resistance. EGCG promotes cisplatin-induced apoptosis in resistant oral cancer cells by targeting Akt, signal transducer and activator of transcription (STAT3), and multi-drug resistance-related proteins [124].

Curcumin has been reported to inhibit P-glycoprotein function in KB-C2 cells by increasing intracellular accumulation of drugs and preventing efflux of P-gp substrates [72]. Curcumin acts as a chemosensitizer in cancer by modulating various pathways that may lead to drug resistance. Curcumin induces apoptosis, regulates drug-target interaction, blocks survival pathways, and targets cancer stem cells [92]. Curcumin shows cytotoxicity in vincristine-resistant human colon cancer cells and inhibits their proliferation. Curcumin and vincristine, when used together, resulted in reduced tumor growth. Curcumin shows an inhibitory effect on MRP1 and survivin expression and helps in reversal of multidrug resistance in colon cancer [60]. Curcumin increased sensitivity of MDR cancer cells to cisplatin by modulating proteins involved in the PI3K/Akt/mTOR pathway [121]. Curcumin has been reported to facilitate tumor necrosis factor-related apoptosis-inducing ligand (TRAIL)-induced killing of drug-resistant ovarian cancer cells. It also induces apoptosis in cisplatin-resistant ovarian cancer cells. Curcumin has been reported to reverse multidrug resistance in in vivo models as well [47]. Curcumin helps to overcome resistance in chronic myeloid leukemia cells resistant to imatinib mesylate by modulation of various cytokines (like TNF-α and interleukins) and their associated downstream pathways (MAP kinase JAK-STAT and Akt) [69]. Epigenetic regulators histone deacetylase (HDACs) and viral oncoproteins (E6/E7) are overexpressed in cervical cancer. Curcumin prevents expression of multidrug-resistant proteins (P-gp and MRP1), HDACs, and enhanced acetylation and increase of p53 in cervical cancer cells resistant to cisplatin. All these modulations lead to arrest of cell cycle at G1-S-phase, and modulation of cell cycle regulatory proteins has been observed. Thus, curcumin acts as a chemoenhancer and helps to revert drug resistance in cervical cancer cells [88].

Capsaicin and [6]-gingerol are the active phytochemicals present in chili peppers and ginger, respectively. They inhibit P-glycoprotein and help to increase the concentration of P-glycoprotein substrates like vinblastine in drug-resistant cancer cells [72]. Rosemary or *Rosmarinus officinalis* contains a plethora of phytochemicals having anticancer potential. Out of these, carnosic acid, carnosol, and ursolic acid enhanced intracellular drug accumulation in multidrug-resistant cancer cells [76]. Citrus phytochemicals show anticancer potential and have been reported to aid in reversal of drug resistance via inhibition of P-glycoprotein [73]. Plant sterols like guggulsterone increase the intracellular accumulation of chemotherapeutic drugs in cancer cells that overexpress P-gp [73]. Guggulsterone does so by increasing the ATPase activity of P-glycoprotein and also inhibits MRP1 [73].

Resveratrol found in grapes, sesamin obtained from sesame seeds, matairesinol from soybean, and ginsenosides from the roots of *Panax ginseng* inhibit P-glycoprotein and may help to reverse drug resistance [46, 74, 75]. MRP1, another multidrug-resistant protein, is inhibited by phytochemicals like EGCG, curcumin, and citrus phytochemicals [71]. Glycyrrhetinic acid inhibits P-glycoprotein and MRP1. It is a substrate of MRP1 and acts by binding at drug-binding site of MRP1. This phytochemical interacts with the ATP hydrolytic site of P-glycoprotein, thereby altering its activity [73].

Ginger is a popular spice in Asian countries and is rich in gingerol. Gingerol downregulates MRP1 and GSTπ expressions in docetaxel-resistant human prostate cancer cells PC3 and reduces proliferation of drug-resistant cells [57]. Lapatinib is a drug commonly used in breast cancer; however, development of resistance to the drug is often encountered. Cruciferous vegetables are rich in isothiocyanates. Lapatinib in combination with isothiocyanates like sulforaphane, erucin, etc. reduces viability of resistant breast cancer cells [39]. Isothiocyanates reverse cytotoxicity of resistant cells to drugs by modulating breast cancer-resistant proteins (BCRP) [37]. Reduced oxygen levels or hypoxia is common in the tumor microenvironment of resistant cells that renders the tumors more aggressive and resistant to chemotherapy. Sulforaphane, found in broccoli, possesses chemopreventive potential by modulation of pathways that are involved in development of carcinogenesis. The anticancer effects of sulforaphane have been investigated on ovarian cancer cells resistant to adriamycin and cisplatin. Sulforaphane regulates hypoxic condition by activating p53, ARE, IRF-1, Pax-6, and XRE genes and repressing the expression of AP-1 and HIF-1. Sulforaphane also modulates targets of HIF-1. Treatment of the cells with this phytochemical thus reduces migration of cancer cells and helps to reverse resistance development in cells [80].

Resveratrol alters levels of MRP1, LRP, Topo II, and apoptotic proteins and aids in reversing drug resistance in bladder cancer by inducing cell cycle arrest at S-phase [116]. Resveratrol also reverses 5-FU-mediated drug resistance in colon cancer cells and induces apoptosis in resistant cells [31]. Lycopene, a carotenoid found in abundance in tomatoes, may reverse drug resistance and induce apoptosis in malignant cells resistant to chemotherapy [29, 70].

Cisplatin, gemcitabine, and docetaxel are widely used drugs in cancer therapy. They act via activation of transcription factor NF-κB in cancer cells which often leads to development of drug resistance. Both in vitro and in vivo studies reveal that administration of genistein (a phytochemical from soya bean) to cells in conjunction with chemotherapeutic drugs inhibits growth of resistant cells and induces apoptosis in these cells. Activity of NF-κB increases after drug treatment, which is reduced by genistein. Involvement of the Akt pathway has also been reported in his regard [3]. Multidrug resistance in breast cancer occurs due to enhanced expression of ABC drug transporters leading to therapeutic failure and recurrence of the disease. Genistein shows efficacy in drug-resistant breast cancers by modulation of these transporters [87].

Naringin is found in citrus fruits especially grapefruits and helps to reverse resistance of ovarian cancer cells to anticancer drug cisplatin via suppression of NF-κB induced expression of P-gp [133]. Baicalin, found in *Scutellaria baicalensis* and *Scutellaria lateriflora*, reduces cisplatin resistance in lung cancer cells by modulating MAPK and Akt pathways, which often get deregulated during drug resistance [122]. Butein, obtained from the bark of *Semecarpus anacardium*, sensitizes cisplatin-resistant HeLa cells to the drug via ERK/p38 pathway [127, 128]. Cepharanthine is an isoquinoline alkaloid obtained from *Stephania cepharantha*. It helps to revert cisplatin resistance in esophageal carcinoma by downregulating P-gp. Oridonin acts on cisplatin resistance in human gastric cancer via suppression of multidrug-resistant proteins and P-gp-mediated drug efflux [104, 105].

miRNAs are posttranscriptional regulators of genes that play key roles in drug resistance. Quercetin, a natural flavonoid, increases cisplatin sensitivity of human osteosarcoma cells by increasing levels and activity of miRNA-217 [127, 128].

Quercetin, commonly found in apples and onions, helps to overcome ovarian cancer drug resistance [90]. Shikonin, isolated from *Lithospermum erythrorhizon*, has been reported to induce cell death in drug-resistant breast cancer in both in vitro and in vivo studies [24]. Silibinin, a flavonolignan, is extracted from milk thistle and has found use in traditional medicine since ancient ages. It acts as a chemosensitizer in various cancer models. It prevents growth of breast cancer cells resistant to doxorubicin. It does so by negatively modulating expression of key oncogenic molecules and pathways like Akt, STAT, and MAP kinase pathways. Thus, silibinin helps to revert chemoresistance by sensitizing drug-resistant cells to drugs [68].

Small-cell lung carcinoma (SCLC) often shows poor prognosis due to development of multidrug resistance (MDR). Silibinin has been reported to induce programmed cell death and overcome MDR in drug-resistant SCLC cells. This molecule acts synergistically with chemotherapeutic drugs, thereby increasing their efficacy [91]. Luteolin belongs to the group of flavonoids and exhibit potent chemosensitizing efficiency in breast cancer. Luteolin targets expression of CCNE2 and helps to prevent TAM resistance in patients suffering from breast cancer [110]. Oxaliplatin is the most common drug for colorectal cancer therapy, but development of drug resistance challenges its efficacy. Nuclear factor erythroid-2 p45-related factor 2 (Nrf2) and NADPH quinone oxidoreductase1 (NQO1) activation is responsible for development of resistance. Luteolin helps to overcome this problem by inhibiting Nrf2 and NQO1. Luteolin acts as an enhancer of chemotherapeutic drugs in resistant cancer cell lines [6].

Apigenin, a dietary flavonoid, has been reported to have anticancer potential based on in vitro and in vivo studies. Apigenin reduces cell growth and colony formation and induces apoptosis in MCF 7 cells resistant to adriamycin. Apigenin reduced multidrug resistance 1 (MDR1) and multidrug resistance-associated proteins (MRPs) in resistant cells. It also reduced efflux of drugs from cells. It exerts its action by repressing JAK-STAT signaling pathway, which is deregulated during drug resistance development. Apigenin reduced activation and nuclear translocation of STAT3 and inhibited target genes of STAT3 [95]. Kaempferol, a flavonoid, has been reported to be effective in ovarian cancer and sensitizes resistant cells to cisplatin. It does so by inhibiting multidrug-resistant genes and cMyc and promoting apoptosis in resistant cells. This phytochemical acts synergistically with anticancer drugs to overcome drug resistance in cancer cells [23]. Kaempferol reverses drug resistances in human glioblastoma cells by preventing efflux of calcein, a MRP substrate. MRP levels are also reduced by kaempferol [77]. Stilbenes are plant secondary metabolites and are widely found in plants like grapes, mulberries, and peanuts. They are well known for their anticarcinogenic potential. Phytoalexins resveratrol and piceatannol are examples of stilbenes. These plant products show efficacy in reversing drug resistance of human adenocarcinoma cells resistant to the drug doxorubicin by modulating MDR genes and inducing apoptosis in resistant cells [119].

6.7 Inhibition of Metastasis Using Phytochemicals

Various genes and their associated signaling pathways are targeted by phytochemicals. As a result, various stages of carcinogenesis like proliferation of cancer cells, invasion, angiogenesis, and metastasis get modulated by phytochemicals in breast cancer. Cancer stem cells are also affected by phytochemicals that may help in overcoming problems of drug resistance in breast cancer. Phytochemicals may increase the efficacy of chemotherapeutic drugs and thus serve as adjuvants in therapy. Some phytochemicals that may aid in chemoprevention of breast cancer include curcumin, resveratrol, epigallocatechin gallate, silibinin, genistein, secoisolariciresinol, thymoquinone, apigenin, etc.

Metastasis, as discussed earlier, is the deadliest stage of cancer. Phytochemicals can impede metastasis. Overexpression of several genes is implicated in metastasis, and these genes get suppressed by these natural compounds. Many phytochemicals regulate genes epigenetically, i.e., via alteration of DNA methylation and histone acetylation and methylation [12, 66]. Mitogen-activated protein kinase kinase (MKK) expression is reduced by phenethyl isothiocyanates, whereas another isothiocyanate benzyl isothiocyanate enhances the expression [64]. Ethanol has been found to increase the expression of MKK4. However, nonmetastatic gene 23 (NM23) gets downregulated in breast cancer cells, but NM23 is upregulated in B16-BL6 melanoma cells due to ethanol [48]. The effect of a phytochemical may be dependent on cancer types as well as cell phenotypes. The phytochemical from tea, epigallocatechin-3-gallate (EGCG), shows different effects on four oral squamous cell cancer cell lines [41]. Effect of EGCG on metastasis suppressor gene, reversion-inducing cysteine-rich protein with Kazal motifs (RECK), was followed in this study. EGCG increases transcription of RECK only in two cell lines, out of four. Many plant-derived molecules show anti-metastatic potential by targeting downstream signaling pathways. One such protein is p38 mitogen-activated protein kinase (MAPK). Metastasis suppressor genes MKK4 and MKK6 target MAAPKs. Again, different phytochemicals have different effects on these genes. The same protein p38MAPK gets enhanced or diminished depending on the type of plant product. Green tea and its constituents have been found to have anti-angiogenic properties, among many others [5, 17].

Polyphenolic extract from red grape and genistein from soy product have been found to inhibit p38 MAPK. Quercetin, found in many fruits, vegetables, and grains, and β-caryophyllene, from rosemary, black pepper, etc., have been found to increase p38. Genistein, the active molecule from soy product, prevents activation of p38 and inhibits cell invasion [30]. This molecule inhibits MKK4 expression and metastasis. Quercetin shows anticancer activity in several types of cancer cells including metastatic colorectal cancer. It prevents metastasis by regulating expression of

MAP kinases and EMT markers like cadherins, β-catenin, and SNAIL. Migration and invasion of CT26 cells were prevented by this phytochemical via regulation of matrix metalloproteinases (MMPs) and tissue inhibitor of metalloproteinases (TIMPs). Quercetin is capable of suppressing colorectal and lung metastasis in in vivo models [42].

Osteosarcoma is a form of malignancy characterized by high mortality rate and very few effective therapies. Quercetin induces arrest of cell cycle, promotes apoptosis, causes differentiation, and prevents cell adhesion, metastasis, and angiogenesis in osteosarcoma cells [49]. Celestrol, a naturally occurring triterpenoid, activates the heat shock transcription factor and is employed to treat neurodegenerative diseases. It activates p38 phosphorylation and impedes migration and invasion of murine melanoma and lung cancer cells of human origin [132]. Curcumin is a good anticancer agent that shows its action on cell cycle regulation, programmed cell death, metastasis, etc. [120]. Carcinogenesis is a multistep process, whose all stages can be regulated by curcumin. Antiproliferative potential of curcumin is an established fact; it also inhibits transcription factors and modulates growth factor receptors and cell adhesion molecules. Cell adhesion molecules are implicated in tumor growth, in formation of blood vessels, and also in metastasis [120]. Curcumin, by regulation of NF-κB, reduced angiogenesis and tumor growth appreciably [4]. Upregulation of tumor suppressors is warranted. Curcumin, by upregulation of some of the tumor suppressor genes like p16, has been found to repress JAK2 and STAT3 pathways, leading to inhibition of invasion and migration. Matrix metalloproteinases that are important for ECM are inhibited by curcumin. In a study it was noted that the ingestion of curcumin was accompanied by a decrease in lymphocytic glutathione S-transferase (GST) [99]. The GSTs are a family of phase II detoxification enzymes. This family has been shown to be involved in the development of resistance to chemotherapy drugs [109].

Resveratrol, a stilbenoid, is found in peanuts, grapes, red wine, and some berries. This polyphenolic compound has a number of health benefits. It is also a good anticancer agent. It acts in all stages of carcinogenesis [103]. In case of colon cancer, it was found that resveratrol is effective in the initiation step and metastasis. Initiation, growth of tumor, cell death, and metastasis are regulated by Wnts and downstream proteins [2], which are regulated by resveratrol [131]. Resveratrol is also a regulator of EMT and metastasis. As a potent chemoenhancer, it can reduce the side effects of chemotherapy [123]. It upregulates PTEN expression (a natural inhibitor of this pathway) and at the same time downregulates Akt phosphorylation, hence inhibiting the P13K/Akt/mTOR pathway [43]. Invasion is also prevented by resveratrol by inactivation of RhoA [44].

Silibinin is a constituent of silymarin, isolated from the milk of thistle seeds. It is known for its hepato-protective nature and anticancer activities. It has been found to inhibit chemokine receptor 4 (CXCR4), thereby preventing metastasis in breast cancer cells [115]. By inhibition of ERK, silibinin decreases the expression of MMP9 and also migration is inhibited [78]. Other important markers associated with metastasis, namely, EGFR and VEGF, are also inhibited. Apigenin, a flavonoid found in parsley and celery, has been found to suppress TNF-α induced chemokines.

The chemokines play a vital role in migration and metastasis. MMP9 and VEGF are reduced by this compound. Therefore, metastasis and angiogenesis are inhibited by apigenin [95].

Isoliquiritigenin, obtained from the licorice root, inhibits metastasis and angiogenesis [81]. Isoliquiritigenin has been shown to downregulate miR-374a and upregulate PTEN, which in turn diminishes the expression of Akt [81, 82]. Some other proteins that aid in migration, like NF-κB, P13K/Akt, and p38, are downregulated by isoliquiritigenin; also matrix metalloproteinases 2 and 9, VEGF, and HIF are downregulated [114]. Cruciferous vegetables are rich in isothiocyanates which are known to possess potent anticancer properties. Benzyl isothiocyanate (BITC), phenethyl isothiocyanate (PEITC), and sulforaphane (SFN) are common isothiocyanates. Sulforaphane from cruciferous vegetables is a suppressor of metastatic proteins like MMP2 and 9 and NF-κB, thereby preventing invasion [53].

PEITC and BITC reduce activity of MMP9 induced by 12-O-tetradecanoylphorbol-13-acetate (TPA) in cancer cell lines. They do so by suppressing expression of AP-1 and NF-κB in osteosarcoma cell lines. Phosphorylation of ERKs, Akt, and FAK is reduced by isothiocyanates. All these events cumulatively contribute to inhibition of metastatic proteins. PEITC has been found to be more effective in suppressing metastasis than BITC or sulforaphane [35]. Epithelial ovarian cancer is an aggressive form of cancer which often undergoes metastasis even at early stages. PEITC exerts its action as anti-metastatic agent by binding to CRM1, a nuclear exportin, and inhibiting its function. PEITC has been reported to prevent migration and invasion of SKOV3 and HO8910 cells. m-TOR, STAT3, MMP2, and MMP9 are also targeted by PEITC in this form of ovarian cancer [98].

One of the leading causes of death due to cancer globally is breast cancer metastasis. Progression of tumor and metastasis are facilitated by hypoxia-inducible factor 1α (HIF-1α) in carcinoma of the breast. It has been reported that PEITC induces accumulation of Nrf2 in the nucleus; increases activities of several antioxidant enzymes; reduces ROS PEITC, by acting as an indirect antioxidant agent; and prevents adhesion, migration, and invasion of cells. PEITC shows its anti-metastatic and anti-angiogenic potential by modulating MMPs and VEGF [89, 93]. Breast tumor cells, having high expression of HER2, often leave their site of origin and metastasize to the brain, which leads to severe therapeutic complications and poor prognosis. In vivo studies demonstrate that PEITC can reduce occurrence of metastatic brain tumors and improve the survival rate of mice. Expressions of HER2, EGFR, and VEGF in the brain of PEITC-treated mice were also found to be less compared to control mice [22]. PEITC reduces risk of prostate cancer invasion and metastasis by modulating microRNAs and reducing expression of MMPs [129].

Indole-3-carbinol (I3C) plays an effective role in chemoprevention of breast cancer. It prevents cell adhesion, migration, and invasion by elevating levels of a tumor suppressor gene PTEN and E-cadherin, which is a regulator of EMT in cancer cells. I3C also reduces growth of breast tumor cells and inhibits metastasis [85]. Apigenin is a flavonoid found in parsley, celery, celeriac, and chamomile tea and shows anticarcinogenic potential. In vivo studies revealed that apigenin reduced metastasis from intestinal adenocarcinomas induced by azoxymethane in male Wistar rats by preventing phosphorylation of mitogen-activated protein kinase (MAPK) [108]. Apigenin also shows efficacy in ovarian cancer metastasis by inhibiting MMPs, p-Akt, and p-p70S6K1 [26].

Kaempferol is a natural flavonoid and shows efficacy as an anticancer agent in various forms of malignancy, including non-small-cell lung cancer cells by targeting epithelial-mesenchymal transition (EMT). This polyphenol reduces proliferation and migration of cells, modulates EMT proteins E-cadherin and vimentin, and suppresses migration of cells and metastasis [25]. Metastatic cholangiocarcinoma or cancer of the bile duct has very poor survival rates. Kaempferol has been reported to inhibit proliferation of HCCC9810 and QBC939 cells. In vitro studies reveal that colony formation ability of cells get reduced by kaempferol. Kaempferol also modulates expression of molecules involved in apoptosis like Bcl-2, Bax, Fas, caspases, and PARP leading to induction of programmed cell death in these cells [84]. Kaempferol also reduced expressions of Akt, TIMP2, and MMP2. In vivo studies show that this phytochemical is capable of reducing tumor volume and metastasis in lung cancer model [84]. Kaempferol also shows anti-inflammatory, antimicrobial, and antioxidant potential. Kaempferol prevents the invasion and migration of renal cell carcinoma by reducing levels and activities of MMPs. It does so by reducing phosphorylation of Akt and focal adhesion kinase (FAK) [32].

Ginsenoside is a member of the group of steroid glycosides and triterpine saponins. It is found in the plant Panax, and has been found to inhibit NF-κB and antiapoptotic protein Bcl-2, besides facilitating upregulation of several apoptotic proteins like Bax and caspases among others. Inhibition of NF-κB by ginsenoside is achieved through inactivation of Akt and ERK. This biomolecule has been found to inhibit cell invasion and angiogenesis by decreasing MMP2, MMP9, VEGF, P13K, mTOR, Akt, and JNK [56].

Luteolin is a flavonoid that is isolated from *Salvia tomentosa*. It is mainly found in the leaves, barks, rind, and pollen of celery, broccoli, thyme, parsley, peppermint, and rosemary. It is an antioxidant, anti-allergic, antitumor, and anti-inflammatory agent. Migration of cells in case of triple-negative breast cancer to the lungs is inhibited by luteolin, through inhibition of VEGF [9]. Reversal of EMT is facilitated by luteolin, thus preventing lung metastasis. Polyphenol extracted from the bark of *Rhus verniciflua* is a traditional medicine, used for the remedy of abdominal pain, gastritis, etc. in Korea. Its anticancer, antioxidant, antimicrobial, and anti-inflammatory activities are known [94]. A clinical study reported the promising effect of *R. verniciflua* extract in improving the overall survival rate in patients with metastatic pancreatic and colorectal cancer [51, 52]. In case of metastatic renal carcinoma, its efficacy has been indicated [50].

NF-κB is a transcription factor and controls DNA transcription, cytokine production, and cell survival. Tumor necrosis factor (TNF-α) is a regulator of the inflammatory process in tumor development. NF-κB is highly implicated in carcinogenesis; besides, it plays a lead role in tumor promotion, thereby influencing cell invasion and metastasis. Control of metastasis is very important for better prognosis, and therefore targeting NF-κB is important. An indole alkaloid hirsutine, isolated from *Uncaria rhynchophylla*, has been found to check inflammation-mediated neurotoxicity and microglial activation [38]. Hirsutine has been found to inhibit NF-κB activity. Metastatic potential has been negatively regulated by hirsutine, by controlling migration and invasion. Two important metastatic proteins, namely, MMP2 and MMP9, have been found to be reduced. Therefore, hirsutine may be considered as a compound that can inhibit metastasis by downregulation of NF-κB [59].

Blueberries are low in calories and good in lowering blood sugar and protecting the heart and brain as well. These are excellent source of vitamin and are good antioxidants. Blueberry decreases phosphatidylinositol 3-kinase (PI3K)/Akt and NF-κB activation in metastatic breast cancer cells (MDA-MB-231 cells). Results of wound healing and migration assay reveal that metastatic potential of triple-negative breast cancer cells gets inhibited by blueberry. Activity of MMP9 was diminished, along with urokinase-type plasminogen activator secretion. Concomitantly MMP1 and plasminogen activator inhibitor-1 were upregulated [1].

As discussed previously, EMT and MET are related to the metastatic potential. Isoliquiritigenin, a phenolic compound from licorice, possesses a number of biological properties. They are good antioxidant, anti-inflammatory, antiviral, antidiabetic, antispasmodic, and above all antitumor agents [81]. Programmed cell death that happens in anchorage-dependent cells when they detach from the surrounding ECM is known as anoikis. Isoliquiritigenin has been found to prevent metastasis in human breast cancer cells through activating anoikis [102, 106, 130].

Active ingredient in coffee is caffeine, which is responsible for the stimulating effect. Coffee imparts protection against liver disease, liver cancer, heart disease, and type 2 diabetes among others [112]. Chlorogenic acids and melanoidins present in coffee deactivate oxidants [28, 67]. Coffee has been found to diminish the activities of MEK1 and TOPK. ERK and NF-κB are also inhibited. Coffee targets MEK1 and TOPK and has been found to reduce metastasis in colon cancer. Chlorogenic acid present in coffee has been found to diminish expression of COX-2 and matrix metalloproteinases, which support cell invasion. Ras is in a mutated form in most colon cancer cells, and they promote abnormal proliferation by activation of ERKs. Chlorogenic acid inhibited phosphorylation of ERKs, thereby inhibiting metastasis [118].

Patients suffering from idiopathic pulmonary fibrosis are highly prone to lung cancer, and prognosis of the disease is very poor. Naringenin, a flavanone found in grapefruit, shows anticancer and anti-metastatic potential as revealed from in vivo studies. C57BL/6 and BALB/c mice with lung fibrosis showed higher incidences of lung metastasis and shorter survival rate. Naringenin prevents lung metastases in mice model with pulmonary fibrosis and increases their lifespan by reducing levels of TGF-β1 and regulatory T cells [58]. Lycopene is bright red in color and belongs to the family of carotenoids. It is found in abundance in tomatoes. Lycopene shows anticancer potential in both in vitro and in vivo models of lung cancer by modulating redox status, inducing cell cycle arrest and consequently apoptosis, and regulating signaling of growth factors. Apart from these, lycopene also prevented invasion, angiogenesis, and metastasis of cancer cells [79].

NADPH oxidase 4 (NOX4) produces reactive oxygen species (ROS) and leads to cancer metastasis. High expression and activity of NOX4 are mediated by TGF-β in cancer cells. Lycopene prevents expression and activity of NOX4 in liver adenocarcinoma SK-Hep-1 cells. This leads to reduced ROS formation and lowered invasion and migration of cells. Metastatic proteins are also downregulated by lycopene [36].

The mangosteen fruit has long been used in Chinese and Ayurvedic medicines. α-Mangostin, an active phytochemical from the fruit, has been reported to induce apoptosis in different cancer cells. In vivo mammary cancer models were developed to study the anti-metastatic potential of α-mangostin. It suppresses volume of the tumor, and lymph node metastasis in breast cancer mode was reduced. Apoptosis of cancer cells mediated by the mitochondrial pathway and expression of caspases were found to be increased after administration of this phytochemical. α-Mangostin diminished phospho-Akt-threonine 308 (Thr308) in both in vitro and in vivo studies [101].

6.8 Conclusion

We have reviewed the roles of phytochemicals in overcoming drug resistance and preventing metastasis in this chapter. These two are the main causes for poor prognosis and the main challenges in cancer therapy. We have examined the reasons for poor prognosis and how phytochemicals act on the key genes and pathways for mitigating drug resistance and preventing metastasis. Along with our discussion on the molecular mechanisms of phytochemicals in chemoprevention and chemosensitization in Chap. 5, a comprehensive view of the roles of phytochemicals in cancer therapy is presented in this chapter. However, many of the beneficial effects of phytochemicals are difficult to reproduce in real disease conditions due to poor bioavailability of the phytochemicals. We discuss the issues related to bioavailability in details in Chap. 8.

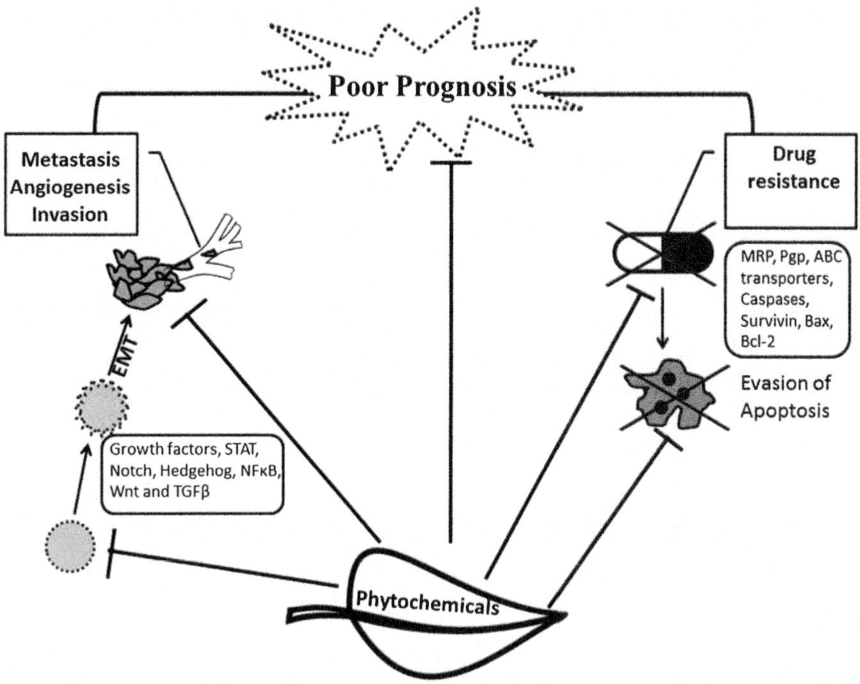

References

1. L.S. Adams, S. Phung, N. Yee, N.P. Seeram, L. Li, S. Chen, Blueberry phytochemicals inhibit growth and metastatic potential of MDA-MB-231 breast Cancer cells through modulation of the phosphatidylinositol 3-kinase pathway. Cancer Res. **70**(9), 3594 (2010)
2. J.N. Anastas, R.T. Moon, WNT signalling pathways as therapeutic targets in cancer. Nat. Rev. Cancer **13**(1), 11–26 (2013)
3. S. Banerjee, Y. Li, Z. Wang, F.H. Sarkar, Multi-targeted therapy of cancer by genistein. Cancer Lett. **269**(2), 226–242 (2008)
4. S. Bimonte, A. Barbieri, G. Palma, D. Rea, A. Luciano, M. D'Aiuto, C. Arra, F. Izzo, BioMed Res. Int. **2015**, 878134 (2015)
5. S.M. Chacko, P. Thambi, R. Kuttan, I. Nishigaki, Beneficial effects of green tea: A literature review. Chin. Med. **5**, 13 (2010)
6. S. Chian, Y.Y. Li, X.J. Wang, X.W. Tang, Luteolin sensitizes two oxaliplatin-resistant colorectal cancer cell lines to chemotherapeutic drugs via inhibition of the Nrf2 pathway. Asian Pac. J. Cancer Prev. **15**(6), 2911–2916 (2014)
7. A.G. Clark, D.M. Vignjevic, Modes of cancer cell invasion and the role of the microenvironment. Curr. Opin. Cell Biol. **36**, 13–22 (2015)
8. K.M. Cook, W.D. Figg, Angiogenesis inhibitors – Current strategies and future prospects. CA Cancer J. Clin. **60**(4), 222–243 (2010)
9. M.T. Cook, Y. Liang, C. Besch-Williford, S.M. Hyder, Luteolin inhibits lung metastasis, cell migration, and viability of triple-negative breast cancer cells. Breast Cancer **9**, 9–19 (2017)
10. D.C. Dale, Poor prognosis in elderly patients with cancer: The role of bias and undertreatment. J. Support. Oncol. **1**(4 Suppl 2), 11–17 (2003)

11. O. Dreesen, A.H. Brivanlou, Signaling pathways in cancer and embryonic stem cells. Stem Cell Rev. **3**(1), 7–17 (2007)
12. S.J. Duthie, Epigenetic modifications and human pathologies: Cancer and CVD. Proc. Nutr. Soc. **70**(1), 47–56 (2011)
13. J.A. Engelman, K. Zejnullahu, T. Mitsudomi, Y. Song, C. Hyland, J.O. Park, N. Lindeman, C.M. Gale, X. Zhao, J. Christensen, T. Kosaka, A.J. Holmes, A.M. Rogers, F. Cappuzzo, T. Mok, C. Lee, B.E. Johnson, L.C. Cantley, P.A. Jänne, MET amplification leads to gefitinib resistance in lung cancer by activating ERBB3 signaling. Science **316**(5827), 1039–1043 (2007)
14. J.A. Engelman, K. Zejnullahu, T. Mitsudomi, Y. Song, C. Hyland, J.O. Park, N. Lindeman, C.M. Gale, X. Zhao, J. Christensen, T. Kosaka, A.J. Holmes, A.M. Rogers, F. Cappuzzo, T. Mok, C. Lee, B.E. Johnson, L.C. Cantley, P.A. Jänne, Principles of cancer therapy: Oncogene and non-oncogene addiction. Cell **136**(5), 823–837 (2009)
15. M.T. Frost, Q. Wang, S. Moncada, M. Singer, Hypoxia accelerates nitric oxide-dependent inhibition of mitochondrial complex I in activated macrophages. Am. J. Physiol. Regul. Integr. Comp. Physiol. **288**(2), R394–R400 (2005)
16. H. Fujiki, E. Sueoka, A. Rawangkan, M. Suganuma, Human cancer stem cells are a target for cancer prevention using (−)-epigallocatechin gallate. J. Cancer Res. Clin. Oncol. **143**(12), 2401–2412 (2017)
17. A. Ghorbani, Best herbs for managing diabetes: A review of clinical studies. Braz. J. Pharm. Sci. **49**, 413–422 (2013)
18. J.P. Gillet, M.M. Gottesman, Mechanisms of multidrug resistance in cancer. Methods Mol. Biol. **596**, 47–76 (2010)
19. D. Gioeli, *Targeted Therapies Mechanisms of Resistance* (Humana Press, New York, 2011)
20. V. Gouaze-Andersson, M.C. Cabot, Glycosphingolipids and drug resistance. Biochim. Biophys. Acta **1758**(12), 2096–2103 (2006)
21. M. Greaves, Darwinian medicine: A case for cancer. Nat. Rev. Cancer **7**, 213–221 (2007)
22. P. Gupta, C. Adkins, P. Lockman, S.K. Srivastava, Metastasis of breast tumor cells to brain is suppressed by Phenethyl Isothiocyanate in a novel in vivo metastasis model. PLoS One **8**(6), e67278 (2013)
23. L. Haitao, M.K. Daddysman, G.O. Rankin, B.-H. Jiang, Y.C. Chen, Kaempferol enhances cisplatin's effect on ovarian cancer cells through promoting apoptosis caused by down regulation of cMyc. Cancer Cell Int. **10**, 16 (2010)
24. W. Han, L. Li, S. Qiu, Q. Lu, Q. Pan, Y. Gu, J. Luo, X. Hu, Shikonin circumvents cancer drug resistance by induction of a necroptotic death. Mol. Cancer Ther. **6**(5), 1641–1649 (2007)
25. M. Hang, F. Zhao, S.-B. Chen, Q. Sun, C.-X. Zhang, Kaempferol modulates the metastasis of human non-small cell lung cancer cells by inhibiting epithelial-mesenchymal transition. Bangladesh J. Pharmacol. **10**(2), 267–270 (2015)
26. J. He, Q. Xu, M. Wang, C. Li, X. Qian, Z. Shi, L.-Z. Liu, B.-H. Jiang, Oral Administration of Apigenin Inhibits Metastasis through AKT/P70S6K1/MMP-9 pathway in Orthotopic ovarian tumor model. Int. J. Mol. Sci. **13**(6), 7271–7282 (2012)
27. M.A. Hediger, M.F. Romero, J.B. Peng, A. Rolfs, H. Takanaga, E.A. Bruford, The ABCs of solute carriers: Physiological, pathological and therapeutic implications of human membrane transport proteins introduction. Pflugers Arch. **447**(5), 465–468 (2004)
28. C. Hoelzl, S. Knasmuller, K.H. Wagner, L. Elbling, W. Huber, N. Kager, F. Ferk, V. Ehrlich, A. Nersesyan, O. Neubauer, A. Desmarchelier, M. Marin-Kuan, T. Delatour, C. Verguet, C. Bezençon, A. Besson, D. Grathwohl, T. Simic, M. Kundi, B. Schilter, C. Cavin, Instant coffee with high chlorogenic acid levels protects humans against oxidative damage of macromolecules. Mol. Nutr. Food Res. **54**(12), 1722–1733 (2010)
29. N.P. Holzapfel, B.M. Holzapfel, S. Champ, J. Feldthusen, J. Clements, D.W. Hutmacher, The potential role of lycopene for the prevention and therapy of prostate cancer: From molecular mechanisms to clinical evidence. Int. J. Mol. Sci. **14**(7), 14620–14646 (2013)

30. X. Huang, S. Chen, L. Xu, Y. Liu, D.K. Deb, L.C. Platanias, R.C. Bergan, Genistein inhibits p38 map kinase activation, matrix metalloproteinase type 2, and cell invasion in human prostate epithelial cells. Cancer Res. **65**(8), 3470–3478 (2005)
31. L. Huang, S. Zhang, J. Zhou, X. Li, Effect of resveratrol on drug resistance in colon cancer chemotherapy. RSC Adv. **9**, 2572–2580 (2019)
32. T.-W. Hung, P.-N. Chen, H.-C. Wu, S.-W. Wu, P.-Y. Tsai, Y.-S. Hsieh, H.R. Chang, Kaempferol inhibits the invasion and migration of renal Cancer cells through the downregulation of AKT and FAK pathways. Int. J. Med. Sci. **14**(10), 984–993 (2003)
33. L.C. Huysentruyt, Z. Akgoc, T.N. Seyfried, Hypothesis: Are neoplastic macrophages/microglia present in glioblastoma multiforme? ASN Neuro. **3**(4), e00064 (2011)
34. M. Inukai, S. Toyooka, S. Ito, H. Asano, S. Ichihara, J. Soh, H. Suehisa, M. Ouchida, K. Aoe, M. Aoe, K. Kiura, N. Shimizu, H. Date, Presence of epidermal growth factor receptor gene T790M mutation as a minor clone in non-small cell lung cancer. Cancer Res. **66**(16), 7854–7858 (2006)
35. Y.J. Jeong, H.J. Cho, F.L. Chung, X. Wang, H.S. Hoe, K.K. Park, C.H. Kim, H.W. Chang, S.R. Lee, Y.C. Chang, Isothiocyanates suppress the invasion and metastasis of tumors by targeting FAK/MMP-9 activity. Oncotarget **8**(38), 63949–63962 (2017)
36. B.Y. Jhou, T.Y. Song, I. Lee, M.L. Hu, N.C. Yang, Lycopene inhibits metastasis of human liver adenocarcinoma SK-Hep-1 cells by downregulation of NADPH oxidase 4 protein expression. J. Agric. Food Chem. **65**(32), 6893–6903 (2017)
37. Y. Ji, M.E. Morris, Effect of organic Isothiocyanates on breast Cancer resistance protein (ABCG2)-mediated transport. Pharm. Res. **21**(12), 2261–2269 (2004)
38. H.Y. Jung, K.N. Nam, B.C. Woo, K.P. Kim, S.O. Kim, E.H. Lee, Hirsutine, an indole alkaloid of Uncaria rhynchophylla, inhibits inflammation-mediated neurotoxicity and microglial activation. Mol. Med. Rep. **7**(1), 154–158 (2013)
39. A. Kaczyńska, A. Herman-Antosiewicz, Combination of lapatinib with isothiocyanates overcomes drug resistance and inhibits migration of HER2 positive breast cancer cells. Breast Cancer **24**(2), 271–280 (2017)
40. M.C. Karlsson, S.F. Gonzalez, J. Welin, J. Fuxe, Epithelial-mesenchymal transition in cancer metastasis through the lymphatic system. Mol. Oncol. **11**(7), 781–791 (2017)
41. K. Kato, N.K. Long, H. Makita, M. Toida, T. Yamashita, D. Hatakeyama, A. Hara, H. Mori, T. Shibata, Effects of green tea polyphenol on methylation status of RECK gene and cancer cell invasion in oral squamous cell carcinoma cells. Br. J. Cancer **99**(4), 647–654 (2008)
42. J.Y. Kee, Y.H. Han, D.S. Kim, J.G. Mun, J. Park, M.Y. Jeong, J.Y. Um, S.H. Hong, Inhibitory effect of quercetin on colorectal lung metastasis through inducing apoptosis, and suppression of metastatic ability. Phytomedicine **23**(13), 1680–1690 (2016)
43. A. Khan, A.N. Aljarbou, Y.H. Aldebasi, S.M. Faisal, M.A. Khan, Resveratrol suppresses the proliferation of breast cancer cells by inhibiting fatty acid synthase signaling pathway. Cancer Epidemiol. **38**(6), 765–772 (2014)
44. T.N. Kim, S.R. Choe, K.H. Cho, D.Y. Cho, J. Kang, C.G. Park, H.Y. Lee, Resveratrol suppresses breast cancer cell invasion by inactivating a RhoA/YAP signaling axis. Exp. Mol. Med. **49**, e296 (2017)
45. S. Kitagawa, T. Nabekura, S. Kamiyama, Inhibition of P-glycoprotein function by tea catechins in KB-C2 cells. J. Pharm. Pharmacol. **56**(8), 1001–1005 (2004)
46. S. Kitagawa, T. Takahashi, T. Nabekura, E. Tachikawa, H. Hasegawa, Inhibitory effects of ginsenosides and their hydrolysed metabolites on daunorubicin transport in KB-C2 cells. Biol. Pharm. Bull. **30**, 1979–1981 (2007)
47. A.B. Kunnumakkara, S. Guha, S. Krishnan, P. Diagaradjane, J. Gelovani, B.B. Aggarwal, Curcumin potentiates antitumor activity of gemcitabine in an orthotopic model of pancreatic cancer through suppression of proliferation, angiogenesis, and inhibition of nuclear factor-kappaB-regulated gene products. Cancer Res. **67**(8), 3853–3861 (2007)
48. K. Kushiro, N.P. Nunez, Ethanol inhibits B16-BL6 melanoma metastasis and cell phenotypes associated with metastasis. In Vivo **26**(1), 47–58 (2012)

49. H. Lan, W. Hong, P. Fan, D. Qian, J. Zhu, B. Bai, Quercetin inhibits cell migration and invasion in human osteosarcoma cells. Cell. Physiol. Biochem. **43**, 553–567 (2017)
50. S.H. Lee, K.S. Kim, W.C. Choi, S.W. Yoon, The concurrent use of Rhus verniciflua stokes as complementary therapy with second or more line regimen on advanced non-small-cell lung cancer: Case series. J. Korean Oriental Med. **30**, 112–117 (2009)
51. C.W. Lee, R.C. Wu, J.-T. Hsu, C.-N. Yeh, T.-S. Yeh, T.-L. Hwang, Y.-Y. Jan, M.-F. Chen, Isolated pancreatic metastasis from rectal cancer: A case report and review of literature. World J. Surg. Oncol. **8**, 26 (2010)
52. S. Lee, K. Kim, H. Jung, S. Lee, S. Cheon, S. Kim, W. Eo, W. Choi, Efficacy and safety of standardized allergen-removed Rhus verniciflua Stokes extract in patients with advanced or metastatic pancreatic cancer: A Korean single-center experience. Oncologia **81**, 312–318 (2011)
53. Y.R. Lee, E.M. Noh, J.H. Han, J.M. Kim, B.M. Hwang, B.S. Kim, S.H. Lee, S.H. Jung, E.Y. Chung, J.S. Kim, Sulforaphane controls TPA-induced MMP-9 expression through the NF-kappa B signalling pathway, but not AP-1, in MCF-7 breast cancer cells. BMB Rep. **46**, 201–206 (2013)
54. H. Li, S. Krstin, M. Wink, Modulation of multidrug resistant in cancer cells by EGCG, tannic acid and curcumin. Phytomedicine **50**, 213–222 (2018)
55. W.-T. Liao, Y.-P. Ye, Y.-J. Deng, X.-W. Bian, Y.-Q. Ding, Metastatic cancer stem cells: From the concept to therapeutics. Am. J. Stem Cells. **3**(2), 46–62 (2014)
56. R.H. Liu, Health benefits of fruit and vegetables are from additive and synergistic combinations of phytochemicals. Am. J. Clin. Nutr. **78**(3), 517S–520S (2003)
57. C.M. Liu, C.L. Kao, Y.T. Tseng, Y.-C. Lo, C.-Y. Chen, Ginger phytochemicals inhibit cell growth and modulate drug resistance factors in docetaxel resistant prostate Cancer cell. Molecules **22**(9), 1477 (2017)
58. C. Lou, F. Zhang, M. Yang, J. Zhao, W. Zeng, X. Fang, Y. Zhang, C. Zhang, W. Liang, Naringenin decreases invasiveness and metastasis by inhibiting TGF-β-induced epithelial to mesenchymal transition in pancreatic Cancer cells. PLoS One **7**(12), e50956 (2012)
59. C. Lou, K. Takahashi, T. Irimura, I. Saiki, Y. Hayakawa, Identification of Hirsutine as an anti-metastatic phytochemical by targeting NF-κB activation. Int. J. Oncol. **45**(5), 2085–2091 (2014)
60. W.D. Lu, Y. Qin, C. Yang, L. Li, Effect of curcumin on human colon cancer multidrug resistance *in vitro* and *in vivo*. Clinics (Sao Paulo) **68**(5), 694–701 (2013)
61. J. Luo, N.L. Solimini, S.J. Elledge, Principles of Cancer therapy: Oncogene and non-oncogene addiction. Cell **138**(4), 807 (2009)
62. M.B. Maniecki, A. Etzerodt, B.P. Ulhøi, T. Steiniche, M. Borre, L. Dyrskjøt, T.F. Orntoft, S.K. Moestrup, H.J. Møller, Tumor-promoting macrophages induce the expression of the macrophage-specific receptor CD163 in malignant cells. Int. J. Cancer **131**(10), 2320–2331 (2012)
63. A. Marusyk, K. Polyak, Tumor heterogeneity: Causes and consequences. Biochim. Biophys. Acta **1805**(1), 105–117 (2010)
64. G.G. Meadows, Diet, nutrients, phytochemicals, and cancer metastasis suppressor genes. Cancer Metastasis Rev. **31**, 441–454 (2012)
65. M.B. Meads, R.A. Gatenby, W.S. Dalton, Environment-mediated drug resistance: A major contributor to minimal residual disease. Nat. Rev. Cancer **9**, 665–674 (2009)
66. S.M. Meeran, A. Ahmed, T.O. Tollefsbol, Epigenetic targets of bioactive dietary components for cancer prevention and therapy. Clin. Epigenetics **1**(3–4), 101–116 (2010)
67. M. Misik, C. Hoelzl, K.H. Wagner, C. Cavin, B. Moser, M. Kundi, T. Simic, L. Elbling, N. Kager, F. Ferk, V. Ehrlich, A. Nersesyan, M. Dušinská, B. Schilter, S. Knasmüller, Impact of paper filtered coffee on oxidative DNA-damage: Results of a clinical trial. *Mutat. Res.***692**(1–2), 42–48 (2010)
68. O. Molavi, F. Narimani, F. Asiaee, S. Sharifi, V. Tarhriz, A. Shayanfar, M. Hejazi, R. Lai, Silibinin sensitizes chemo-resistant breast cancer cells to chemotherapy. Pharm. Biol. **55**(1), 729–739 (2017)

69. A. Mukherjee, S. Mukherjee, J. Biswas, M. Roy, Imatinib mesylate resistance in chronic Myelogenous leukemia cells may be overcome by Indian spice Curcumin. Int. J. Med. Sci. Innov. Res. **2**(6), 535–543 (2017)
70. J. Molnár, N. Gyémánt, I. Mucsi, A. Molnár, M. Margaret Szabó, T. Körtvélyesi, A. Varga, P. Molnár, G. Tóth, Modulation of multidrug resistance and apoptosis of cancer cells by selected carotenoids. University of Szeged University of Pécs. In Vivo **18**(2), 237–244 (2004)
71. T. Nabekura, Overcoming multidrug resistance in human cancer cells by natural compounds. Toxins (Basel). **2**(6), 1207–1224 (2010)
72. T. Nabekura, S. Kamiyama, S. Kitagawa, Effects of dietary chemopreventive phytochemicals on P-glycoprotein function. Biochem. Biophys. Res. Commun. **327**(3), 866–870 (2005)
73. T. Nabekura, T. Yamaki, S. Kitagawa, Effects of chemopreventive citrus phytochemicals on human P-glycoprotein and multidrug resistance protein 1. Eur. J. Pharmacol. **600**, 45–49 (2008a)
74. T. Nabekura, T. Yamaki, K. Ueno, S. Kitagawa, Effects of plant sterols on human multidrug transporters ABCB1 and ABCC1. Biochem. Biophys. Res. Commun. **369**(2), 363–368 (2008b)
75. T. Nabekura, T. Yamaki, K. Ueno, S. Kitagawa, Inhibition of P-glycoprotein and multidrug resistance protein 1 by dietary phytochemicals. Cancer Chemother. Pharmacol. **62**(5), 867–873 (2008c)
76. T. Nabekura, T. Yamaki, T. Hiroi, K. Ueno, S. Kitagawa, Inhibition of anticancer drug efflux transporter P-glycoprotein by rosemary phytochemicals. Pharmacol. Res. **61**(3), 259–263 (2010)
77. A. Nakatsuma, T. Fukami, T. Suzuki, T. Furuishi, K. Tomono, S. Hidaka, Effects of kaempferol on the mechanisms of drug resistance in the human glioblastoma cell line T98G. Pharmazie **65**, 379–383 (2010)
78. S.J. Oh, S.P. Jung, J. Han, S. Kim, J.S. Kim, S.J. Nam, J.E. Lee, J.H. Kim, Silibinin inhibits TPA-induced cell migration and MMP-9 expression in thyroid and breast cancer cells. Oncol. Rep. **29**, 1343–1348 (2013)
79. P. Palozza, R.E. Simone, A. Catalano, M.C. Mele, Tomato lycopene and lung Cancer prevention: From experimental to human studies. Cancers (Basel). **3**(2), 2333–2357 (2011)
80. M. Pastorek, V. Simko, M. Takacova, M. Barathova, M. Bartosova, L. Hunakova, O. Sedlakova, S. Hudecova, O. Krizanova, F. Dequiedt, S. Pastorekova, J. Sedlak, Sulforaphane reduces molecular response to hypoxia in ovarian tumor cells independently of their resistance to chemotherapy. Int. J. Oncol. **47**(1), 51–60 (2015)
81. F. Peng, Q. Du, C. Peng, N. Wang, H. Tang, X. Xie, J. Shen, J. Chen, A review: The pharmacology of Isoliquiritigenin. Phytother. Res. **29**(7), 969–977 (2015)
82. F. Peng, H. Tang, P. Liu, J. Shen, X. Guan, X. Xie, J. Gao, L. Xiong, L. Jia, J. Chen, Isoliquiritigenin modulates miR-374a/PTEN/Akt axis to suppress breast cancer tumorigenesis and metastasis. Sci. Rep. **7**(1), 9022 (2017)
83. M. Prieto-Vila, R. Takahashi, W. Usuba, I. Kohama, T. Ochiya, Drug resistance driven by cancer stem cells and their niche. Int. J. Mol. Sci. **18**(12), 2574 (2017)
84. Y. Qin, W. Cui, X. Yang, B. Tong, Kaempferol inhibits the growth and metastasis of cholangiocarcinoma in vitro and in vivo. Acta Biochim. Biophys. Sin. **48**(3), 238–245 (2016)
85. M. Qinghui, D.G. Itzhak, R. Eliot, F. Saijun, Inhibitory effects of Indole-3-carbinol on invasion and migration in human breast cancer cells. Breast Cancer Res. Treat. **63**, 147–152 (2000)
86. T. Reya, S.J. Morrison, M.F. Clarke, I.L. Weissman, Stem cells, cancer, and cancer stem cells. Nature **414**(6859), 105–111 (2001)
87. J.P. Rigalli, G.N. Tocchetti, M.R. Arana, S.S. Villanueva, V.A. Catania, D. Theile, M.L. Ruiz, J. Weiss, The phytoestrogen genistein enhances multidrug resistance in breast cancer cell lines by translational regulation of ABC transporters. Cancer Lett. **376**(1), 165–172 (2016)
88. M. Roy, S. Mukherjee, Reversal of resistance towards cisplatin by curcumin in cervical Cancer cells. Asian Pac. J. Cancer Prev. APJCP **15**(3), 1403–1410 (2014)

89. M. Roy, R. Sarkar, S. Mukherjee, A. Mukherjee, J. Biswas, Sulforaphane inhibits metastatic events in breast cancer cells through genetic and epigenetic regulation. J. Carcinog Mutagen **6**, 4 (2015)
90. M. Russo, P. Nigro, R. Rosiello, R. D'Arienzo, G.L. Russo, Quercetin enhances CD95- and TRAIL-induced apoptosis in leukemia cell lines. Leukemia **21**(5), 1130–1133 (2007)
91. D. Sadava, S.E. Kane, Silibinin reverses drug resistance in human small-cell lung carcinoma cells. Cancer Lett. **339**(1), 102–106 (2013)
92. S. Saha, A. Adhikary, P. Bhattacharyya, T. Das, G. Sa, Death by design: Where curcumin sensitizes drug-resistant tumours. Anticancer Res. **32**(7), 2567–2584 (2012)
93. R. Sarkar, S. Mukherjee, J. Biswas, M. Roy, Phenethyl isothiocyanate, by virtue of its antioxidant activity, inhibits invasiveness and metastatic potential of breast cancer cells: HIF-1α as a putative target. Free Radic. Res. **19**, 1–48 (2015)
94. Z. Selamoglu, Biotechnological approaches on anticancer activity of flavonoids-mini review. Mod. Appro. Drug Des. **1**(2) (2017). MADD.000510.2017
95. H.S. Seo, J.M. Ku, H.S. Choi, J.K. Woo, B.H. Lee, D.S. Kim, H.J. Song, B.H. Jang, Y.C. Shin, S.G. Ko, Apigenin overcomes drug resistance by blocking the signal transducer and activator of transcription 3 signaling in breast cancer cells. Oncol. Rep. **38**(2), 715–724 (2017)
96. T.N. Seyfried, *Cancer as a Metabolic Disease: On the Origin, Management, and Prevention of Cancer* (Wiley, Hoboken, 2012), p. 432
97. T.N. Seyfried, L.C. Huysentruyt, On the origin of cancer metastasis. Crit. Rev. Oncog. **18**(1–2), 43–73 (2017)
98. W.Y. Shao, Y.L. Yang, H. Yan, Q. Huang, K.J. Liu, S. Zhang, Phenethyl isothiocyanate suppresses the metastasis of ovarian cancer associated with the inhibition of CRM1-mediated nuclear export and mTOR-STAT3 pathway. Cancer Biol. Ther. **18**(1), 26–35 (2017)
99. R.A. Sharma, C.R. Ireson, R.D. Verschoyle, K.A. Hill, M.L. Williams, Leuratti, M.M. Manson, L.J. Marnett, W.P. Steward, A. Gescher, Effects of dietary curcumin on glutation S-transferase and malondialdehyde-DNA adducts in rat liver and colon mucosa: relation with drug levels. Clin. Cancer Res. **7**, 1452–1458 (2001)
100. S.V. Sharma, J. Settleman, Oncogene addiction: setting the stage for molecularly targeted cancer therapy. Genes Dev. **21**(24), 3214–3231 (2007)
101. M.-A. Shibata, M. Iinuma, J. Morimoto, H. Kurose, K. Akamatsu, Y. Okuno, Y. Akao, Y. Otsuki, α-Mangostin extracted from the pericarp of the mangosteen (*Garcinia mangostana* Linn) reduces tumor growth and lymph node metastasis in an immunocompetent xenograft model of metastatic mammary cancer carrying a p53 mutation. BMC Med. **9**, 69 (2011)
102. L. Si, X. Yang, X. Yan, Y. Wang, Q. Zheng, Isoliquiritigenin induces apoptosis of human bladder cancer T24 cells via a cyclin-dependent kinase-independent mechanism. Oncol. Lett. **14**, 241–249 (2017)
103. J.M. Smoliga, J.A. Baur, H.A. Hausenblas, Resveratrol and health–a comprehensive review of human clinical trials. Mol. Nutr. Food Res. **55**(8), 1129–1141 (2011)
104. C.-Y. Sun, Q.-Y. Zhang, G.-J. Zheng, B. Feng, Autophagy and its potent modulators from phytochemicals in cancer treatment. Cancer Chemother. Pharmacol. **83**(1), 17–26 (2019a)
105. C.-Y. Sun, Q.-Y. Zhang, G.-J. Zheng, B. Feng, Phytochemicals: Current strategy to sensitize cancer cells to cisplatin. Biomed. Pharmacother. **110**, 518–527 (2019b)
106. K. Tanagornmeatar, C. Chaotham, B. Sritularak, K. Likhitwitayawuid, P. Chanvorachote, Cytotoxic and anti-metastatic activities of phenolic compounds from Dendrobium ellipsophyllum. Anticancer Res. **34**(11), 6573–6579 (2014)
107. H. Tang, L. Zeng, J. Wang, X. Zhang, Q. Ruan, J. Wang, S. Cui, D. Yang, Reversal of 5-fluorouracil resistance by EGCG is mediate by inactivation of TFAP2A/VEGF signaling pathway and down-regulation of MDR-1 and P-gp expression in gastric cancer. Oncotarget **8**(47), 82842–82,853 (2017)
108. A. Tatsuta, H. Iishi, M. Baba, H. Yano, K. Murata, M. Mukai, H. Akedo, Suppression by apigenin of peritoneal metastasis of intestinal adenocarcinomas induced by azoxymethane in Wistar rats. Clin. Exp. Metastasis **18**(8), 657–662 (2000)

109. D.M. Townsend, K.D. Tew, The role of glutathione-S-transferase in anti-cancer drug resistance. Oncogene **22**(47), 7369–7375 (2003)
110. S.H. Tu, C.T. Ho, M.F. Liu, C.S. Huang, H.W. Chang, C.H. Chang, C.H. Wu, Y.S. Ho, Luteolin sensitises drug-resistant human breast cancer cells to tamoxifen via the inhibition of cyclin E2 expression. Food Chem. **141**(2), 1553–1561 (2013)
111. M.F. Ullah, Cancer multidrug resistance (MDR): a major impediment to effective chemotherapy. Asian Pac. J. Cancer Prev. **9**(1), 1–6 (2008)
112. R.M. van Dam, Coffee consumption and risk of type 2 diabetes, cardiovascular diseases, and cancer. Appl. Physiol. Nutr. Metab. **33**(6), 1269–1283 (2008)
113. E. Vrdoljak, N. Marschner, C. Zielinski, J. Gligorov, J. Cortes, F. Puglisi, M. Aapro, L. Fallowfield, A. Fontana, M. Inbar, Z. Kahan, A. Welt, C. Lévy, E. Brain, X. Pivot, C. Putzu, A. González Martín, S. de Ducla, V. Easton, G. von Minckwitz, Final results of the TANIA randomised phase III trial of bevacizumab after progression on first-line bevacizumab therapy for HER2-negative locally recurrent/metastatic breast cancer. Ann. Oncol. **27**(11), 2046–2052 (2016)
114. K.L. Wang, S.M. Hsia, C.J. Chan, F.Y. Chang, C.Y. Huang, D.T. Bau, P.S. Wang, Inhibitory effects of isoliquiritigenin on the migration and invasion of human breast cancer cells. Expert Opin. Ther. Targets **17**(4), 337–349 (2013)
115. Y. Wang, W.C. Liang, W.L. Pan, W.K. Law, J.S. Hu, D.T.M. Ip, M.M.Y. Waye, T.B. Ng, D.C.C. Wan, Silibinin, a novel chemokine receptor type 4 antagonist, inhibits chemokine ligand 12-induced migration in breast cancer cells. Phytomedicine **21**, 1310–1317 (2014)
116. S. Wang, Q. Meng, Q. Xie, M. Zhang, Effect and mechanism of resveratrol on drug resistance in human bladder cancer cells. Mol. Med. Rep. **15**(3), 1179–1187 (2017)
117. O. Warburg, On the origin of cancer cells. Science **123**(3191), 309–314 (1956)
118. M. Wei, Z. Zheng, L. Shi, Y. Jin, L. Ji, Natural polyphenol Chlorogenic acid protects against acetaminophen-induced hepatotoxicity by activating ERK/Nrf2 antioxidative pathway. Toxicol. Sci. **162**(1), 99–112 (2018)
119. O. Wesołowska, J. Wiśniewski, A. Bielawska-Pohl, M. Paprocka, N. Duarte, M. Ferreira, D. Duś, K. Michalak, Stilbenes as Multidrug Resistance Modulators and Apoptosis Inducers in Human Adenocarcinoma Cells. Anticancer Res. **30**(11), 4587–4593 (2010)
120. R. Wilken, M.S. Veena, M.B. Wang, E.S. Srivatsan, Curcumin: A review of anti-cancer properties and therapeutic activity in head and neck squamous cell carcinoma. Mol. Cancer **10**, 12 (2011)
121. L. Xiao-ai, W. Bei, X. Xiao-hong, P. Lei, W. Bin, X. Dong, C. Zheng, Q. Du, Curcumin re-sensitizes multidrug resistant (MDR) breast cancer to cisplatin through inducing autophagy by decreasing CCAT1 expression. RSC Adv. **7**, 33572–33579 (2017)
122. Z. Xu, J. Mei, Y. Tan, Baicalin attenuates DDP (cisplatin) resistance in lung cancer by downregulating MARK2 and p-Akt. Int. J. Oncol. **50**(1), 93–100 (2017)
123. M. Younas, C. Hano, N. Giglioli-Guivarc'h, B.H. Abbasi, Mechanistic evaluation of phytochemicals in breast cancer remedy: current understanding and future perspectives. RSC Adv. **8**, 29714 (2018)
124. C.H. Yuan, C.T. Horng, C.F. Lee, N.N. Chiang, F.J. Tsai, C.C. Lu, J.H. Chiang, Y.M. Hsu, J.S. Yang, F.A. Chen, Epigallocatechin gallate sensitizes cisplatin-resistant oral cancer CAR cell apoptosis and autophagy through stimulating AKT/STAT3 pathway and suppressing multidrug resistance 1 signaling. Environ. Toxicol. **32**(3), 845–855 (2017)
125. H. Zahreddine, K. Borden, Mechanisms and Insights into drug resistance in cancer. Front. Pharmacol. **4**, 28 (2013). https://doi.org/10.3389/fphar.2013.00028
126. J. Zhang, F. Visser, K.M. King, S.A. Baldwin, J.D. Young, C.E. Cass, The role of nucleoside transporters in cancer chemotherapy with nucleoside drugs. Cancer Metastasis Rev. **26**(1), 85–110 (2007)
127. L. Zhang, X. Yang, X. Li, C. Li, L. Zhao, Y. Zhou, H. Hou, Butein sensitizes HeLa cells to cisplatin through the AKT and ERK/p38 MAPK pathways by targeting FoxO3a. Int. J. Mol. Med. **36**(4), 957–966 (2015a)

128. X. Zhang, Q. Guo, J. Chen, Z. Chen, Quercetin enhances cisplatin Sensitivity of human osteosarcoma Cells by modulating microRNA-217-KRAS axis. Mol. Cell **38**(7), 638–642 (2015b)

129. C. Zhang, L. Shu, H. Kim, T.O. Khor, R. Wu, W. Li, A.N.T. Kong, Phenethyl isothiocyanate (PEITC) suppresses prostate cancer cell invasion epigenetically through regulating microRNA-194. Mol. Nutr. Food Res. **60**(6), 1427–1436 (2016)

130. H. Zheng, Y. Li, Y. Wang, H. Zhao, J. Zhang, H. Chai, T. Tang, J. Yue, A.M. Guo, J. Yang, Downregulation of COX-2 and CYP 4A signaling by isoliquiritigenin inhibits human breast cancer metastasis through preventing anoikis resistance, migration and invasion. Toxicol. Appl. Pharmacol. **280**(1), 10–20 (2014)

131. H. Zhou, L. Shang, X. Li, X. Zhang, G. Gao, C. Guo, B. Chen, Q. Liu, Y. Gong, C. Shao, Resveratrol augments the canonical Wnt signaling pathway in promoting osteoblastic differentiation of multipotent mesenchymal cells. Exp. Cell Res. **315**(17), 2953–2962 (2009)

132. H. Zhu, X.W. Liu, T.Y. Cai, J. Cao, C.X. Tu, W. Lu, Q.J. He, B. Yang, Celastrol acts as a potent antimetastatic agent targeting beta1 integrin and inhibiting cell-extracellular matrix adhesion, in part via the p38 mitogen-activated protein kinase pathway. J. Pharmacol. Exp. Ther. **334**(2), 489–499 (2010)

133. H. Zhu, J. Gao, L. Wang, K.-J. Qian, L.-P. Cai, *In vitro* study on reversal of ovarian cancer cell resistance to cisplatin by naringin via the nuclear factor-κB signaling pathway. Exp. Ther. Med. **15**(3), 2643–2648 (2018)

Pharmacogenomics and Phytochemicals 7

Abstract

It is well known for a long time that the same drugs have different efficacy in different patients due to variations in metabolism of the drugs. One of the reasons for these variations is the differences in genetic makeup of individuals. There may be variations in key inherited genes through germline cells from parents due to differences in copy numbers, variations due to single nucleotide polymorphism, or presence of different alleles of the same gene. There may also be variations in genetic makeup of somatic cells due to mutations in diseases like cancer. It is important to personalize medicines considering these genetic variations in individuals in order to decide on appropriate dosage for better outcome in treatment and also to reduce side effects of drugs. We have discussed the details of pharmacogenomics both due to genetic variations in germline cells and somatic cells. In particular we have discussed the effects of such genetic variations in cancer and how phytochemicals can mitigate the effects of some of these variations in cancer.

7.1 Introduction

It is well known for a long time that the same drug has different effects on different individuals due to differences in genetic makeup among individuals. Though the same genes are present in the genome of all humans, there may be copy number variations of genes, as well as other differences in expression levels of genes among individuals. The effects of drugs on individuals may vary significantly if such genes are targets of drugs or related to targets of drugs. However, it was not easy to explore these variations in genetic makeup of individuals until the turn of the century. Large-scale sequencing of genome became both feasible and affordable with the advent of the Human Genome Project. The cost of deep sequencing has progressively dropped, and we are in an era when sequencing is becoming a diagnostic tool for

M. Roy, A. Datta, *Cancer Genetics and Therapeutics*,
https://doi.org/10.1007/978-981-13-9471-3_7

clinicians. Hence there is an increasing opportunity to explore the effects of specific drugs on individuals, which may culminate in the era of personalized medicine in the not-so-distant future.

There are two possible causes for genetic variations among individuals. Variations in the germline cells are responsible for inherited variations in the genotype, including copy number variations, whereas variations in the genetic makeup of somatic cells are not inherited, but caused by disease conditions. Genetic variations in somatic cells are most common in almost all types of cancers. Moreover, the genetic variations in somatic cells may be across individual cancer patients or even in the same tumor of a single patient. This makes the efficacy of chemotherapy challenging as the same drug has different effects and responses in different patients. We will discuss both of these types of variations in this chapter, genetic variations in both germline and somatic cells. We will also discuss the applications of phytochemicals in pharmacogenomics and how these plant-derived chemicals can help in treating individuals with different genetic makeups.

7.2 Pharmacogenomics of Genetic Variations of Germline Cells

Patients having similar health issues have been found to respond to drugs differently. Response to drugs may be due to various factors, which not only include genetics but also epigenetics, environmental factors, traits of an individual including gender, age, and other medications given to that individual at the same time. Therefore, the reason why the same treatment protocol does not work equally well for all individuals depends on a number of factors, which are pictorially represented in Fig. 7.1.

Genetic trait of an individual is decided by more than one gene. Often abnormal or aberrant genes are inherited from parents through germline cells due to mutation, resulting in different traits. Difference in genetic constitution is the cause of different traits and characteristics in individuals. Every person has a unique variation of the human genome. In an ideal case, the medicines prescribed for an individual should take into account that person's unique genetic traits; hence it is called "personalized medicine." Therefore, personalized medicine may render much better diagnosis and correct therapy.

Behavior of the genome and its variability in response to a drug are pharmacogenetics. The suffix "omics" has been incorporated later on, and the term pharmacogenomics has gained in importance in the last two decades. The purpose of pharmacogenomics is to scrutinize the genetic makeup of an individual and choose the drug and dosage based on genetic traits. Pharmacogenetics or pharmacogenomics together may leave a great impact on the development of drugs and their marketing in the future [15]. To be effective, genetic records of individuals need to be maintained, and treatment strategies should be guided by these. High-throughput sequencing techniques like RNAseq have become affordable recently, and such techniques are very important for the success of pharmacogenomics.

Fig. 7.1 The factors that determine the efficacy of a treatment using drugs

Variation within the human genome is common for every 500–1000 bases on an average. Genetic polymorphisms in drug-metabolizing enzymes have been found to influence drug response [31]. Focus is also given on single nucleotide polymorphisms (SNPs), which is a substitution of a single nucleotide that occurs at a specific position in the genome. Appreciable variation in SNP is present within a population. SNPs within the coding regions of genes can give indications for drug response. In particular it is possible to explore genome-wide associations tying multiple SNPs and haplotypes to drug response [13]. With the help of SNP profiling, it is possible to modulate drug dosage, in order to achieve maximum efficacy and minimum toxicity.

We now discuss the important genes that are particularly relevant in pharmacogenomics. Our discussion is based on the recent review article by Daly [6].

Certain genes are responsible for the difference in drug metabolism and its response. These genes are cytochrome P450 2D6, vitamin K epoxide reductase complex subunit 1 (VKORC1), thiopurine methyltransferase (TPMT), etc.

7.2.1 Cytochrome P450 (CYP)

The *cytochrome P450 (CYP)* family of enzymes are associated with the phase I metabolism of drugs. Phase I metabolism includes bioactivation and detoxification. The CYP family of proteins are membrane-associated proteins with heme as a cofactor [14]. In order to reduce iron, electrons are required. Cytochromes need a protein partner for delivery of the electron to the heme iron from NADH or NADPH. Four different CYP families of proteins are particularly important for phase I metabolism of drugs, namely, CYP2D6, CYP2C9, CYP3A4, and CYP2C19 [6]. However, these four proteins show significant polymorphism among humans. Even a significant fraction of the population completely lack the CYP2D6 and CYP2C19 proteins. This happens when there are inactivating genetic polymorphisms in both the copies of these genes [55]. These alleles code for inactive versions of these proteins and compromise the ability for drug metabolism. Another serious problem is the overexpression of some of the CYP family of proteins, as the rate of drug metabolism is too high, and the drugs may not reach the intended targets without high dosage, causing adverse side effects. Many individuals have higher expression levels of CYP2D6 and CYP2C19 proteins, either due to the presence of multiple copies of the gene or due to overexpression of the gene [41].

There are several other important members of the P450-containing systems that we discuss below. A membrane-bound hemoprotein cytochrome b5, also called Cyb5, acts as an electron carrier for several membrane-bound oxygenases. Depending on the nature of electron transfer, these enzymes can be broadly classified into the following categories [17]:

(i) Bacterial P450: They catalyze oxidations of hydrocarbons and are good biocatalysts that use ferredoxin reductase and ferredoxin for electron transfer. The turnover rates are high and bacterial P450 are attractive biocatalysts for fine chemical production [42].
(ii) $CYB_5R/cyb_5/P450$: The required electrons come from CYP b5.
(iii) FMN/Fd/P450: These are enzymes with a flavin mononucleotide (FMN)-binding domain, produced from riboflavin. It is fused to the heme-binding domain.
(iv) Microsomal P450: Cytochrome P450 reductase transfers the electron from NADPH. Cytochrome b5 also helps in electron transfer after getting reduced by cytochrome b5 reductase.
(v) Mitochondrial P450: The enzyme adrenodoxin reductase is recruited, which transfers electrons from NADPH to P450.
(vi) P450 only: External reducing power is not required. Also, molecular oxygen is not required for the catalytic action.

Drugs regulate the activity of various CYP isozymes (enzymes that have different amino acid sequences but catalyze the same reactions; they are also known as isoenzymes), sometimes negatively or positively. Biosynthesis of an isoenzyme may be induced, where activity is increased, or it may be inhibited. Metabolism of drugs and their elimination depends on CYP enzymes. Suppose there are two drugs, A and B. Drug A inhibits CYP-mediated metabolism of drug B, and then B may be accumulated within our body and cause toxicity. Therefore, it is important to have knowledge on drug interactions. It is better to select a drug that does not affect CYP, so that the metabolism of the second drug is not affected.

In *Homo sapiens* there are 18 families of CYP450 genes divided among a further 43 subfamilies [2]. We discuss below a summary of the genes and of the proteins they encode.

Some drugs, which are cytotoxic, may get inactivated by the action of CYP; on the other hand, some pro-drugs may get activated. Pro-drugs are medications that get activated only after metabolism. Activation of pro-drugs renders them effective as chemotherapeutic agents. Many different cytotoxic drugs are inactivated by the action of CYP, whereas several pro-drugs are activated by the action of CYP, rendering them cytotoxic and effective in cancer chemotherapy [30]. CYPs are therefore important enzymes, which play important roles in activation and inactivation of carcinogens, as well as the anticancer drugs. Therefore, they play a vital role in the etiology of cancer. Also, the outcomes of chemotherapy are also guided by this enzyme [36, 38].

Most of the CYP genes are polymorphic. Among them are *CYP2B6, having* 48 alleles, *CYP2C9* with 32 alleles, *CYP2D6* with 92 alleles, and *CYP3A4* having 34 alleles. Functional polymorphisms are found in the *CYP2A6*, *CYP2B6*, *CYP2C9*, *CYP2C19*, and *CYP2D6* genes. Changes or mutation in CYP genes leads to enhanced or reduced expression of enzymes, total absence of enzyme, and enzyme with altered specificity for the substrate [37]. Alterations in CYP gene may be deletion mutation, missense mutation, mutation in stop codon, or those causing splicing defects.

Polymorphic metabolizing CYP enzymes may be of different classes, namely, class I and class II. Of these two, class I is well conserved, without an important functional polymorphism. Those belonging to class I have low molecular weight and contain a single heme group. They aid in the metabolism of precarcinogens and drugs. Class II is polymorphic and active in drug metabolism, but inactive for precarcinogens. Heme group in this class is attached to a C-terminal binding motif. There are two other classes, class III and class IV. Enzymes of class III have multiple covalently attached heme groups with low redox potential. Class IV enzymes contain heme groups besides other prosthetic groups [4].

Some of the cytochrome P450s which are of importance for the metabolism of drugs and carcinogens are CYP1A1, CYP1A2, CYP1B1, CYP2A6, CYP2B6, CYP2C8, CYP2C9, CYP2C19, CYP2D6, CYP2E1, CYP3A4, CYP3A5, and CYP3A7. Of these 13 types, carcinogens are the substrates for CYP1A1, CYP1A2, CYP1B1, CYP2A6, CYP2E1, CYP3A4, CYP3A5, and CYP3A7, while drugs act as substrates for CYP1A2, CYP2A6, CYP2B6, CYP2C8, CYP2C9, CYP2C19, CYP2D6, CYP2E1, CYP3A4, CYP3A5, and CYP3A7.

In the context of carcinogenesis, CYP1A1 and CYP2E1 are most studied. CYP1A1 is not expressed in the liver, whereas CYP2A2 is expressed in the liver; therefore regulations are different. Induction of these two enzymes occurs via aryl hydrocarbon receptor [16]. CYP1A may activate or detoxify a number of cancer-causing agents, like cigarette smoke, heterocycle amines, polycyclic aromatic hydrocarbons, etc. This explains the fact why there lie inter-individual differences in susceptibility of cancer risk. CYP1B1, an extrahepatic enzyme, is highly expressed in cancer tissues. Susceptibility to hormone-dependent cancers is guided by this enzyme. Overexpression of CYP1B1 is inversely related to the sensitivity of the anticancer drug docetaxel. CYP2A6 is expressed in mucosa. This enzyme metabolizes some drugs like nicotine, coumarin, methoxyflurane, halothane, valproic acid, disulfiram, etc. Another CYP which is expressed in the liver is CYP2B6. A change in CYP2B6 affects the pharmacokinetics of cyclophosphamide. This enzyme is also responsible for the metabolism of ifosfamide, thiotepa, and procarbazine. CYP2C18, CYP2C19, CYP2C8, and CYP2C9 are expressed in the liver. CYP2C8 works on antidiabetic and antimalarial drugs, though its polymorphism is still controversial. CYP2C9 metabolizes phenytoin, warfarin, and nonsteroidal anti-inflammatory drugs (NSAIDs).

CYP2D6 is an important enzyme and has a wide variation among individuals. It aids in the metabolism of tamoxifen. It is expressed in the liver and is encoded by the CYP 2D6 gene [35]. Expression is found to be high in the central nervous system as well. CYP 2D6, which is a part of the cytochrome P450 mixed-function oxidase enzyme family, plays an important role in the metabolism of xenobiotics. Many of the drugs are metabolized and eliminated by CYP 2D6. This is achieved by addition and removal of some functional groups, specifically hydroxylation, demethylation, and dealkylation. Explanation of the nomenclature of CYP 2D6 is a cytochrome which belongs to family 2, subfamily D, and gene number 6. In humans, CYP consists of a total of 57 genes with 18 families and 44 subfamilies. Common CYPs that are important in clinical perspectives are CYP2D6, CYP2C19, CYP2C9, CYP3A4, and CYP3A5 [32]. CYP2D6 shows the maximum phenotypic variability among CYPs, mainly because of genetic polymorphism. CYP2D6, the most important CYP gene, is responsible for the metabolism of debrisoquine, metoprolol, sparteine, propranolol, encainide, codeine, dextromethorphan, clozapine, desipramine, haloperidol, amitriptyline, imipramine, etc. CYP2C19 aids in the metabolism of antidepressants and proton pump inhibitors like S-mephenytoin, amitriptyline, diazepam, omeprazole, proguanil, hexobarbital, propranolol, and imipramine. Drugs like tolbutamide, ibuprofen, mefenamic acid, tetrahydrocannabinol, losartan, diclofenac, etc. have been found to be metabolized by CYP2C9 [18]. CYP3A4 and CYP3A5 are abundantly found in the liver, and both help in the metabolism of erythromycin, ethinylestradiol, nifedipine, triazolam, cyclosporine, amitriptyline, imipramine, and aldosterone.

Pharmacogenomically it is possible to identify an individual with variations in the CYP2D6 allele. In a particular individual, function of CYP 2D6 may be of various types. Genotype of a patient may be divided in several categories [20, 21]:

(i) Poor metabolizer: when there is scanty or absence of CYP 2D6 function
(ii) Intermediate metabolizer: as the name says, can metabolize drugs at an intermediate rate, in between the poor and extensive category
(iii) Extensive metabolizer: with normal CYP 2D6 activity and ultra-rapid metabolizer, due to multiple copies of the gene abnormal function of the CYP is seen

7.2.2 VKORC1

VKORC1, an enzymatic protein complex, helps to reduce vitamin K 2,3-epoxide to the active form and is therefore important for clotting. This protein is encoded by the VKORC1 gene. It is an important membrane protein linked with the endoplasmic reticulum. Vitamin K is essential for coagulation. VKORC1 protein reduces vitamin K 2,3-epoxide to the active form (Owen et al. [34]). This form is the key factor for blood coagulation or clotting. Mutations in this VKORC1 gene lead to deficiency of vitamin K and other clotting factors. Therefore, this protein is the main enzyme in vitamin K cycle.

In case of several health disorders like pulmonary embolism and thrombosis or to avoid stroke, the drug warfarin is used. In acute case of deep vein thrombosis or pulmonary embolism, anticoagulant therapy is recommended [22]. VKORC1 is responsible for the pharmacodynamics of warfarin, which is a VKOR inhibitor [7, 26]. Bleeding during warfarin administration can be identified by VKORC1, as well as CYP2C9. Polymorphism in VKORC1 leads to varied response to warfarin, a drug given as anticoagulant or blood thinner.

In case of VKORC1 single nucleotide polymorphism (SNP), in place of wild-type G allele, A allele is found. Those having A allele produce less VKORC1 than those having G allele. This variation is seen among various ethnic groups [51]. As a consequence clotting defect results [53]. Those with this defect should be given lower dose of warfarin.

7.2.3 TPMT

Thiopurine methyltransferase (TPMT) is an enzyme encoded by the TPMT gene; the pseudogene is located on the chromosome 18q [23]. TPMT catalyzes S-methylation of thiopurine drugs, e.g., azathioprine, 6-mercaptopurine, and 6-thioguanine, and metabolizes them [25]. Alteration in this gene results in decreased methylation. This may cause anemia, leukopenia, myelosuppression, etc. [11]. 6-Mercaptopurine, a thiopurine drug, is a chemotherapeutic and immunosuppressive agent. Polymorphisms in TPMT alter the activity of the enzyme and hence increase sensitivity to the drug.

TPMT regulates the balance between thioguanine and inactive metabolites by aiding in the S-methylation of thiopurines [45]. Mercaptoethanol is an anticancer drug, whose metabolism is regulated by TPMT. This drug is highly cytotoxic as it causes severe myelosuppression and myelotoxicity. TPMT is responsible for the pharmacodynamics of several drugs like codeine, clopidogrel, tamoxifen, warfarin, etc.

7.2.4 An Example of Allele Variation in Drug Metabolism

We have discussed some of the enzymes that are responsible for drug metabolism and show variations among individuals due to over or under expressions of the enzymes. We refer to the article by Daly [6] for a more detailed review of such enzymes. Efficacy of a drug depends not only on the metabolism but also on the drug to be used. Drugs may be active drug or pro-drug. Active drugs are those that get inactivated by metabolism, whereas pro-drugs need metabolic activation for their action. Suppose two patients having the same health issues are given the same pro-drug at the same dose. Suppose the genotype of the first patient is *1/*1, i.e., this patient has received 1 allele of a gene from mother and 1 from father. This patient has an extensive metabolizer phenotype, and metabolic activity will be normal. The second person has inherited allele 1 from mother and allele 4 from father, having a genotype *1/*4. This patient will belong to the category of intermediate metabolizer, who cannot show the efficacy of the pro-drug to the fullest extent. The drug cannot be broken down completely in this case, and the second patient needs a much lower dose. In case of the second patient, unmetabolized drug is accumulated in the body, resulting in side effects. This has been depicted in Fig. 7.2. Though prescribed the same medicine at similar dose, the drug will be more efficacious in case of the first patient. Therapeutic benefit for the second patient will be much less. An intermediate-metabolizing enzyme is considered to be less active. This category of metabolizing enzymes fails to break down a drug completely as a normal metabolizer. Therefore, the same therapeutic effect may be obtained at a lower dose, which prevents the accumulation of unmetabolized drug and hence reduces side effects.

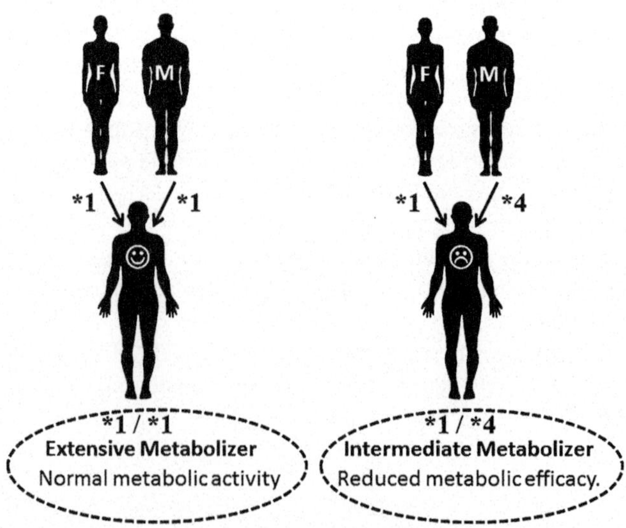

Fig. 7.2 Examples of normal and reduced metabolic efficacy

Though this was observed for a long time, the use of such alleles as valid and predictive biomarkers has been possible after the advent of cheap sequencing technologies. These biomarkers help to minimize side effects and enhance therapeutic efficacy. Therefore, pharmacogenomics is an important area because it can improve the safety of the drug for individual patients, by proper selection of drug dose. Treatment protocols can be tailor-made as per genetic disposition of the patient.

7.3 Pharmacogenomics of Genetic Variations in Somatic Cells

We have discussed pharmacogenomics for germline variations or inherited traits in the previous section. Genetic variations in somatic cells of tumors are an important hallmark of cancer. Moreover, there are significant heterogeneity in the genetic makeup of cells due to mutation within the same tumor and tumors of the same kind across patients. Hence pharmacogenomics has a major role to play in chemotherapy of cancer patients.

Pharmacogenomics is highly beneficial for cancer treatment [28]. Chemotherapeutic drugs are usually highly toxic. Hence, several issues need to be kept in mind for prescribing medication, including patient safety, minimizing health-care costs, enhancement of efficiency, and determination of optimum dosage. These are the salient points which pharmacogenomics can address. Pharmacogenomics helps to identify the population, who are susceptible to adverse effect of the drugs. Having a prior idea, these patients may be rendered extra care during treatment. The dose of the drug may be adjusted, or an alternate treatment may be given to them in case of any adverse effects [48]. Chemotherapeutic drugs are very expensive; therefore, administration of the proper drug that may help in controlling the disease is of enormous importance. Trying out various drugs without knowing much about the genetic status of an individual may cause problems and increase the burden of expenditure. Generally, dosage is guided by the body weight and age of an individual. But genetic makeup of an individual should be the key deciding factor for dose determination, so as to avoid over dosage and the associated toxicity. During treatment immune status of an individual is impaired, which is another causative factor for poor prognosis. Pharmacogenomics may guide to develop vaccines by knowing the genetic makeup, and application of proper vaccines may boost up the immune system of an individual, thereby reducing the risk of infection [29, 44].

Pharmacogenomics enables us to understand how the genetic variability has impact on drug efficiency and toxicity. Variation in the genetic material influences the pharmacokinetics and pharmacodynamics of the drugs. Morbidity and mortality rate in case of cancer is much higher compared to other diseases. Failure to treatment modality is also a common and major problem. Therefore, it is of much importance to predict the response of a patient to the drug. This is personalized oncology. Cancer treatment has to be personalized for better outcome. Treatment strategy is decided under the supervision of an oncologist, but in order to undertake the regimen, not only oncologists but specialists from various medical fields take the decision. In a tumor, not all cells are cancerous; rather a mixture of normal and cancer cells is found. In cancer pharmacogenomics candidate genes are taken into consideration. Metabolizing enzymes that are responsible for the activation of drugs are tested for relation with phenotypes of interest.

Cancer subtypes are driven by somatic mutations. Many of the genes are mutated, and these mutations are heterogeneous in a cancer cell. This is because a full-grown tumor may be derived from different clonal expansions. Targeted therapy is applied against some of the proteins which are activated by somatic mutations [52]. Most of the cancers are treated with chemotherapeutic drugs, which are cytotoxic. They may not target the somatic mutations. Response of any therapeutic drug is influenced by the response of an individual. It is of great value to understand the reason for this variation, in order to make treatment a success.

Genetic variations in each individual leave an impact on the efficacy and toxicity. Receptors and drug targets get altered due to enzymes that metabolize cancer drugs and transporters that carry the drugs. Therefore, all the anticancer drugs may not work equally well on all individuals. Making drugs tailor-made for individual patients may be of help. In pharmacogenomics of cancer treatment, often candidate genes are considered. Most of the drugs used for the treatment are in the pro-drug form; therefore metabolic activation is necessary for their action. Variation in drug-metabolizing enzymes needs to be considered. Hence the importance of genotypic arrays containing SNPs will give an idea about the absorption (A), distribution (D), metabolism (M), and elimination (E) of the drugs to be used. ADME of individual metabolic enzymes and the corresponding genes have gained in importance in pharmacogenomics. Polymorphisms of drug-metabolizing enzymes are important parameters to get an idea about the active metabolites in the blood of an individual, other organs, and tumor tissues.

Drug-metabolizing enzymes are various isoforms of cytochrome P450. If the conversion of pro-drug to active drug form is not enough, then the drug may fail to show its efficacy. If too much of active form is produced, therapeutic toxicity will be pronounced. In many cancers pro-drug response and severity of toxicity may occur. Frequency and severity of side effects at therapeutic doses for most of the anticancer drugs are very high. These drugs target highly proliferating tumor cells. ADME analysis for the pro-drugs given to the patient is being given more and more emphasis in cancer therapy. For other anticancer drúgs that do not need metabolic activation, ADME still needs consideration. Most of the anticancer drugs, though they are aimed at curing tumors, themselves are carcinogens. They are causative factors for secondary cancers. Patients having aberrant wild-type human metabolizing enzymes due to polymorphisms are unable to transform antineoplastic pro-drugs into active metabolites. They also fail to reduce active metabolite elimination rate. These often culminate in accumulation of active metabolites in the body, causing severe toxicity. Harmful effects are caused due to the treatment protocol, including serious adverse effects like death. Initially, much focus was given on maximizing the efficacy, besides reducing toxicity. In this context ADME was the main consideration. But, later on more attention was given on drug response and toxicity-related pathways. The main stay of failure to therapy is development of multidrug resistance. After chemotherapy intervention, a number of genes and proteins start working hand in hand. These are drug transporters, DNA repair molecules, P-glycoprotein, multidrug resistance proteins, etc. All these dampen the efficacy of therapy, resulting in therapeutic failure. Therefore, detection of polymorphism is of much relevance to improve therapy.

For metastatic cancers, treatment is different from that given for the primary tumor site. Polymorphisms and genotyping of metastasis-related and cancer stem cell-related genes may help to predict response of drugs. Development of drug resistance has a major contribution in this regard. After chemotherapy intervention, DNA transporter molecules, along with repair molecules, start working concertedly to dampen the therapeutic efficacy. These molecules include ATP-binding cassette transporters, P-glycoproteins, and multidrug-related proteins among others. Detection of polymorphism of these genes may give an indication of development of drug resistance. Therefore, pharmacogenomics finds a very important niche in cancer therapy. Forecasting the response of the drug against cancer growth and associated toxicity caused thereof is important to consider in cancer therapy. We have already discussed that chemotherapeutic outcome of a drug depends on polymorphisms of the genes that are targeted by the drug. Also, factors, such as polymorphisms in the molecules at tumor microenvironment, stem cell molecules, and metabolizing enzymes, play vital roles. Identification of genetic or epigenetic information of drug response genes may guide to select the treatment regime for each individual patient and, therefore, may aid in better therapeutic outcome. Efficacy of drugs used for control of metastasis is dependent on the stage. Often the oncologists use more than one drug to achieve best results of therapy. For other diseases, generally one single drug is used. This is an important difference between the pharmacogenomics of anticancer drugs and that of other diseases.

Several examples of usefulness of pharmacogenomics that are associated with specific gene polymorphisms in drug-metabolizing enzymes are TPMT, DPD, and UGT1A1. Some more examples are drug transporters MDR1 and drug target enzymes TS after treatment with 5-fluorouracil and irinotecan [24]. For the treatment of leukemia, 6-mercaptopurine is used, which shows its action by inhibiting DNA and RNA synthesis [56]. The enzyme TMPT catalyzes the formation of active drug from its pro-form. It was found that genetic variability of this gene compromises the bioavailability of the drug and enhances toxicity. Therefore, for those having genetic variation in TPMT gene, hematological toxicities may be a probable problem. Genetic studies may predict deficiency of this gene, and oncologists may prescribe a safe dose for the patient.

Another chemotherapeutic drug, irinotecan, is used to treat a number of solid tumors [12]. This, being a pro-drug, needs metabolic activation to the active form SN-38. Irinotecan works by inhibition of topoisomerase I. Like all other chemotherapeutic drugs, irinotecan has its toxicities like diarrhea and myelosuppression. Polymorphism in UGT1A1 is responsible for this.

5-Fluorouracil (5-FU) is also a chemotherapy drug used for a number of cancers like colorectal, stomach, pancreatic, breast, etc. [46]. Most of this drug undergoes anabolism to elicit its cytotoxicity to the cancer cells, while a major part is excreted from human body. DPD is the enzyme that is responsible for the catabolism.

Genetic variation in this enzyme leads to different degrees of catabolism and as a consequence causes adverse effects. A pharmacogenomic study may help to detect this defect in the gene of a patient, helping the doctors to adjust the dose. 5-FU acts by the inhibition of the enzyme thymidylate synthase, which is required for DNA synthesis and repair. Genetic changes in TS lead to resistance.

One of the proteins responsible for multidrug resistance is P-glycoprotein, which is encoded by the gene MDR1. P-gp aids in the transport of a number of drugs, which are hydrophobic in nature [33]. Some of these drugs are paclitaxel, adriamycin, etc. MDR1 polymorphism has been observed, which affects the pharmacokinetics and drug response.

We have given an overview of the application of pharmacogenomics in cancer. However we recommend the reader the recent review article by Hyman et al. [19] for further details and recent progresses in this area.

7.4 Phytochemicals and Pharmacogenomics

Plant-derived products are always beneficial in therapy. Anticancer drugs that are conventionally used in chemotherapy kill the cells, irrespective of whether they are cancerous or normal. This affects normal cells, causing severe side effects. To cope with the toxicity, sub-optimal doses are given, and resistant tumor cells escape. This is a cause of recurrence of the disease. Phytochemicals may come to the rescue, as we discuss below.

For the resistant population of cells in a tumor, therapy causes toxicity to the healthy organs, whereas cancer is not fully cured. If the cells are resistant to a particular drug, oncologists change the treatment protocol. Oncologists try to identify other drugs that are sensitive to the resistant population of cells that survive after the first treatment. Therapeutic protocols have undergone enormous advancement in recent years, and a plethora of treatment options are now available. Antibody therapy, immunotherapy, high-dose therapy, hematopoietic stem cell transplantation, gene therapy, and hyperthermia are many such options which may help. There is one more option, and that is phytotherapy. Concept of chemosensitivity is gaining in importance recently. These plant molecules have been found to be used by many countries around the world throughout history. Phytochemicals were quite popular for several remedies of a number of ailments in Chinese medicine, Kampo, Ayurveda, Unani, and Korean medicine. Many phytochemicals are suitable in cancer therapy, including DNA topoisomerase inhibitor camptothecin, microtubule inhibitor vinca alkaloids, paclitaxel, podophyllotoxin, etoposide, and others. These are now established anticancer agents.

As we have discussed in other chapters, these active biomolecules from plant sources are secondary metabolites, having a defensive action against herbivores and other pathogens. They raise signals to the insects and thus help in reproduction. They are important for the plants, as plants cannot fight against enemies. Phytochemicals elicit a number of pharmacological features, for which they are used in therapy. However, many other phytochemicals are not recognized as drugs and, therefore, cannot be prescribed or sold in the pharmacy. At the most they can be given as dietary supplements. From safety point of view, as well as quality, they are not proven through extensive clinical trials. Therefore, in order to establish these in the medical fraternity, emphasis must be given on quality control. This includes ethnobotany, which means documentation of knowledge. Botanical authentication comes next, which includes taxonomy, fingerprinting, and DNA sequencing. These plant products need to be standardized, and they must be collected and isolated in a proper manner, so that there are no contaminations and adulterations. Most importantly, in order to establish their role, modes of action need to be studied, both at the cellular and molecular level. Finally, without proper clinical trial, no molecule can be accepted, as safety is the priority.

The unique property of phytochemicals is that, unlike conventional chemotherapy drugs, they can discriminate normal and cancer cells; therefore, they can act preferentially on the cancer cells, without harming the normal cells. Targeted drugs specifically attack the target site, thus killing the cancer cells. Normal cells are devoid of these target genes, and hence these cells escape the assault of the drugs. This is one way of getting rid of side effects using the phytochemicals. Amplification of genes is often observed in tumors, but not in normal cells. Overexpression of proteins is facilitated by gene amplification. Chromosomal translocations, resulting in fusion of genes, are also seen in cancer cells, but not in healthy cells. Genomic instability is also seen in tumors. All these events drive carcinogenesis. A patient may have one or more of these changes in the genomes of cancer cells; therefore, design and development of drugs that can target and address these problems may render a better therapeutic outcome. Phytochemicals may be employed for this purpose, and precision or personalized medicine may be developed.

Targeted drugs may be categorized as monoclonal antibodies or small molecules that can enter the tumor cells. Monoclonal antibodies can be employed for cell surface proteins, like EGFR, VEGF, etc. Small molecules may cross the membrane and enter the cells and find their target. However, these targeted therapy drugs are not free from problems. Tumor cells may develop resistance against these drugs. Often target gets altered by mutations, or there may be SNPs. Signaling pathways may get altered, cell cycle arrest may be perturbed, and all these hamper the efficacy of the targeted drug. These targeted drugs may show some side effects on normal cells due to non-specific off-target effect. Liver toxicity, skin rash, and many other side effects are possible.

Phytochemicals may be a respite to all these problems. They can be used in targeted therapy.

Plant-derived molecules can be modified to generate derivatives having more pharmacological attributes. Phytochemicals can show their action against development of drug resistance as well. Some of the plant molecules have been found to inhibit ATPase activity of ABC transporter family of proteins. In this way, they may aid in overcoming multidrug resistance, which is a major problem in cancer therapy. Toxicity of conventional anticancer drugs may be well tackled by phytochemicals.

Phytomedicines have been considered to have a beneficial effect for a better health. It was thought that for a particular disease, there should be a unique drug, which will have one target. But now the idea is changing. Scientists and physicians start believing that one disease may be controlled by one drug but that drug needs to have multiple targets. In network pharmacology, all the nodes are not inhibited by the drugs. Phytochemicals may be employed in this respect. Remediation with plant-derived molecules with known chemical composition may be considered. A single phytochemical can have multiple targets.

Signal transducer and activator of transcription 3 (STAT3), one of the targets of cancer therapy, is an important molecule in signal transduction of tumorigenesis. Ligands like epidermal growth factors, interferons, and interleukins (5 and 6) bind with their specific receptors and activate receptor-associated Janus kinases (JAK), finally resulting in activation of STAT3. Nuclear translocation of STAT3 is facilitated upon phosphorylation, and it binds to the DNA, acting as a transcription factor. As a result of this, the transcription of a number of genes intricately associated with cell growth, programmed cell growth, invasion, and metastasis and angiogenesis gets affected. Mitogen-activated protein kinases (MAPK) and c-SRC non-receptor tyrosine kinase phosphorylate STAT3 and activate it. STAT 3 has been found to be active in many cancers. This molecule sometimes may act as tumor suppressor, particularly when it is mutated. Therefore, STAT3 is an important target of cancer therapy, and molecules that can target STAT3 are important. Inhibiting the upstream molecules in STAT3 pathways is a good approach. Molecules that inhibit tyrosine kinases inhibit downstream signaling pathways. This is also applicable to JAK, which is in the upstream of STAT3. STAT3 inhibitors bind to the SH2 domain and disrupt dimerization. Some molecules bind to the DNA and thereby inhibit transcriptional potential of STAT3. STAT3 inhibitors consist of SH2 ligands. Oligodeoxynucleotides (ODN) can bind STAT3 and cause death of cancer cells [10]. Small molecular inhibitors of STAT3 may be helpful in cancer therapy. In case of NSCLC with EGFR mutation, where EGFR TKIs fail, these small inhibitors may be of help [50]. Some of the phytochemicals have been found to inhibit STAT3, by blocking phosphorylation of STAT3 and its nuclear translocation, thereby inhibiting its role as a transcription factor. Some of the phytochemicals may have a number of targets. Therefore, it can be inferred that a number of genes or proteins need to be targeted for drug action. Phytochemicals like curcumin, butein, etc. have been found to inhibit STAT3 via a plethora of ways. They have been found to alter phosphorylation of STAT 3; or they may dimerize, acetylate, and change the DNA-binding ability of this molecule [49].

The plant *Ambrosia maritima* L. has two phytochemicals, namely, neoambrosin and damsin, which have anticancer properties. Besides targeting STAT3, these two plant molecules have been found to target and modulate many other genes including in leukemia cells. This is the concept of polypharmacology which is gradually emerging, and it is the application of drugs that have multiple targets, instead of a single target. Phytochemicals that inhibit phosphorylation of STAT3 are curcumin from *Curcumin longa* L.; epigallocatechin-3-gallate from green tea (*Camellia sinensis* L.); honokiol from *Magnolia officinalis*; resveratrol from the skin of red grapes, pea nut, and berries; cucurbitacin I from *Cucurbita andreana*; guggulsterone from *Commiphora mukul*; withaferin A from *Withania somnifera*

L.; quercetin from fruits, vegetables, leaves, grains, red onions, and kale. Some phytochemicals having anti-angiogenic potential are Acer mono Maxim., whose leaves are irritant, but the bark is a good astringent with quercetin and naringenin as active constituents. *Adenophora racemosa* and *Cinnamomum japonicum* also can protect against angiogenesis. *Schisandra chinensis*, an antihepatotoxic and anti-inflammatory plant product containing lignans, is anti-angiogenic. Some other phytochemicals that inhibit angiogenesis are apigenin and resveratrol.

Carcinogenesis is a complex process, and drugs that can attack more than one target are desirable. This avenue may be put to use against drug-resistant cancer cells. Understanding the complex signaling networks may untangle new and unique targets for development of better drugs. Network-based approaches for drug development are coined as network pharmacology. These propositions may able to unravel the mode of action of drugs. All the pathways that lead to carcinogenesis may be targeted by drugs.

A major problem of cancer chemotherapy is the development of drug resistance, during treatment. In particular, the development of multidrug resistance is a serious problem, due to which treatment outcome becomes less effective. ATP-binding cassette transporters are responsible for efflux of chemotherapeutic drugs out of the cells, so that they are no longer available within the cell for their action. Any novel compound that can work against multidrug resistance will be beneficial. There is some indication that traditional Chinese medicines may overcome this problem. Red sage, also known as Chinese sage (*Salvia miltiorrhiza*), a perennial plant is not involved in P-gp-mediated drug resistance. Three *S. miltiorrhiza* compounds that are not involved in P-gp-mediated multidrug resistance may be helpful in case of multidrug-resistant tumors [8, 9].

Genetic variations in *CYP2D6*17*, *CYP2D6*10*, and *CYP2D6*2N* have been observed among Africans and Asians and in other populations with varying frequencies. Other diversities in genes have been found in the African population. A number of drugs that can modulate ADME are *Cassia siamea*, *Phyllanthus amarus*, and *Lactuca taraxacifolia* [43].

Cytotoxic chemotherapy has several side effects. Therefore, it is better to address targeted therapy, which is tumor specific. These targeted chemotherapeutic drugs also have side effects. Also, resistance may develop in course of treatment. In order to avoid all these problems, phytochemicals may be considered to be good candidates for targeted therapy. Phytochemicals cannot target all the genes in a genetic pathway, as only a small percentage of these nodes can be targeted by drugs [8]. Several strategies need consideration in this respect. Phytochemicals may be considered if the bioactive chemical constituents are known. Multi-target therapeutic approaches may be explored with one phytochemical with selective polypharmacological approaches. There are subtle differences between a normal and a cancer cell. A protein which is not essential in a normal cell may have some therapeutic importance in cancer cells, if they are in a relevant pathway of carcinogenesis. Target of therapy should be to eliminate or at least inhibit such proteins. Suppression of a single gene may not have any effect on tumor growth. Therefore, one needs to focus not on suppression of a single gene, but to perturb

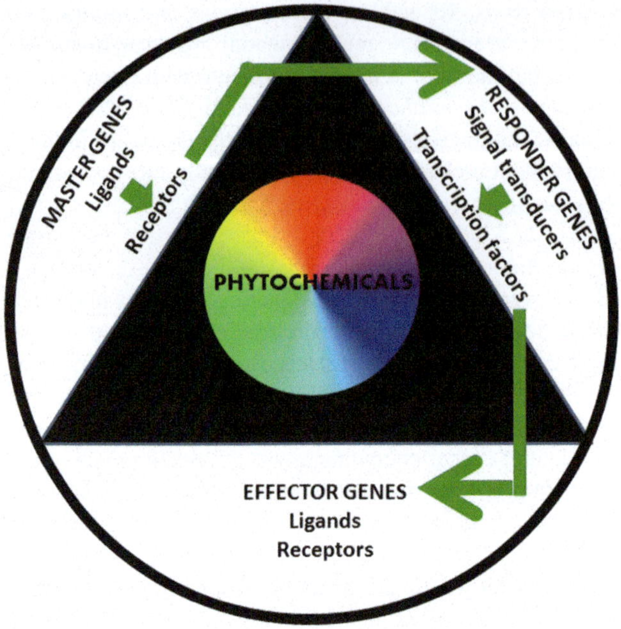

Fig. 7.3 Possible actions of phytochemicals on different types of genes

the network implicated in carcinogenesis. It is also important to know how these biomolecules act on human body at the molecular level.

The molecular mechanisms that can be targeted by phytochemicals are master genes, responder genes, and effector genes. This is illustrated in Fig. 7.3.

(i) Master genes, including ligands (EGF, WNT, chemokines, hormones, etc.) and receptors (EGFR, HER2, etc.)
(ii) Responder genes, including signal transducers (JAK/STAT, RAS/RAF/ERK, PI3K/AKT, etc.) and transcription factors (NF-κB, Myc, mTOR, etc.)
(iii) Effector genes, including those responsible for drug resistance (ABCB1, etc.), invasion and metastasis (MMPs, etc.), proliferation and cell cycle regulatory genes (cyclins, CDKs, CKIs, etc.), DNA repair genes (MGMT, ERCC, XRCC, etc.), genes responsible for angiogenesis (VEGF, etc.), and apoptosis (Bcl-2, caspases, etc.)

Implementation of precision medicine into regular cancer therapy is one of the predominant tasks of the future. It is necessary to establish the scientific outcome of precision medicine in the clinic and at the same time to develop integrated cost-effective models, which will promote personalized medicine.

Several phytochemicals have been shown to modulate cytochrome P450 (CYP) enzymes. Maidenhair tree, *Ginkgo biloba* or ginkgo, *Allium* vegetables (*Allium sativum* and *Allium cepa*), *Camellia sinensis*, *Zingiber officinale*, *Piper nigrum*, rosemary, curcumin, *Panax ginseng*, etc. can interact with CYP isoforms [9].

Phytochemicals have been shown to affect the expression or activity of specific CYP in in vivo and in vitro systems. A number of phytochemicals have been found to induce or inhibit CYP. CYP helps in carcinogen activation on the one hand and can metabolize a number of substrates on the other hand. Induction of CYP by a substrate may enhance metabolism of another. Grapefruit is one such plant product. It inhibits CYP3A4-mediated metabolism of a number of drugs. This may lead to enhanced bioavailability and may cause overdosing. Therefore, for those drugs, it is always advisable to stay away from grapefruit [57].

Carambola or star fruit inhibits CYP2A6. Watercress inhibits CYP2E1, and CYP2E1 may affect the metabolism of other drugs like chlorzoxazone, a muscle relaxant used in the treatment of muscle spasm, pain, etc. The dried roots of goldenseal, a herb, are used for preparation of medicines. This herb contains several alkaloids like hydrastine, berberine, canadine, etc. [47]. Goldenseal affects enzyme activities of CYP2C9, CYP2D6, CYP3A4, etc. [3]. Some of the CYP enzymes that are modulated by phytochemicals have been depicted in Fig. 7.4.

Traditional *Chinese medicine* is popular in China to treat and prevent a number of health problems. It includes different forms of herbal medicines. Tianqi Tianqi Jiangtang is popularly used to control blood sugar. This formulation comprises Radix Astragali, Radix Trichosanthis, Fructus Ligustri Lucidi, Dendrobii Caulis, Radix Ginseng, Cortex Lycii Radicis bone, Rhizoma Coptidis, Asiatic Cornelian cherry fruit, Ecliptae Herba, and Chinese gall. Of these ten components, action of Rhizoma Coptidis and Radix Astragali are similar to that of diformin, a drug used for treating diabetes. Rhizoma Coptidis contains berberine, which aids in improvement of glycemic parameters. Radix Ginseng contains ginsenoside which also elicits antihyperglycemic effect. Diabetic vascular problems are well tackled by Asiatic Cornelian cherry fruit. Single nucleotide polymorphism in TPMT gene is correlated with the hypoglycemic effect. Tianqi Tianqi has a modulatory effect on TPMT [27].

Warfarin (Coumadin) is an anticoagulant, whose effect may be modulated by a number of plant biomolecules. Many factors are found to influence the efficacy of warfarin. Efficacy of warfarin may be increased by a Chinese herb Shu-Jing-Hwo-Shiee-Tang (SJHST) [40]. This Chinese formulation is used for treatment of degenerative joint diseases. Warfarin is also used in cardiac diseases. Therefore, during treatment, close monitoring of patients is a must. Warfarin has been reported to interact with a number of medications. Selective serotonin reuptake inhibitors (SSRIs) [5]; antibiotics [1] like sulfamethoxazole/trimethoprim, fluconazole, azithromycin ciprofloxacin, etc.; nonsteroidal anti-inflammatory drugs (NSAIDs) [39]; antiplatelet agents like aspirin; etc. have been found to increase the chances of bleeding in susceptible individuals. A number of herbal formulations also interact with warfarin. Garlic, ginger, chamomile, clove, fenugreek, grapefruit, mango, onion, turmeric, gingko biloba, anise, asafoetida, etc. may enhance the chances of bleeding when taken during warfarin administration, whereas St. John's wort, ginseng, *Cannabis*, chamomile, cranberry, garlic, ginkgo, grapefruit, *Lycium*,

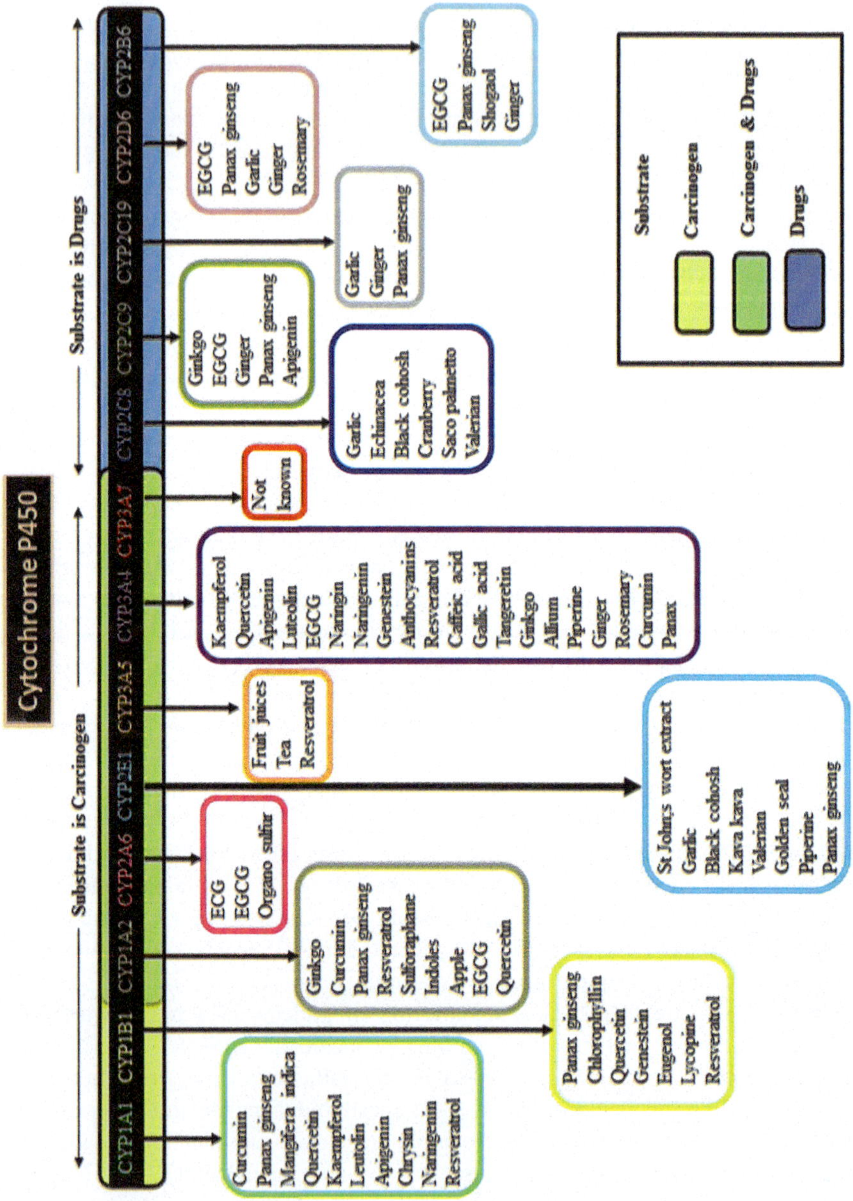

Fig. 7.4 CYP enzymes that are modulated by phytochemicals

and red clover may reduce the effect of warfarin. Reduction of efficacy of warfarin may enhance the risk of clotting. Chances of bleeding may also be enhanced when taken together with high doses of vitamins A and E.

7.5 Conclusion

We have discussed pharmacogenomics in details in this chapter. First, we have introduced pharmacogenomics in a general setting, as a method for prescribing drugs taking into account the individual genetic variations. We have discussed how pharmacogenomics has become a reality due to the availability of low-cost whole genome sequencing technologies, and we are entering the era of personalized medicine. We have discussed the pharmacogenomics of inherited genetic variations and how these variations can be used for prescribing appropriate drugs and dosage. Next, we have discussed the pharmacogenomics of cancer, and finally we have discussed the roles phytochemicals can play in pharmacogenomics of cancer.

References

1. J. Baillargeon, H. Holmes, Y. Lin, M.A. Raji, G. Sharma, Y.F. Kuo, Concurrent use of warfarin and antibiotics and the risk of bleeding in older adults. Am. J. Med. **125**(2), 183–189 (2012)
2. S. Bak, F. Beisson, G. Bishop, B. Hamberger, R. Höfer, S. Paquette, D. Werck-Reichhart, Cytochromes p450. Arabidopsis Book. **9**, e0144 (2011)
3. P. Chatterjee, M.R. Franklin, Human cytochrome p450 inhibition and metabolic-intermediate complex formation by goldenseal extract and its methylenedioxyphenyl components. Drug Metab. Dispos. **31**(11), 1391–1397 (2003)
4. M.A. Correia, P.R. Sinclair, F. De Matteis, Cytochrome P450 regulation: The interplay between its heme and apoprotein moieties in synthesis, assembly, repair, and disposal. Drug Metab. Rev. **43**(1), 1–26 (2010)
5. M.F. Crader, J.K. Arnold, Warfarin drug interactions. [Updated 2019 Jan 17], in *StatPearls [Internet]*, (Treasure Island, StatPearls Publishing, 2012).; 2019 Jan–. Available from: https://www.ncbi.nlm.nih.gov/books/NBK441964/
6. A.K. Daly, Pharmacogenetics: A general review on progress to date. Br. Med. Bull. **124**, 65–79 (2017)
7. L. Dean, Warfarin therapy and the genotypes CYP2C9 and VKORC1, medical genetics summaries, in *Medical Genetics Summaries [Internet]*, ed. by V. Pratt, H. McLeod, L. Dean, et al., (National Center for Biotechnology Information (US), Bethesda, 2012)
8. T. Efferth, S. Kahl, K. Paulus, M. Adams, R. Rauh, H. Boechzelt, X. Hao, B. KAina, R. Bauer, Phytochemistry and pharmacogenomics of natural products derived from traditional Chinese medicine and chinese materia medica with activity against tumor cells. Mol. Cancer Ther. **7**(1), 152 (2008)
9. T. Efferth, M. Saeed, E. Mirghani, A. Alim, Z. Yassin, E. Saeed, H.E. Khalid, S. Daak, Integration of phytochemicals and phytotherapy into cancer precision medicine. Oncotarget **8**(30), 50284–50304 (2017)
10. R. Fagard, V. Metelev, I. Souissi, F. Baran-Marszak, STAT3 inhibitors for cancer therapy: Have all roads been explored? JAKSTAT **2**(1), e22882 (2013)
11. K. Fujita, Y. Sasaki, Pharmacogenomics in drug-metabolizing enzymes catalyzing anticancer drugs for personalized cancer chemotherapy. Curr. Drug Metab. **8**(6), 554–562 (2007)

12. K. Fujita, Y. Kubota, H. Ishida, Y. Sasaki, Irinotecan, a key chemotherapeutic drug for metastatic colorectal cancer. World J. Gastroenterol. **21**(43), 12234–12248 (2015)
13. K.M. Giakomini, C.M. Brett, R.B. Altman, N.L. Benowitz, M.E. Dolan, D.A. Flockhart, J.A. Johnson, D.F. Hayes, T. Klein, R.M. Krauss, D.L. Kroetz, H.L. McLeod, A.T. Nguyen, M.J. Ratain, M.V. Relling, V. Reus, D.M. Roden, C.A. Schaefer, A.R. Shuldiner, T. Skaar, K. Tantisira, R.F. Tyndale, L. Wang, R.M. Weinshilboum, S.T. Weiss, I. Zineh, The Pharmacogenetics research network: From SNP discovery to clinical drug response. Clin. Pharmacol. Ther. **81**(3), 328–345 (2007)
14. F.J. Gonzalez, H.V. Gelboin, Human cytochromes P450: Evolution and cDNA-directed expression. Environ. Health Perspect. **98**, 81–85 (1992)
15. D. Gurwitz, M.M. Hopkins, S. Gaisser, D. Ibarreta, Pharmacogenetics in Europe: Barriers and opportunities. Public Health Genomics **12**, 134–141 (2009)
16. O. Hankinson, The aryl hydrocarbon receptor complex. Annu. Rev. Pharmacol. Toxicol. **35**, 307–340 (1995)
17. I. Hanukoglu, Electron transfer proteins of cytochrome P450 systems. Physiological functions of cytochrome P450 in relation to structure and regulation. Adv. Mol. Cell Biol. **14**, 29–56 (1996)
18. J.A. Hasler, R. Estabrook, M. Murray, I. Pikuleva, M. Waterman, J. Capdevila, V. Holla, C. Helvig, J.R. Falck, G. Farrell, L.S. Kaminsky, S.D. Spivack, E. Boitier, P. Beaune, Human cytochromes P450. Mol. Asp. Med. **20**(1–2), 1–137 (1999)
19. D.M. Hyman, B.S. Taylor, J. Baselga, Implementing genome-driven oncology. Cell **168**(4), 584–599 (2017)
20. M. Ingelman-Sundberg, C. Rodriguez-Antona, Pharmacogenetics of drug-metabolizing enzymes: Implications for a safer and more effective drug therapy. Philos. Trans. R. Soc. Lond. Ser. B Biol. Sci. **360**(1460), 1563–1570 (2005)
21. M. Ingelman-Sundberg, Pharmacogenetics of cytochrome P450 and its applications in drug therapy: The past, present and future. Trends Pharmacol. Sci. **25**(4), 193–200 (2004)
22. C. Kearon, E.A. Akl, A.J. Comerota, P. Prandoni, H. Bounameaux, S.Z. Goldhaber, M.E. Nelson, P.S. Wells, M.K. Gould, F. Dentali, M. Crowther, S.R. Kahn, Antithrombotic therapy for VTE disease: Antithrombotic therapy and prevention of thrombosis, 9th ed: American College of Chest Physicians Evidence-Based Clinical Practice Guidelines. Chest **141**(2 Suppl), e419S–e496S (2012)
23. D. Lee, C. Szumlanski, J. Houtman, R. Honchel, K. Rojas, J. Overhauser, E.D. Wieben, R.M. Weinshilboum, Thiopurine methyltransferase pharmacogenetics. Cloning of human liver cDNA and a processed pseudogene on human chromosome 18q21.1. Drug Metab. Dispos. **23**(3), 398–405 (1995)
24. W. Lee, A.C. Lockhart, R.B. Kim, M.L. Rothenberg, Cancer pharmacogenomics: Powerful tools in cancer chemotherapy and drug development. Oncologist **10**, 104–111 (2005)
25. X. Li, F.M. Lian, D. Guo, L. Fan, J. Tang, J.-B. Peng, H.-W. Deng, Z.-Q. Liu, X. Xiao, Y. Wang, K. Qu, S. Deng, Q. Zhong, Y. Sha, Y. Zhu, Y. Bai, X. Chen, Q. Zhou, H. Zhou, X. Tong, W. Zhang, The rs1142345 in TPMT affects the therapeutic effect of traditional hypoglycemic herbs in Prediabetes. Evid. Based Complement. Alternat. Med. **2013**, 327629 (2013). 8 pages
26. N.A. Limdi, D.L. Veenstra, Warfarin Pharmacogenetics. Pharmacotherapy **28**(9), 1084–1097 (2008)
27. M.Z. Liu, Y.L. Zhang, M.Z. Zeng, F.Z. He, Z.Y. Luo, J.Q. Luo, J.G. Wen, X.P. Chen, H.H. Zhou, W. Zhang, Pharmacogenomics and herb-drug interactions: Merge of future and tradition. Evid. Based Complement. Alternat. Med. **2015**, 321091 (2015). 8 pages
28. D.Y. Lu, T.R. Lu, B. Xu, J. Ding, Pharmacogenetics of cancer therapy: Breakthroughs from beyond? Future Sci. OA **1**(4), FSO80 (2015)
29. M.H. Manjili, Therapeutic cancer vaccines. J. Clin. Cell Immunol. **2**, e101 (2011)
30. M.C. McFadyen, W.T. Melvin, G.I. Murray, Cytochrome P450 enzymes: Novel options for cancer therapeutics. Mol. Cancer Ther. **3**, 363–371 (2004)
31. R. Nagasubramanian, F. Innocenti, M.J. Ratain, Pharmacogenetics in cancer treatment review of medicine. Annu. Rev. Med. **54**(1), 437–452 (2003)

32. D.W. Nebert, D.W. Russell, Clinical importance of the cytochromes P450. Lancet **360**(9340), 1155–1162 (2002)
33. H. Omote, M.K. Al-Shawi, Interaction of transported drugs with the lipid bilayer and P-glycoprotein through a solvation exchange mechanism. Biophys. J. **90**(11), 4046–4059 (2006)
34. R.P. Owen, L. Gong, H. Sagreiya, T.E. Klein, R.B. Altman, VKORC1 pharmacogenomics summary. Pharmacogenet. Genomics **20**(10), 642–644 (2010)
35. R.P. Owen, K. Sangkuhl, T.E. Klein, R.B. Altman, Cytochrome P450 2D6. Pharmacogenet. Genomics **19**(7), 559–562 (2009)
36. T. Oyama, N. Kagawa, N. Kunugita, K. Kitagawa, M. Ogawa, T. Yamaguchi, R. Suzuki, T. Kinaga, Y. Yashima, S. Ozaki, T. Isse, Y.D. Kim, H. Kim, T. Kawamoto, Expression of cytochrome P450 in tumor tissues and its association with cancer development. Front. Biosci. **9**, 1967–1976 (2004)
37. S.C. Preissner, M.F. Hoffmann, R. Preissner, M. Dunkel, A. Gewiess, S. Preissner, Polymorphic cytochrome P450 enzymes (CYPs) and their role in personalized therapy. PLoS One **8**(12), e82562 (2013)
38. M. Rooseboom, J.N. Commandeur, N.P. Vermeulen, Enzyme-catalyzed activation of anticancer prodrugs. Pharmacol. Rev. **56**(1), 53–102 (2004)
39. R.I. Shorr, W.A. Ray, J.R. Daugherty, M.R. Griffin, Concurrent use of non-steroidal anti-inflammatory drugs and oral anticoagulants places elderly persons at risk for hemorrhagic peptic ulcer disease. Arch. Intern. Med. **153**, 1665–1670 (1993)
40. S.-H. Yang, C.-L. Yu, H.-Y. Chen, Y.-H. Lin, A commonly used Chinese herbal formula, Shu-Jing-Hwo-Shiee-Tang, potentiates anticoagulant activity of warfarin in a rabbit model. Molecules **18**, 11712–11723 (2013)
41. S.C. Sim, C. Risinger, M.L. Dahl, E. Aklillu, M. Christensen, L. Bertilsson, M. Ingelman-Sundberg, A common novel CYP2C19 gene variant causes ultrarapid drug metabolism relevant for the drug response to proton pump inhibitors and antidepressants. Clin. Pharmacol. Ther. **79**, 103–113 (2006)
42. C.Y. Tan, R. Suzuki, T. Haga, F. Iwata, T. Nagamune, Immobilization of a bacterial cytochrome P450 Monooxygenase system on a solid support. Angew. Chem. Int. Ed. **55**(48), 15002–15006 (2016)
43. N.E. Thomford, K. Dzobo, D. Chopera, A. Wonkam, M. Skelton, D. Blackhurst, S. Chirikure, C. Dandara, Pharmacogenomics implications of using herbal medicinal plants on African populations in health transition. Pharmaceuticals **8**, 637–663 (2015)
44. N. Toomula, K. Hima Bindu, D. Sathish Kumar, R. Arun Kumar, Pharmacogenomics-personalized treatment of Cancer, diabetes and cardiovascular diseases. J Pharmacogenom Pharmacoproteomics **3**, 1 (2012)
45. D. Urbančiča, A. Kotar, A. Šmid, M. Jukič, S. Gobec, L.G. Mårtensson, J. Plavec, I. Mlinarič-Raščan, Methylation of selenocysteine catalysed by thiopurine S-methyltransferase. Biochim. Biophys. Acta **1863**(1), 182–190 (2019)
46. M.M. Valenzuela, J.W. Neidigh, N.R. Wall, Antimetabolite treatment for pancreatic cancer. Chemotherapy (Los Angel). **3**(3), 137 (2014)
47. H.A. Weber, M.K. Zart, A.E. Hodges, H.M. Molloy, B.M. O'Brien, L.A. Moody, A.P. Clark, R.K. Harris, J.D. Overstreet, C.S. Smith, Chemical comparison of goldenseal (Hydrastis canadensis L.) root powder from three commercial suppliers. J. Agric. Food Chem. **51**(25), 7352–7358 (2003)
48. M.T. Whitstock, C.M. Pearce, E.J. Eckermann, Randomised controlled trials and 'unexpected' adverse events associated with newly released drugs: Improvements in Pharmacovigilance systems are necessary for real-time identification of patient safety risks. J. Clin. Toxicol. **S2**, 001 (2011)
49. A.L. Wong, J. Hirpara, S. Pervaiz, J.-Q. Eu, G. Sethi, B.C. Goh, Do STAT3 inhibitors have potential in the future for cancer therapy? Expert Opin. Investig. Drugs **26**(8), 883–887 (2017)

50. A.L. Wong, R.A. Soo, D.S. Tan, S.C. Lee, J.S. Lim, P.C. Marban, L.R. Kong, Y.J. Lee, L.Z. Wang, W.L. Thuya, R. Soong, M.Q. Yee, T.M. Chin, M.T. Cordero, B.R. Asuncion, B. Pang, S. Pervaiz, J.L. Hirpara, A. Sinha, W.W. Xu, M. Yuasa, T. Tsunoda, M. Motoyama, T. Yamauchi, B.C. Goh, Phase I and biomarker study of OPB-51602, a novel signal transducer and activator of transcription (STAT) 3 inhibitor, in patients with refractory solid malignancies. Ann. Oncol. **26**, 998–1005 (2015)
51. M. Yamamura, M. Yamamoto, Tumor metastasis and the fibrinolytic system. Gan to Kagaku Ryoho. Cancer Chemother. **16**(4 Pt 2–1), 1246–1254 (1989)
52. B. Yu, S.A. O'Toole, R.J. Trent, Somatic DNA mutation analysis in targeted therapy of solid tumors. Translat. Pediatr. **4**(2), 125–138 (2015)
53. H.Y. Yuan, J.J. Chen, M.T. Lee, J.C. Wung, Y.F. Chen, M.J. Charng, M.J. Lu, C.R. Hung, C.Y. Wei, C.H. Chen, J.Y. Wu, Y.T. Chen, A novel functional VKORC1 promoter polymorphism is associated with inter-individual and inter-ethnic differences in warfarin sensitivity. Hum. Mol. Genet. **14**(13), 1745–1751 (2005)
55. U.M. Zanger, M. Schwab, Cytochrome P450 enzymes in drug metabolism: Regulation of gene expression, enzyme activities, and impact of genetic variation. Pharmacol. Ther. **138**, 103–141 (2013)
56. G. Zaza, M. Cheok, W. Yang, J.C. Panetta, C.-H. Pui, M. Relling, W.E. Evans, Gene expression and thioguanine nucleotide disposition in acute lymphoblastic leukemia after in vivo mercaptopurine treatment. Blood **106**, 1778–1785 (2005)
57. K. Zeratsky, Grapefruit juice: Can it cause drug interactions? Ask a food & nutrition specialist. MayoClinic.com. Retrieved 2009-02-09 (2008)

8 Phytochemicals as Bioenhancers

Abstract

In the previous chapter, we have learned about absorption of drugs. Bioavailability is somewhat related to absorption. In pharmacology this is defined as a category of absorption, the fraction of the given dosage of drug which remains unchanged and goes to the systemic circulation. Absorption of drugs in general and bioavailability in particular are crucial issues in therapy. We will discuss bioavailability of phytochemicals in details in this chapter, along with the factors that influence bioavailability, the processes that improve bioavailability, and the phytochemicals that act as important bioenhancers, agents that improve the bioavailability of drugs.

8.1 Fundamentals of Bioavailability

Bioavailability of a drug is defined as the fraction of a drug that reaches the intended target sites of the drug, i.e., tissues or cells. Understanding the different modes of drug administration is important in order to understand bioavailability. Drugs may be taken by mouth; sometimes may be placed underneath the tongue; may be dropped in the eye, ear, or nose; or may be inserted in the rectum or vagina. Some drugs may be inhaled to be delivered to the lungs. Most effective is the injectable mode; a drug may be injected in the vein, muscle, and skin or into the space between vertebrae. Sometimes drugs may be applied in the form of a patch for a systemic effect. Drugs may be in liquid form or may be in the form of capsules, tablets, or suppositories, which are inserted into the vagina or rectum. Drugs in solid form get dissolved and elicit local or systemic effect. There are certain drugs that are administered by inhalation. These drugs are atomized into tiny drops, smaller the size of the drops, deeper they can go. Different modes of drug delivery are depicted in Fig. 8.1.

M. Roy, A. Datta, *Cancer Genetics and Therapeutics*,
https://doi.org/10.1007/978-981-13-9471-3_8

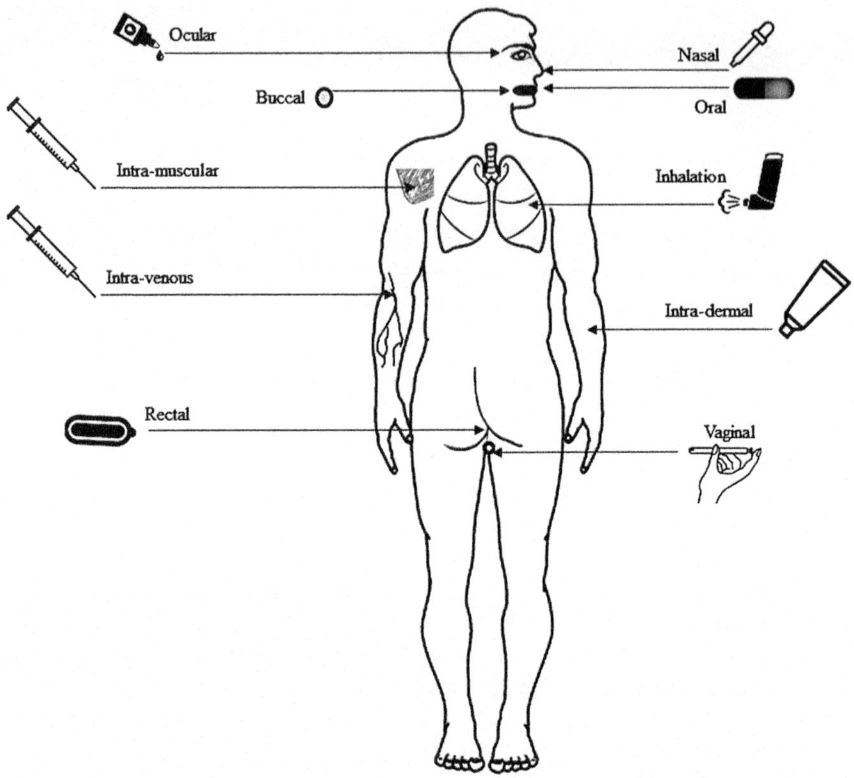

Fig. 8.1 Different modes of administration of drugs

Concentration of a drug decays with time in the human body. Orally administered drugs go to the stomach and get dissolved, and a part is absorbed in the blood. Next it goes to the small intestine, where absorption is completed. Bioavailability may be poor due to poor absorption from the stomach and small intestine. Upon absorption, drugs go to the liver through the hepatic vein. Drugs are metabolized in the liver, which is called first-pass metabolism.

Oral administration of drug is the best way of medication, as it is easy and convenient and does not cause pain due to injection [22]. But in many cases when the oral drug fails to work efficiently, it is given intravenously. This is due to the fact that when delivered intravenously, its bioavailability is very high, almost 100% [17]. Oral absorption is not much effective due to compromised absorption and loss due to first-pass metabolism. Absorption of a drug also depends on the genetic makeup of an individual.

Two components of the circulatory system are the systemic circulation and pulmonary circulation. After penetrating the biological cell membranes, drugs are carried by the systemic circulation to their target sites. Systemic circulation is the mechanism of circulation in which oxygenated blood from the left ventricle is carried to the capillaries through the arteries. The cell membranes are lipophilic as they have phospholipid bilayer [1]. In order to permeate through the membrane, the drug molecule needs to have enough hydrophobic properties. Besides, lipophilic properties are another requisite, so that passive absorption of drug molecules can be achieved.

The amount of drug that reaches the intended site is important for better treatment outcomes. The measure of bioavailability is denoted by the letter **f**. Since intravenous administration of a drug is most efficacious, it defines the absolute bioavailability. When injected into the vein, the drug goes through the bloodstream and reaches its target site. The time when the highest concentration of a drug is reached in the blood is known as T_{max}, and the highest concentration of the drug found in the blood is known as C_{max}. While travelling through the blood stream, a part of the administered drug may get metabolized and also may be excreted from our body.

Though oral administration of a drug is the most convenient, its bioavailability is often compromised. For a drug molecule to be efficacious, it has to be bioavailable, and for that purpose the molecules must be able to cross the gastrointestinal (GI) membrane. Regarding absorption of drugs in the body, another term, "bioequivalence," is popularly used. In pharmacokinetics, bioequivalence is used to assess if two drugs have similar bioavailability. Two drugs which are bioequivalent should be at par so far their efficacy is concerned. It is therefore a comparison between two drugs and can give an idea whether one can be used in place of the other.

8.2 Factors Affecting Bioavailability

There are inter-individual variations among patients, and sometimes, the same patient may show different characteristics at different times. Bioavailability is affected due to a number of factors, e.g.:

1. If the drug is hydrophobic and its pKa and solubility
2. If the drug is given in full stomach
3. If there is interaction with other drugs like antacids, alcohol, nicotine, etc. or foods like grapefruit, cranberry, etc.
4. If the drug formulation is immediate release or modified release, e.g., delayed, extended, or sustained release
5. If there is enzyme induction or inhibition by other drugs or foods
6. If there is individual variation in metabolism, like age and differences in phenotypes
7. If there is any circadian differences
8. If there is any other disease state, e.g., hepatic or renal problem, stress, and any gastrointestinal surgery has been done before
9. If there is substrate of an efflux transporter

Bioavailability of a drug also depends on the pH of the GI fluid, interaction with other drugs, food, etc. There are other routes of administration of drug to an individual as mentioned before. Nasal administration is neither invasive nor painful. The epithelial surface in the nasal cavity is permeable and drugs can easily enter. This method is applicable to the drugs which are meant for nose and brain. If applied through the nasal cavity, these drugs can avoid first-pass metabolism in the liver. Drugs that are not stable in gastric fluid and are harmed during the first-pass metabolism in the liver may be administered by buccal route. In this way, absorption is slow. This is suitable for drugs given in low dose [10]. Drugs administered by this way are given beneath the tongue or may be between gums and teeth. Drugs are not swallowed, rather they get absorbed directly. Sometimes drugs are given via pulmonary or inhalation route. They are breathed in the lungs and delivery of drug by this method is rapid. Nebulizer is an example.

There are some other ways of drug administration. These are for injectable drugs. These may be intravenous, intramuscular, subcutaneous, or around the spinal cord, which is called intrathecal. Intravenous is the most effective way of drug delivery. Drugs are also delivered through rectum, as suppository. In case of nausea, drugs are given by this route. For the treatment of vaginal infections, often drugs are applied to the vagina, in the form of tablet, cream, suppository, and others. The applied drug gets absorbed through the vaginal wall, but the absorption rate is slow. There also may be ocular and dermal routes of administration [37].

Absolute bioavailability is therefore defined as the ratio of bioavailability of the drug following non-intravenous administration to that following intravenous administration. Non-intravenous administration may be oral, ocular, rectal, subcutaneous, sublingual, and transdermal [41].

Measure of bioavailability **f** for routes other than intravenous is fractional, while for intravenous it is unity. There are a number of factors that govern bioavailability. Some of the drugs act better if taken with food; some show the best efficacy when taken in empty stomach. Intestinal mobility may have an effect. Physical properties of the drug to be administered are vital factors too. Hydrophobicity (if the molecule is repelled from a mass of water) of the drug compound and its solubility (the property to dissolve in a solvent) and pK (acid dissociation constant, which gives a quantitative measure of the strength of an acid in solution) values are of extreme importance. The drugs may be of different types, depending on the time of release. Drugs may be categorized as:

(i) Modified release
(ii) Immediate release
(iii) Extended release
(iv) Sustained release
(v) Delayed release

If a drug after administration is delivered with a delay, it falls under the category of delayed release. If the drug is active for a long time, it is said to be an extended release drug. Sustained-release ones maintain a constant drug concentration for a specific time, and side effects with these drugs are minimum. Immediate release drugs need to be taken at a much frequent interval.

Interaction with other drugs or foods also contributes to the bioavailability. Nicotine, antacid, alcohol, etc. may affect bioavailability. Some of the foods that are consumed in our daily diet like fruits and vegetables may also affect bioavailability of the drugs. Fruits like cranberry and grape fruit juice and brassica vegetables fall under this category. They may inhibit CYP3A (a cytochrome isoform that is responsible for drug metabolism), whereas anti-seizure medicine formulation phenytoin may induce CYP1A2, CYP2C9, etc. Metabolism of drugs also changes with time. For infants and geriatric people, metabolism rate may be very slow. Bioavailability of drugs may not be enough for patients with poor hepatic and renal functions.

The nutrients that we consume are not taken up by our body completely. Human body can use only a part of the nutrients consumed. This is due to poor bioavailability of that nutrient. Vitamins and other phytochemicals are therefore not fully utilized in our body. Vitamins and minerals behave in a very complex way. Sometimes they work hand in hand; sometimes they work differently. Vitamins C and D and non-heme iron help in absorption of drugs. Zinc and iron oppose each other. Some anti-nutrients like oxalates found in dark green leafy vegetables hinder absorption of minerals like calcium, zinc, iron, etc. Absorption and retention of nutrients depend on the need of the body, which again varies with age, gender, health condition, etc.

Many of the drugs are poorly absorbed; therefore, newer strategies need to be evolved to enhance the bioavailability and hence efficacy. Improvement in the methodology of drug synthesis could solve the problem. A number of physico-chemical properties are responsible for the poor availability of the drugs. Complex structure of the drug, size of the molecule, its molecular weight, ionic charges, pH, high lipophilic nature of the drug, and others are the causative factors [28]. Rule of five by Lipinski, also called Pfizer's rule of five, is a model which can predict and assess the oral absorption of drugs. Lipinski observed that mostly the drugs that are given orally have lower size and they are lipophilic. The rule describes the properties which are essential for pharmacokinetics, like absorption, distribution, metabolism, and excretion. It states that for a drug to be orally active, there are certain criteria that need to be fulfilled. Activity of the drug diminishes if more than one of these criteria is violated. Criteria of an ideal drug are:

(i) Should have lower than five hydrogen bond donors
(ii) Should have lower than ten acceptors of hydrogen bonds
(iii) Should have molecular weight lower than 500 Dalton with high lipophilicity
(iv) Should have Log P less than 5
(v) Should have the molar refractivity in the range 40–130

P stands for partition coefficient, which is the proportion of concentration of a compound in a mixture of two immiscible phases at equilibrium. Molar refractivity is a measure of polarizability [19].

Oral bioavailability depends on several factors, like permeability of the drug, solubility of the drug in water, dissolution rate, first-pass metabolism, and susceptibility to efflux. Among other factors, susceptibility to efflux pumps leads to poor bioavailability for certain drugs. These include P-glycoprotein or multi-drug resistance proteins. An agent that increases the efficacy and bioavailability of a drug when administered together is called a "bioenhancer." Usually bioenhancers do not have any pharmacological activity at the dose they are administered. Therefore, they are not therapeutic agents, but in combination with an active drug result in higher potential of the pharmacological effect of the drug.

8.3 Improvements in Bioavailability

Enhancement of bioavailability is a current challenge in pharmacology. Poor solubility of the drugs in water and poor bioavailability are the main issues that hinder development of many drug molecules at the initial stage. Several approaches have been developed for improving the oral bioavailability of these compounds, mainly by increasing their dissolution rates and aqueous solubility.

Conventional ways to enhance the breakup of drugs are salt formation, taking help of solubilizing agents, etc. But these methods have some problems. Formation of salts may render side effects. Inclusion of an additional solvent may pose problem. Inclusion of microemulsion, emulsion, and liposomes may increase bioavailability. Reduction of size to submicron size, which is known as nanosizing, of drugs is also becoming popular.

Before diffusion a drug needs to be dissolved. Hence solubility in the gut is an important determinant. When the size of the drug particles is reduced, they have greater chances of interaction with the solvent, and hence there is a chance of increased solubility. Therefore, enhancement of solubility of drugs may be an ideal approach for increasing bioavailability.

Drugs that are poorly bioavailable can never show their fullest activity. Only a small proportion of the dose reaches the target organ through circulation. Hence a very high dose of the drug needs to be administered to mitigate poor bioavailability. Such a high dose of the drug leads to severe side effects. An example is danazol for treating endometriosis. Being a poorly bioavailable drug, a very high dose of danazol is prescribed. This causes various side effects. Also, prescribing such a high dose is a financial burden for the patients [4].

Therefore, improvement of bioavailability of a drug is of enormous importance.

Poorly absorbed drugs may be made bioavailable, employing a number of approaches:

(i) Using enhancers that increase absorption and hence bioavailability
(ii) Using pro-drugs
(iii) Pharmaceutical approaches, i.e., drug delivery systems
(iv) Using P-gp inhibitors

8.3.1 Using Enhancers that Increase Absorption

Bile salts, surfactants, chelating agents, polymers, fatty acids, etc. may aid in intestinal absorption of drugs [40]. Chitosan is another option. All of these enhance the solubility of hydrophobic drugs. Calcium chelators like EGTA and EDTA enhance absorption by decreasing calcium level.

8.3.2 Pro-drug

Often drugs are in their pro-drug form. In order to improve bioavailability, these pro-drugs are covalently attached to a drug, which modifies the physicochemical properties. The covalent bond is so designed that they can be cleaved very easily, liberating the active drug. The pro-drug gets absorbed in the circulatory system. Sometimes, these pro-drugs are not very soluble. Another demerit is seen when nonpolar moieties mask the pro-drug. In that case the drug is accumulated within the cell, where the pro-drug gets biotransformed into an active drug. Highly charged moieties like phosphate and carboxyl groups are masked by nonpolar groups through an ester bond, this increases the lipophilicity of the drug to be delivered, and hence its permeability is increased. The ester bond gets cleaved very easily.

Sometimes pro-drugs are used to increase transporter action. Here the attached moiety works as a recognition site for the transporter proteins. Bile acid transporters also improve bioavailability. The attachment site and the size of the drug are determinants of efficiency of the drug. The pro-drugs, besides increasing bioavailability, can impart long-acting pharmacological profiles. Half-life of the drugs can be increased this way.

8.3.3 Pharmaceutical Approaches

Dosage forms that can increase the permeability are a very good way to enhance intestinal absorption of drugs, which are poorly absorbed. These formulations are liposomes and emulsions, which increase intestinal absorption [15, 35]. Another possibility is to reduce the size of drug particles [47]. By particle size reduction, we mean micronization, using nanoparticular carriers, etc., which can facilitate drug absorption. Nanosizing will be discussed in details later.

8.3.4 Inhibition of P-gp

P-glycoprotein 1 (P-gp) is a multi-drug resistance protein 1 (MDR1) or ATP-binding cassette subfamily B member 1 (ABCB1). This is an important protein that pumps out any unwanted foreign substances from a cell. Phytochemicals that act as bioenhancers inhibit this protein. Bioenhancers can act as inhibitor of the substrate P-gp in several modes, competitive inhibitor, uncompetitive inhibitor, noncompetitive inhibitor, or irreversible inhibitor. These inhibitors restrict active drug transport. As a consequence bioavailability of the drug in the gut is enhanced [21].

This glycoprotein is encoded by the *ABCB1* gene [11]. The normal excretion of xenobiotics gets back to gut lumen by P-gp, and due to this, the efficacy of some pharmaceutical drugs gets reduced. These drugs are supposedly substrates of P-gp. Expression of P-gp in cancer cells is high. Cancer cells develop multidrug resistance due to this. P-gp helps in the transportation of various drugs including several anticancer drugs like etoposide, doxorubicin, vinblastine, gefitinib, sunitinib, etc. Cardiac glycosides digoxin and glucocorticoids like dexamethasone are some more drugs transported by P-gp. P-gp efflux lowers intracellular concentrations of drugs to such a concentration when they no longer show therapeutic efficacy. Some transcriptional factors like p53 [16] and NF-κB [48] regulate P-gp by binding to the promoter region of the P-gp gene.

8.4 Enhancement of Bioavailability

Drugs prescribed for a patient may be delivered by alternate routes. Oral route is the most desired and convenient one; however, to enhance systemic bioavailability, it may be delivered by parenteral, nasal, vaginal, rectal, or transdermal way. But, considering convenience, it is always desirable to improve oral bioavailability of the drug. These can be achieved by a number of ways [46]:

1. By enhancing the solubility and dissolution rate
2. By modification of partition coefficient
3. By avoiding hepatic first-pass metabolism
4. By avoiding degradation of the drug in the gastrointestinal tract
5. By inhibiting drug efflux
6. By devising novel drug delivery system (Fig. 8.2).

8.4.1 Details of Techniques that Improve Bioavailability

Nanosizing is a pharmaceutical process which reduces the size of the drug to nanometer size. Availability of oral drugs depends on the absorption at the gastrointestinal tract. Absorption refers to the movement of the administered *drug* from the site of administration to the circulatory system. Absorption is dependent on the properties of the active ingredients, on the dosage, and also on the characteristics of the absorption site. Drugs may get absorbed by two ways, passive and active transport. The concentration gradient across the membrane facilitates passive diffusion, where drugs drift from a higher to a lower concentration without the aid of a carrier or extra energy. Active transport, on the other hand, needs carriers and

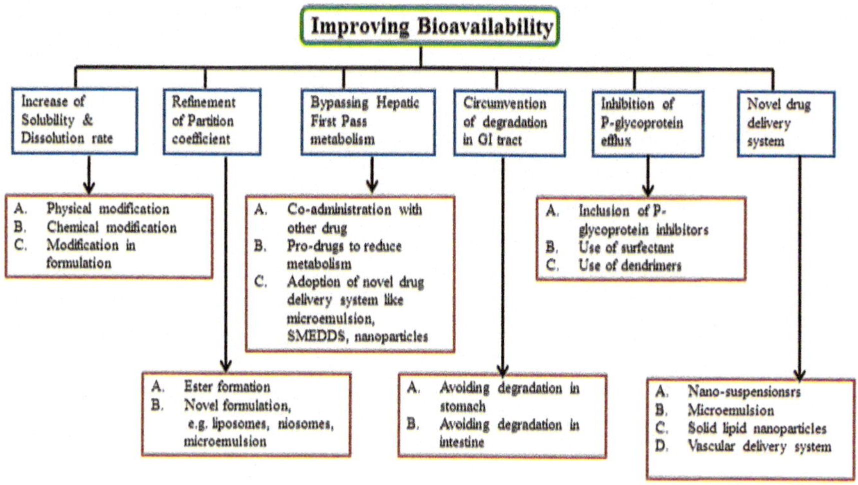

Fig. 8.2 Different ways of improving bioavailability of drugs

ATP as energy source. This method of transport can transport the drug from lower to higher concentration. There is another way to transport drugs, which is known as endocytosis. In this method the drug is engulfed by the membrane, which is then released into the cell. Absorption is affected mainly by three factors, pH, blood flow, and movement. Most of the drugs are either weakly basic or weakly acidic. If uncharged, they pass through the membrane at ease. Relative concentration of charged or uncharged atoms determines the effective concentration of the drug at the site of absorption. Intestine is a site of absorption and blood flow to the intestine is high. Also, the large surface area of intestine supports better absorption. Movement of the drug is another vital factor. A fast-moving drug has less time of contact with the site of absorption. Hence absorption rate will be less. Nanosizing is a good technology to be followed for sparingly soluble drugs. By reducing the size of the drug, surface area is increased and hence the dissolution. Absorption of the drug is greatly enhanced.

Reduction of particle size is of enormous importance. This can be explained by Noyes-Whitney equation [12], which is:

$$\frac{dC}{dt} = \frac{DA\,(C_S\text{-}C)}{h}, \text{ where}$$

dC/dt is the rate of dissolution of drug.
D is the diffusion coefficient of the drug.
A is the effective surface area.
H is the thickness of diffusion layer around the drug molecule.
C_S is the saturation solubility of the drug in solution.
C is the concentration of the drug in GI fluid.

All the parameters except A and C are constants. Therefore, only A and C can be changed. If the particle size in diminished, total surface area is increased and hence A gets increased, hence the rate of dissolution is increased. [12]. In this equation, as the thickness of the diffusion layer is decreased, due to reduction in size, dC/dt increases. Therefore higher dissolution of the drug can be obtained [42].

Nanosizing of a drug means reducing the size to submicron size [27]. Nanosizing of drugs can be done by several methods; these are top-down, bottom-up [32], and combination approaches.

Size reduction by employing high-energy approaches is followed in the top-down technique. High-pressure homogenization is required in this technique. All the steps in this technique are carried out in liquid media, and nanosuspensions are formed. Later on they are processed in capsule or tablet forms or may remain as a suspension. The suspension form refers to a colloidal dispersion, which is further stabilized by a suitable polymer. The second technique, called the bottom-up, is a precipitation method. Here the nanosized particles are obtained from a drug solution that is supersaturated. This technique has certain merits. It is a low-energy technique and more economical. Bottom-up in combination with top-down method has been followed to get much smaller-sized particles. Attempts were made to achieve particle size less than 100 nm, as particles of this dimension can penetrate through various biological barriers. The bottom-up technique can improve bioavailability much better. The drawbacks of this method are the long time for processing and low drug loading efficiencies. The top-down method has been used for commercial purpose.

Physical modification can be achieved by (i) reducing the size of the particle, i.e., through micronization and nanonization; (ii) modifying the crystal form, i.e., polymorphs and pseudopolymorphs; and (iii) modifying the drug dispersion carriers. Chemical modifications like change in pH and salt formation may affect solubility and dissolution rate. Co-crystallization, addition of solubilizer, ultra-rapid freezing, and other formulation-based approaches may affect this. Some unique drug delivery systems like microemulsion, solid lipid nanoparticles, and self-microemulsifying drug delivery system (SMEDDS) may help to avoid hepatic first-pass metabolism. Microemulsions are thermodynamically stable, and isotropic mixtures of oil, water, and surfactant are often mixed with a cosurfactant. Microemulsion is a promising way of drug delivery, which aids in controlled release of drugs [31]. The extent and rate of absorption of lipophilic drugs are greatly improved by this delivery method [44]. Solid lipid nanoparticles are another effective means of pharmaceutical delivery system. The solid lipid core matrix can solubilize lipophilic molecules. Solid lipid nanoparticles can overcome the limitations of other carriers, like emulsions, polymeric nanoparticles, and liposomes. They have very good release profile and can deliver a drug to its target with good efficiency [33]. SMEDDS can improve the solubility and bioavailability of poorly soluble drugs and hence are good delivery means [36].

Degradation in the stomach can be avoided by putting an enteric coating and that in the intestine by floating drug delivery system or with the help of bioadhesive polymers. Targeted drug delivery systems can also help in this regard. Many of the drugs are delivered by floating drug delivery system, where drugs are released at a slower rate. Sustained release and avoiding release of the drug to the stomach are the two advantages of this delivery method [39]. Bioadhesive delivery systems can localize the drug delivery and release the drug at a slower rate. Binding of a natural or synthetic polymer to a biological system is called "bioadhesion." Taking consideration of patient compliance, this mode of delivery is a good approach for drug delivery [6]. The plasma membrane protein P-glycoprotein (P-gp), also a multidrug resistant protein, helps in drug transport mechanism, by expelling drugs out of the cell. P-gp inhibitors can alter pharmacokinetics of a drug [29]. P-gp inhibitors increase the oral bioavailability of its substrates. Some of the P-gp inhibitors include colchicine, felodipine, diltiazem, erythromycin, omeprazole, ketoconazole, lansoprazole, and other proton-pump inhibitors, tamoxifen, nifedipine, quinidine, verapamil, etc.

Solubility of a drug is influenced by particle size, molecular size, temperature, pressure, nature of the solute and solvent, polarity, and polymorphs. Particle size is an important determinant. As the particle size is reduced, the effective surface area is higher. The ratio of the surface area to the volume is increased. The more the surface area, the better the interaction with the solvent and hence the solubility. Molecular size is inversely proportional to the solubility. Temperature is directly related to the solubility. Increase in pressure enhances solubility. Solubility also depends on the nature of the solute. Polarity is also an important factor. Nonpolar molecules dissolve in nonpolar solvents; polar molecules have an affinity for polar solvents. Polymorphs are crystalline materials having same chemical composition, but they vary in molecular packing. They have impact on the bioavailability and solubility. For drugs with poor solubility, polymorphism is very important [7].

By definition, solubility is the amount of a substance called the solute which is dissolved in a unit volume of a liquid substance known as the solvent, to get a saturated solution under a particular temperature and pressure. It was mentioned previously that solubility can be influenced by many factors and therefore, in order to improve solubility, several techniques can be adopted. These are traditional techniques, newer and novel techniques, and solid dispersion techniques. Use of cosolvents, changing the dielectric constant of the solvent, chemical modification of the drug, use of surfactants, change of pH of solvent, etc. can increase solubility. Among the newer and novel technologies, size reduction technologies, nanoparticle technology, cryogenic technology, lipid-based delivery system, microemulsion, etc. are very important. The third technique is solid dispersion technology, which is a solubilization technology for poorly water-soluble drugs. Depending on the source, bioenhancers can be classified into two categories, one of plant origin and the second one of animal origin. Those from plant sources will be discussed elaborately. Cow urine distillate is a good bioenhancer from animal source.

8.4.2 Ideal Properties of the Bioenhancers

An ideal bioenhancer must fulfil certain criteria [20]. They should be:

(a) Nontoxic
(b) Non-allergenic and nonirritating
(c) Acting rapidly and predictably and reproducible activity
(d) Unidirectional in action
(e) Compatible with other active pharmaceutical ingredients
(f) Stable with time and environment
(g) Can be formulated into various dosage forms
(h) Easily available and cost-effective [8]

One more property essential for an ideal bioenhancer is that it should not have any pharmacological actions of its own.

Enhancement of bioavailability renders much value to the patients and their families. This is a noninvasive approach too. The dose of drug can be minimized and chances of drug resistance are less. These are all ecologically beneficial and safe for the environment [20]. Plant-derived molecules satisfy all the above criteria and are therefore good candidates to be effective bioenhancers.

8.5 Phytochemicals as Enhancers of Bioavailability

In the previous chapters, we have discussed various attributes of phytochemicals in different areas of cancer treatment. Many drugs, including herbal or plant-derived drugs, fail to show their fullest potential in in vivo models. The reason behind this is poor lipid solubility and size of the drug, which results in poor absorption and poor bioavailability. Plant-derived molecules are good enhancers of bioavailability. Therefore, it is important to know about the enhancers of plant origin and their mode of action in increasing bioavailability.

Many of the plant-derived products, though they are potential anticancer agents, fail to be effective in vivo. This may be due to poor solubility, molecular size, or both. These drawbacks lead to poor absorption and hence poor bioavailability. Many of the phytochemicals are phenolics, and they are water soluble, due to which they cannot navigate and enter the intestine crossing the lipid membranes. There are certain plant products, which can aid in increasing bioavailability. The action of a bioavailability enhancer was first discovered by Bose in 1929, who described the effect of long pepper along with adhatoda vasaka leaves and reported the increased activity of vasaka [18]. The term bioavailability enhancer itself was first coined by an Indian scientist at Regional Research Lab. Jammu, who discovered and named piperine as the first enhancer of bioavailability in 1979 [8]. It offers comfortable, convenient, and noninvasive way to administer drugs due to following advantages: a) dose reduction, b) minimization of drug resistance, c) minimization of drug (especially true in case of anticancer drug like Taxol), d) ecological benefit, and e) safety of the environment [20]. In Ayurveda some of the herbal products were mentioned as bioenhancers. In Sanskrit the term Trikatu was used, which means three acrids. The three compounds are black pepper, long pepper, and ginger [8]. These agents facilitate availability of the drugs, but they do not show any adverse side effects. Activity of the drug molecule may be increased when the bioavailability of the drug is increased across the membrane, the drug may get potentiated through conformational interaction and may act as receptors for the drug molecule, or these agents may render the targets more compliant to the drugs. In order to function, the drug has to cross the epithelial barrier of the intestinal mucosa. There are a few anatomical and biological barriers which restrict the entry of the drugs. Some characteristic structures of the epithelium prevent drugs from entering systemic circulation. Also, for the drug molecules above a critical size (0.4 nm), entry is a problem [23].

Bioenhancers can be categorized based on the mode of action:

(i) Those which inhibit P-gp and other efflux pump
(ii) Those which suppress CYP-450
(iii) Those which regulate GIT function to impart more absorption of drugs

To enhance the absorption of oral drugs, there are some enhancers, like bile salts, surfactants, fatty acids, chelating agents, etc. An extract from shells of crustaceans like shrimps, crabs, etc. called chitosan may block absorption of dietary fat and cholesterol. Chitosan, a fiber derived from chitin, aids in the enhancement of drug absorption by redistribution of cytoskeletal F-actin that opens the tight junctions. Surfactants like bile, bile salts, and fatty acids increase the solubility of hydrophobic drugs and enhance absorption. Calcium chelators like ethylene glycol tetra-acetic acid (EGTA) and ethylene diamine tetra-acetic acid (EDTA) reduce the calcium concentration, resulting in cell-cell contacts. By this way they increase absorption of drugs [8].

Ampicillin, an antibiotic employed to treat respiratory and urinary tract infections and infections of other kinds, is not fully absorbed from GI tract due to its hydrophilic nature. Esterification of carboxyl group of ampicillin leads to the synthesis of a number of pro-drugs with increased lipophilicity. Absorption capacity of these synthesized drugs is higher.

Mechanisms by which phytochemicals aid in bioenhancement of drug are not all the same. Bioenhancers may be nutritional or antimicrobial, the former function by acting on the gastrointestinal tract and the latter on drug metabolism processes. The phytochemicals that act as enhancers reduce the secretion of hydrochloric acid and increase blood supply to the gastrointestinal tract [2]. In addition to this, they inhibit the gastrointestinal transit time, gastric emptying time, and intestinal motility [5, 30]. The permeability of the cell membrane of gastrointestinal tract gets modified by these enhancers. First-pass metabolism and drug-metabolizing enzymes get inhibited as well [3].

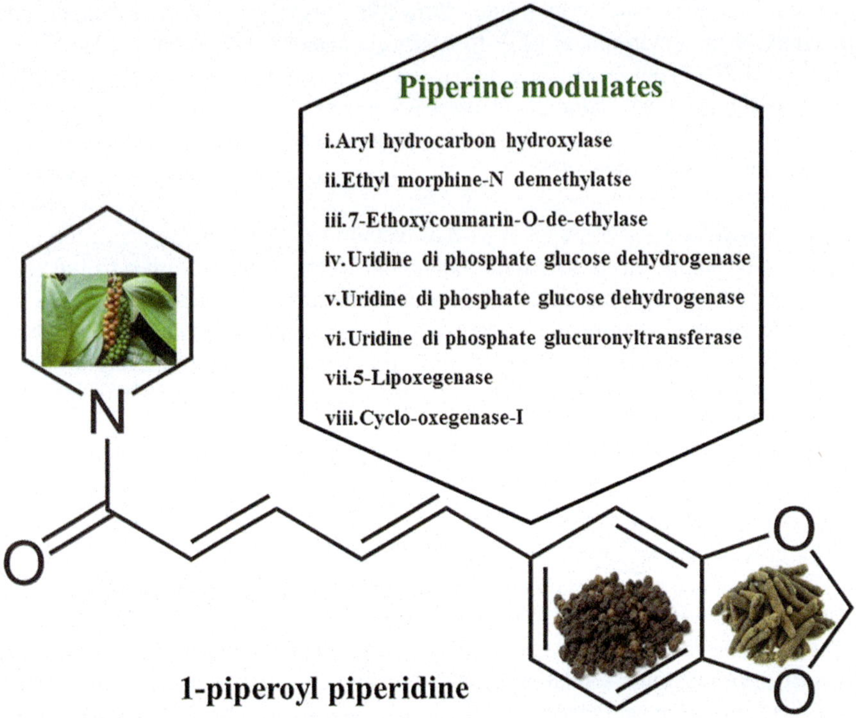

Fig. 8.3 Actions of piperine as a bioenhancer

Piperine, one of the three components of Trikatu, is a well-known enhancer. It is an alkaloid which is responsible for the pungency of *Piper nigrum*. It is highly soluble in alcohol, chloroform, and ether, though its solubility in water is less. It modulates cell signaling pathways, inhibits drug efflux pumps, and aids in DNA receptor binding. Drug-metabolizing enzymes are inhibited by piperine; also the action of drug pump that eliminates cells is decreased [24]. Piperine inhibits P-glycoprotein and cytochrome P450, which have enormous contribution in first-pass elimination of drugs. By inhibition of these drug-metabolizing enzymes, drugs can work better. The mode of action of piperine is depicted in Fig. 8.3.

There are several more natural compounds which show bioenhancing properties. These are quercetin, niazeridine, capmul, genistein, naringin, lysergol, *Callistemon rigidus*, *Carum carvi*, sinomenine, glycyrrhizin, menthol, diosmin, gallic acid ester, nitrile glycoside, etc. Quercetin, a plant flavonol from the flavonoid group of polyphenols, is found in citrus fruits, vegetables, and leaves and is an inhibitor of cytochrome P3A4 (CYP3A4), MRP, and P-gp. Apart from its wide range of biological activities that include radical scavenging, antioxidant, anti-inflammatory, antitumor, antiatherosclerotic, and antiviral effects, it was found that a prior treatment with quercetin significantly alters the pharmacokinetic properties of the drug verapamil [34]. It also alters the efficacy of drugs like diltiazem, paclitaxel, digoxin, doxorubicin, and tamoxifen [38]. Ranolazine, irinotecan, valsartan, fexofenadine, and etoposide are few more drugs whose bioavailability is influenced by quercetin. Quercetin also enhances the efficacy of EGCG, another important phytochemical. Intake of red onion with EGCG increases the absorption of EGCG. Red onion is a good source of quercetin.

Genistein, a phytoestrogen from soy products, inhibits P-gp and MRP efflux function. Genistein also inhibits CYP3A4. Intestinal absorption of paclitaxel is greatly influenced by genistein. A study revealed that genistein can enhance the bioavailability of EGCG, but the tumorigenesis was increased. This finding shows that studies need to be conducted carefully and data interpretation must be done with great care [26]. Naringenin, a bitter, colorless flavanone, is found in grapefruit and in a variety of other fruits and herbs. It has been found to inhibit CYP3A1/2, CYP3A4, and P-gp. Naringenin has been found to increase the efficacy of paclitaxel, diltiazem, tamoxifen, clopidogrel, and verapamil. Curcumin from *Curcuma longa* has been found to suppress drug-metabolizing enzymes, and it suppresses drug transporter P-gp and thereby enhances the efficacy of the drugs celiprolol and midazolam. Curcumin inhibits UDP-glucuronyl transferase, efflux transporters, and CYP3A4. Effect of curcumin on the bioavailability of tamoxifen was studied in rats. Both the compounds have been found to interact with CYP and P-glycoprotein. Curcumin may act as an enhancer for tamoxifen through inhibition of the CYP3A4-mediated metabolism of tamoxifen. It also enhances bioavailability of norfloxacin. P-gp efflux transporter gets inhibited in the small intestine. Curcumin may thus decrease the first-pass metabolism of the drug tamoxifen by inhibiting P-gp or CYP3A4 [9]. Curcumin has been found to inhibit UDP-glucuronyl transferase level in the intestine and hepatic tissues, which is a reason for its potential as a bioenhancer [45].

Ginger from *Zingiber officinale* contains gingerol and is an important ingredient for Indian cuisine. Gingerol aids better absorption by regulating the function of the GI tract. This compound has been found to increase the availability of rifampicin and ethionamide, drugs that are used in pulmonary tuberculosis. Antibiotics like azithromycin, antifungal like ketoconazole, and antiviral like zidovudine are made more bioavailable by gingerol. The activity of the anticancer drug 5-fluorouracil is also enhanced by gingerol [45]. Caraway oil is extracted from dried and crushed seeds of *Carum carvi*, whose active ingredients are carvone and limonene. Caraway is in abundance in Meridian fennel/Persian cumin. It has several beneficial properties, including enhancement of bioavailability of antibiotics, antifungal, antiviral, and anticancer drugs [13]. Caraway/cumin is a P-gp efflux pump inhibitor. It aids in the tissue modulation in local mucosa. It increases the bioavailability of rifampicin, isoniazid, cycloserine, and ethionamide [45].

Peppermint oil from *Mentha piperita* is another bioenhancer that increases the oral bioavailability of cyclosporine by inhibition of CYP3A. Menthol from corn mint enhances the bioavailability of dideoxycytidine, though the mechanism of action of menthol is not well understood. Indian aloe or *Aloe vera* is an important source of phytochemicals and also a promising bioenhancer. The gel in the whole leaf is known for its diverse properties. Two different *Aloe vera* preparations, whole leaf extract and the gel, were studied for the absorption of vitamins C and E. It was observed that aloe improves the absorption of both vitamins C and E. The absorption of vitamins is slower, and vitamins last longer in the plasma with aloe, and it increases the bioavailability of vitamins C and E in humans [43]. *Aloe vera* has effect on intracellular tight junction modulation. The drugs whose availability is enhanced by *Aloe vera* include didanosine, insulin, atenolol, etc.

Diosmin found in citrus fruits inhibits P-gp and enhances the bioavailability of digoxin. The active phytomolecule in garlic that is also a bioenhancer is allicin. It enhances the fungicidal activity of amphotericin B and acts against some pathogenic fungi such as *Candida albicans* and *Aspergillus fumigatus* in addition to yeast *Saccharomyces cerevisiae*. Amphotericin B exhibited enhanced antifungal activity against *S. cerevisiae* when given along with allicin. Other bioenhancers include liquorice, stevia or honey leaf, *Sinomenium acutum*, gallic acid, capsaicin, capmul, and *Ammoniac multiflora*. Drumstick pods contain an ingredient called niaziridin, a glycoside which increases the bioavailability of rifampicin [25]. The biomolecule from morning glory plant called lysergol is a potent bioenhancer. This compound increases the bioavailability of rifampicin, tetracycline, berberine, sulfasalazine, and ampicillin. It also enhances the effects of curcumin and other antibiotics [13]. *Moringa oleifera* acts as a bioenhancer for the drug rifampicin. This compound inhibits CYP450. Lycopene from red fruits and vegetables, particularly tomato, is effective in enhancing the availability of simvastatin. Resveratrol, found in berries and grapes, inhibits CYP2C9, CYP2E1, P-gp, and MRP2 and inhibits renal uptake transporters. This compound shows bioenhancing potential on diclofenac, NSAID, methotrexate, etc. Sinomenine from *Sinomenium acutum* influences the drug paeoniflorin by inhibiting P-gp. Insulin can be potentiated by citric acid, which is found in abundance in citrus fruits. Citric acid has been found to inhibit mucosal tissue modulation metabolism. Gokhru, which is found in the plant *Tribulus terrestris*, is a modulator of local mucosal tissue and has been found to increase the action of metformin. Another plant-derived compound emodin from *Rheum palmatum* inhibits P-gp efflux and modulates MRPs 1, 2, and 3. It shows its action on verapamil [37].

8.6 Conclusion

We have discussed issues related to bioavailability with an emphasis on phytochemicals and their roles as bioenhancers. We have discussed the measures of bioavailability, the factors influencing bioavailability, how bioavailability can be improved through different techniques like nanosizing, and the roles of bioenhancers in improving bioavailability. We have discussed the roles of a broad range of phytochemicals that act as bioenhancers. Some of the bioenhancers work better in presence of other enhancers. The bioenhancers potentiate the efficacy of the drugs, thereby reducing the drug dosage and the associated toxicity and adverse side effects. Treatments may become shortened, drug resistance may be reduced, and treatments may become cost-effective due to bioenhancers.

References

1. B. Alberts, A. Johnson, J. Lewis, M. Raff, K. Roberts, P. Walter. *Molecular biology of the cell*, 4th edn. (Garland Science/The Lipid Bilayer, New York, 2002). https://www.ncbi.nlm.nih.gov/books/NBK26871/
2. A.R. Annamalai, R. Manavalan, Effects of Trikatu and its individual components and piperine on gastrointestinal tracts: Trikatu a bioavailable enhancer. Ind. Drugs **27**(12), 595–604 (1989)
3. C.K. Atal, R.K. Dubey, J. Singh, Biochemical basis of enhanced drug bioavailability by piperine: Evidence that piperine is a potent inhibitor of drug metabolism. J. Pharmacol. Exp. Therap. **232**, 258–262 (1985)
4. S.I.F. Badawy, M.M. Ghorab, C.M. Adeyeye, Characterization and bioavailability of danazolhydroxypropyl-β-cyclodextrin coprecipitates. Int. J. Pharm. **128**(1–2), 45–54 (1996)
5. S. Bajad, K.S. Bedi, A.K. Singla, R.K. Johri, Piperine inhibits gastric emptying and gastrointestinal transit in rats and mice. Planta Med. **67**, 176–179 (2001)
6. D. Brahmbhatt, Bioadhesive drug delivery systems: Overview and recent advances. Int. J. Chem. Life Sci. **6**(3), 2016–2024 (2017)
7. R. Censi, P.D. Martino, Polymorph impact on the bioavailability and stability of poorly soluble drugs. Molecules **20**, 18759–18776 (2015)
8. S.A. Chavhan, S.A. Shinde, H.N. Gupta, Natural bioenhancers. Int. J. Pharm. Sci. Res. **3**(1), 29–37 (2018)
9. Y.A. Cho, W. Lee, J.S. Choi, Effects of curcumin on the pharmacokinetics of tamoxifen and its active metabolite, 4-hydroxytamoxifen, in rats: Possible role of CYP3A4 and P-glycoprotein inhibition by curcumin. Pharmazie **67**, 124–130 (2012)
10. P. Crowley, L. Martini, Optimising drug delivery: The challenges and opportunities. ONDrug Deliv. **59**, 4–11 (2015)
11. S. De Iudicibus, A. De Pellegrin, G. Stocco, F. Bartoli, R. Bussani, G. Decorti, ABCB1 gene polymorphisms and expression of P-glycoprotein and long-term prognosis in colorectal cancer. Anticancer Res. **28**(6B), 3921–3928 (2008)
12. S.M. Dizaj, Z. Vazifehasl, S. Salatin, K. Adibkia, Y. Javadzadeh, Nanosizing of drugs: Effect on dissolution rate. Res. Pharm. Sci. **10**(2), 95–108 (2015)
13. G.B. Dudhatra, S.K. Mody, M.M. Awale, H.B. Patel, C.M. Modi, A. Kumar, D.R. Kamani, B.N. Chauhan, A comprehensive review on Pharmacotherapeutics of herbal bioenhancers. Sci. World J. **2012**, 637953 (2012)
14. I. Elsayed, A. Ahmed, A. Ahmed, H. Elshafeey, Nanosizing of a poorly soluble drug: Technique optimization, factorial analysis, and pharmacokinetic study in healthy human volunteers. Int. J. Nanomedicine **9**(1), 2943–2953 (2014)
15. R.H. Engel, S.J. Riggi, M.J. Fahrenbach, Insulin: Intestinal absorption as water-in-oil-in-water emulsions. Nature **219**, 856–857 (1968)
16. M.E. Goldsmith, J.M. Gudas, E. Schneider, K.H. Cowan, Wild type p53 stimulates expression from the human multidrug resistance promoter in a p53-negative cell line. J. Biol. Chem. **270**(4), 1894–1898 (1995)
17. J.P. Griffin, *The Textbook of Pharmaceutical Medicine*, 6th edn. (BMJ Books/Wiley-Blackwell, New Jersey, 2013)
18. R. Gupta, K. Kesarwani, Bioavailability enhancers of herbal origin: An overview. Asian Pac. J. Trop. Biomed. **3**(4), 253–266 (2013)
19. S. Gupta, R. Kesarla, A. Omri, Formulation strategies to improve, the bioavailability of poorly absorbed drugs with special emphasis on self-emulsifying systems. ISRN Pharm **2013**, 1–16 (2013)., Article ID 848043
20. G. Jain, U.K. Patil, Strategies for enhancement of bioavailability of medicinal agents with natural products. Int. J. Pharm. Sci. Res. **6**(12), 5315–5324 (2015)
21. S. Javed, W. Ahsan, K. Kohli, The concept of bioenhancers in bioavailability enhancement of drugs – a patent review. J. Sci. Lett **1**(3), 143–165 (2016)

22. J. Jin, L. Zhu, M. Chen, H. Xu, H. Wang, X. Feng, X. Zhu, Q. Zhou, The optimal choice of medication administration route regarding intravenous, intramuscular, and subcutaneous injection. Patient Prefer. Adherence **9**, 923–942 (2015)
23. R.L. Juliano, V. Ling, A surface glycoprotein modulating drug permeability in Chinese hamster ovary cell mutants. Biochim. Biophys. Acta **455**(1), 152–162 (1976)
24. K. Kesarwani, R. Gupta, Bioavailability enhancers of herbal origin: An overview. Asian Pac. J. Trop. Biomed. **3**(4), 253–266 (2013)
25. S.P.S. Khanuja, J.S. Arya, T. Santakumar, D. Saikia, H. Kaur, Nitrile glycoside useful as a bioenhancer of drugs and nutrients, process of its isolation from Moringa oleifera. U.S. Patent US 6858588 (2005)
26. J.D. Lambert, S. Kwon, J. Ju, M. Bose, M.-J. Lee, J. Hong, X. Hao, C.S. Yang, Effect of genistein on the bioavailability and intestinal cancer chemopreventive activity of (−)-epigallocatechin-3-gallate. Carcinogenesis **29**(10), 2019–2024 (2008)
27. X. Li, L. Gu, Y. Xu, Y. Wang, Preparation of fenofibrate nanosuspension and study of its pharmacokinetic behaviour in rats. Drug Dev. Ind. Pharm. **35**, 827–833 (2009)
28. C.A. Lipinski, Drug-like properties and the causes of poor solubility and poor permeability. J. Pharmacol. Toxicol. Methods **44**(1), 235–249 (2000)
29. A. Lutful, P-glycoprotein inhibition for optimal drug delivery. Drug Target Insights **7**, 27–34 (2013)
30. M. Majeed, V. Badmaev, R. Rajendran, Use of piperine to increase the bioavailability of nutritional compounds. United States Patent, Number 5536506 (1995)
31. A. Mishra, R. Panola, A.C. Rana, Microemulsions: As drug delivery system. J. Sci. Innov. Res. **3**(4), 467–474 (2014)
32. J. Möschwitzer, Particle size reduction technologies in the pharmaceutical development process. Am. Pharm. Rev. **2010**, 54–59 (2010)
33. N. Naseri, H. Valizadeh, P. Zakeri-Milani, Solid lipid nanoparticles and nanostructured lipid carriers: Structure, preparation and application. Adv. Pharm. Bull. **5**(3), 305–313 (2015)
34. R.J. Nijveldt, E.V. Nood, D.E. van Hoorn, P.G. Boelens, K. Norren, P.A. van Leeuwen, Flavonoids: A review of probable mechanisms of action and potential applications. Am. J. Clin. Nutr. **74**, 418–425 (2001)
35. H.M. Patel, B.E. Ryman, Oral administration of insulin by encapsulation within liposomes. FEBS Lett. **62**, 60–63 (1976)
36. M.J. Patel, S.S. Patel, N.M. Patel, M.M. Patel, A self-microemulsifying drug delivery system (SMEDDS). Int. J. Pharm. Sci. Rev. Res. **4**(3), 29–35 (2010)
37. B. Peterson, M. Weyers, J.H. Steenekamp, J.D. Steyn, C. Gouws, J.H. Hamman, Drug bioavailability enhancing agents of natural origin (Bioenhancers) that modulate drug membrane permeation and pre-systemic metabolism. Pharmaceutics **11**, 33 (2019)
38. G.K. Randhawa, J.S. Kullar, Rajkumar, Bioenhancers from mother nature and their applicability in modern medicine. Int. J. Appl. Basic Med. Res. **1**(1), 5–10 (2011)
39. A. Sarawade, M.P. Ratnaparkhi, S. Chaudhari, Floating drug delivery system: an overview. Int. J. Res. Dev. Pharm. Life Sci. **3**(5), 1106–1115 (2014)
40. N.G. Schipper, S. Olsson, J.A. Hoogstraate, A.G. de Boer, K.M. Varum, P. Artursson, Chitosans as absorption enhancers for poorly absorbable drugs 2: Mechanism of absorption enhancement. Pharm. Res. **14**(7), 923–929 (1997)
41. L. Shargel, A.B.C. Yu, *Applied Biopharmaceutics & Pharmacokinetics*, 7th edn. (McGraw-Hill, New York, 1999). ISBN 0-8385-0278-4
42. K. Sigfridsson, A.J. Lundqvist, M. Strimfors, Particle size reduction and pharmacokinetic evaluation of a poorly soluble acid and a poorly soluble base during early development. Drug Dev. Ind. Pharm. **37**, 243–251 (2011)
43. R. Singha, S. Devib, J.H. Patela, U.D. Patela, S.K. Bhavsar, A.M. Thaker, Indian herbal bioenhancers: A review. Phcog. Rev. **3**(5), 80–82 (2009)
44. S.S. Solanki, B. Sarkar, R.K. Dhanwani, *Microemulsion Drug Delivery System: For Bioavailability Enhancement of Ampelopsin*. ISRN Pharmaceutics. Article ID 108164, 4 pages (2012)

45. D.V. Tatiraju, V.B. Bagade, P.J. Karambelkar, V.M. Jadhav, V. Kadam, Natural bioenhancers: An overview. J. Pharm. Phytochem. **2**(3), 55–60 (2013)
46. H. Thakkar, B. Patel, S. Thakkar, A review on techniques for oral bioavailability enhancement of drugs. Int. J. Pharm. Sci. Rev. Res. **4**(3), 203–223 (2010)
47. F. Veiga, C. Fernandes, F. Teixeira, Oral bioavailability and hypoglycaemic activity of tolbutamide/cyclodextrin inclusion complexes. Int. J. Pharm. **202**(1–2), 165–171 (2000)
48. G. Zhou, M.T. Kuo, NF-kappaB-mediated induction of mdr1b expression by insulin in rat hepatoma cells. J. Biol. Chem. **272**(24), 15174–15183 (1997)

Appendices

Appendix I

Type of cancer	Causes	Signs and symptoms
Acute lymphoblastic leukemia (ALL)	Previous cancer therapeutics like chemotherapeutic drugs or radiation, genetic disorders like Down syndrome, history of a sibling having ALL	Tiredness, weakness, dizziness, shortness of breath, fever, persistent infection, bleeding and bruising, night sweats, anorexia, enlargement of lymph nodes, bone or joint pain
Acute myeloid leukemia (AML)	Smoking; long-term exposure to benzene and chemotherapeutic drugs, particularly alkylating agents or platinum compounds; exposure to radiation; those having myeloproliferative disorders like polycythemia vera, thrombocytopenia, or myelodysplastic syndrome have a higher risk of developing AML. Certain disorders like Fanconi anemia, Bloom syndrome, ataxia-telangiectasia, Down syndrome, or Trisomy 8 impart a higher risk	Tiredness, fever, loss of appetite, weight loss, night sweats, weakness, pallor, arrhythmia, dizziness, dyspnea, cold hands and feet, headaches, weight loss and shortness of breath, anorexia, bruising and bleeding, bleeding from nose or gum, bleeding and clotting problem, bone or joint pain, seizures, facial numbness, blurred vision
Anal cancer	Age, HPV 16 or 18 infection, multiple sexual partners, anal intercourse, smoking, prior history of cancer like cervical and vaginal, medication that causes immunosuppression	Rectal bleeding, pain in anal area, lumps around the anus, sticky discharge from the anus with occasional itching, changes in bowel habit, fecal incontinence, bloating, vaginal dryness and lower back pain for women

(continued)

M. Roy, A. Datta, *Cancer Genetics and Therapeutics*,
https://doi.org/10.1007/978-981-13-9471-3

Type of cancer	Causes	Signs and symptoms
Basal cell carcinoma of the skin	Ultraviolet exposure; those having less melanin are susceptible; sunburn; having more moles; family or personal history of skin cancer; weak immune system; radiation exposure; exposure to the metalloid arsenic	Appearance of new moles; increase in the size of mole; change in color of the mole; a spot which is increased in size due to a lump within it; rough, scaly mole; itchy moles; bleeding from moles; waxy bump; appearance of a scar-like lesion; red nodule
Bladder cancer	Age, race (Caucasian), and gender (male) are more susceptible; history of previous cancer; bladder inflammation; certain chemicals like arsenic, etc.	Hematuria, frequent urination, pain or discomfort during urination, urge of urination even when the bladder is empty, lower back pain, loss of appetite, weakness, swelling of feet, weight loss
Bone cancer	Paget's disease, radiation therapy to treat some other types of cancer, some genetic syndromes like hereditary retinoblastoma	Pain and swelling at the tumor site, stiffness and swelling of joint, fever, anemia, fracture, numbness, fatigue
Brain cancer	Exposure to ionizing radiation, genetic syndromes, and family history	Frequent and severe headache, nausea or vomiting, blurred or double or loss of vision, loss of sensation, difficulty with balance, development of speech and hearing problem, behavioral changes, seizures, loss of control of body function, loss of awareness, drowsiness, forgetfulness, confused state
Breast cancer	Age, family history, early age at first menstruation, prior incidence of breast cancer, exposure to estrogen, having children late or not at all, lack of or inadequate breastfeeding, obesity, alcohol consumption, exposure to radiation, use of hormone replacement therapy (HRT), birth control pills	Thickened tissue in the breast, a lump with/without pain in the breast or in an armpit, reddening of breast skin, discharge from the nipple (sometimes bloody), rash around the nipples, inverted nipple, change in the size or shape of the breast, flaking or scaling of the skin on the breast or nipple
	Old age, estrogen-related drugs (for treatment of prostate cancer), family history, Klinefelter's syndrome, liver disease, obesity, testicle disease or surgery (orchiectomy) may increase the risk of male breast cancer	Signs and symptoms of male breast cancer (1% incidence) are painless lump, thickened breast tissue, redness, scaling of skin of breast, redness of nipple which turns inward, and discharge from nipple
Burkitt lymphoma	A rare but aggressive form of non-Hodgkin lymphoma. It is related to the Epstein-Barr virus (EBV) and chronic malaria. People with weak immune status are most susceptible	Fever, abdominal swelling, distortion of facial bones, night sweats, weight loss, intestinal obstruction, an enlarged thyroid, enlarged tonsils

(continued)

Type of cancer	Causes	Signs and symptoms
Cervical cancer	Human papillomavirus (HPV) infection, multiple sexual partners, smoking, taking birth control pills, engaging in early sexual contact, poor hygiene, immunosuppression due to other reasons, *Chlamydia* infection, obesity, long-term use of oral contraceptives, family history of cervical cancer, multiparity, use of diethylstilbestrol (DES) to control miscarriage, tender age at first child birth	Abnormal vaginal bleeding, increased vaginal discharge, vaginal bleeding after menopause, pain during sexual intercourse, bleeding after having sex, pelvic pain, long duration and heavy menstrual bleeding than normal, longer or heavier menstrual periods than usual, other abnormal vaginal discharges, pain during sexual intercourse
Cholangiocarcinoma (bile duct cancer)	Age; smoking; liver flukes (parasite) from raw or uncooked fish; congenital bile duct problem; chronic liver disease, people having primary sclerosing cholangitis are susceptible; cirrhosis; exposure to certain chemicals like nitrosamines and polychlorinated biphenyls	Jaundice, pale stool, dark urine, itching, abdominal pain proceeding toward the back, chill, fever, loss of appetite, weight loss, fatigue, sense of fullness in the belly
Chronic lymphocytic leukemia (CLL)	Old age, more common among Caucasian, incidence of blood or bone marrow cancer in the family, exposure to certain chemicals like benzene	Enlarged, but painless lymph nodes, fever, fatigue, upper abdominal pain, sweating at night, weight loss, repeated infections, bleeding, bruising, nosebleeds or gum bleeding, shortness of breath
Chronic myelogenous leukemia (CML)	Old age; males are more susceptible; exposure to radiation	Anemia, weakness and fatigue, loss of weight, anorexia, pallor, sweating at night, fever, pain in bone, enlarged spleen, sense of fullness in belly, bleeding and bruising, having infection frequently
Colorectal cancer	Exact cause is not known. Inherited gene mutations may be a risk factor: hereditary nonpolyposis colorectal cancer (HNPCC) or Lynch syndrome increases the risk of colon cancer before the age of 50. Familial adenomatous polyposis (FAP) may be responsible for development of polyps in the lining of your colon and rectum. FAP, if untreated, may cause colon cancer before the age of 40. Diet high in fat and low in fiber may be a risk factor. Old age, inflammatory intestinal condition, family history, sedentary lifestyle, diabetes, smoking, alcohol consumption	Sudden change in bowel habits, such as diarrhea, constipation, rectal bleeding, dark stools, or blood in the stool, cramping or abdominal pain, weakness and fatigue, weight loss, persistent abdominal discomfort, cramps, gas, or pain

(continued)

Type of cancer	Causes	Signs and symptoms
Endometrial cancer	Imbalance of female hormone in the body, onset of menstruation at early age, not becoming pregnant, obesity, old age, inherited colon cancer syndrome, hormone therapy for breast cancer, oral contraceptives, polycystic ovarian syndrome, effect of radiotherapy to treat other cancers, diabetes, use of intrauterine devices	Vaginal bleeding after menopause, feeling a mass
		Bleeding in between periods, discharge from vagina, pelvic pain
Esophageal cancer	Alcohol consumption; smoking; having a reflux disorder, such as gastroesophageal reflux disease (GERD); having Barrett's esophagus, which is a condition characterized by damaged esophageal lining due to GERD; obesity; not eating enough fruits and vegetables; having achalasia, a condition where the muscle at the bottom of the esophagus doesn't relax properly. Men are three times more likely than women to develop esophageal cancer	Weight loss, indigestion, heartburn, painful swallowing, choking while eating, vomiting, chest pain, fatigue, hiccups, chronic cough, pain in the throat or breastbone
	Esophageal cancer is more common in African-Americans than in other ethnicities	
	The chances of developing esophageal cancer increase with age; the risks are particularly higher for people over 45	
Gallbladder cancer	Most common in women; aged; people having gallstones and gallbladder polyps are susceptible	Jaundice, pain in the abdomen, nausea, vomiting, fever, lump in the abdomen, swollen abdomen, bloating, loss of appetite, weight loss
Gastric cancer	*Helicobacter pylori* infection; stomach polyps; tumors in other parts of the digestive system; old age; smokers; family history; consumption of salty or processed foods; overeating; obesity; alcohol use; women are more susceptible; previous stomach surgery; nitrates and nitrites are common causes	Nausea and vomiting, fatigue, stomach pain, heartburn, anorexia, weight loss, bloating, feeling of fullness, bloody stool, indigestion, jaundice, anemia

(continued)

Type of cancer	Causes	Signs and symptoms
Head and neck cancer	Alcohol and tobacco are the main risk factors; HPV infection; betel quid; preserved foods; poor oral hygiene; exposure to wood dust, asbestos, and synthetic fibers. Workers in a construction, metal, formaldehyde; textile, ceramic, and food industries are at high risk. Radiation exposure due to treatment of some other cancers may be another cause. Infection with the Epstein-Barr virus is a risk factor	Red or white patch in the mouth; lump in the head and neck area; sore throat; foul odor from mouth; hoarseness; ear pain and infection; nasal congestion; bleeding or discharge from nose; shortness of breath; double vision; numbness in head and neck region; chewing, swallowing, and jaw movement difficulty; blood in saliva; weight loss; fatigue; sore that does not heal
Hepatocellular carcinoma	More common in male, chronic viral hepatitis (Hep-B or Hep-C) infection, cirrhosis, nonalcoholic fatty liver disease, alcohol consumption, obesity, type 2 diabetes, exposure to aflatoxins, anabolic steroids, arsenic, tobacco use	Weight loss, anorexia, upper abdominal pain, nausea and vomiting, weakness, fatigue, swelling of the abdomen, yellow discoloration of the skin and white portion of the eye, white stools
Kidney cancer	Old age, smoking, obesity, hypertension, long-term dialysis, family history, exposure to some chemicals like arsenic, chromium, etc.	Blood in urine, back pain, anorexia, weight loss, fever, weakness, and fatigue
Laryngeal cancer	Men are at higher risk than female; old age; gastrointestinal reflux disease; tobacco and alcohol use; lack of nutrition; human papillomavirus infection; some genetic disorders like Fanconi anemia and dyskeratosis congenita; exposure to wood dust, paint, and chemicals used in petroleum; textile and plastic industries; asbestos	Cough, change in voice, hoarseness, swallowing difficulty, pain in the ear, sore throat, weight loss
Lung cancer	Smoking; even passive smokers are at high risk. Exposure to radon gas, asbestos, arsenic, chromium, nickel, family history of lung cancer	Persistent cough, blood with cough, shortness of breath, chest pain, bone pain, hoarseness, weight loss, and headache
Lymphoma	Lymphoma may be of two types: Hodgkin lymphoma and non-Hodgkin lymphoma. For non-Hodgkin lymphoma, probable risk factors are old age; women are more susceptible; common in white population; exposure to nuclear radiation and certain chemicals; immunodeficiency; autoimmune disease like rheumatoid arthritis and celiac disease; infections like Epstein-Barr virus (EBV); breast implants; obesity	Lymphoma may be without obvious symptoms. There may be swelling of lymph nodes, with or without pain. Other signs are fever, night sweats, chills, anorexia, weight loss, itching, fatigue, weakness, persistent cough, breathing difficulty, swelling of the abdomen

(continued)

Type of cancer	Causes	Signs and symptoms
	For Hodgkin lymphoma, men are more prone to develop the disease. It is common among people between the age group of 20 and 30 and those above 55. Other risk factors are family history, HIV infection, mononucleosis, and affluence	
Melanoma	Fair skin (less melanin), sunburn, UV light exposure, many moles, unusual mole family history of melanoma, weak immune status	A change in an existing mole, appearance of a new pigmented or unusual-looking growth on the skin
Mouth cancer	Tobacco use like cigarettes, cigars, pipes, chewing tobacco and snuff, alcohol use, exposure of lips to sun, HPV infection, weak immune status	A bleeding sore that doesn't heal, growth, lump, thickening of the lining of the mouth, loose teeth, poor dentures, jaw and tongue pain, difficulty in swelling or chewing, sore throat
Mouth cancer may affect the lips, gums, tongue, inner lining of the cheeks, and roof and floor of the mouth		
Neuroblastoma	Children with a history of neuroblastoma in the family are susceptible to develop the disease. But, cause is not identified till now	Lump or swelling in the child's abdomen or neck; swelling of the legs or upper chest, neck, and face; enlarged belly; breathing and swallowing difficulty; weight loss; sense of being full; toilet problem; drooping eyelid with bulging eyelids; swelling of lymph nodes when the disease has spread
		Neuroblastomas sometimes release hormones (chemicals) that can cause problems with tissues and organs in other parts of the body, even though the cancer has not spread to those tissues or organs. These problems are called paraneoplastic syndromes. Signs are fever; watery diarrhea; increased blood pressure and heart beat; and sweating and reddening of the skin
Ovarian cancer	Family history of ovarian or breast cancer, age (postmenopausal), fertility drugs, having breast cancer, HRT, obesity, and endometriosis are the risk factors. Contraceptive pills, full-term pregnancies before the age of 26, and gynecological surgery reduce the risk	Pain in the pelvis and lower abdomen, back pain, feeling full, frequent urination, pain during sexual intercourse, heartburn, indigestion, changes in bowel habit, nausea, weight loss, shortness of breath, tiredness, anorexia

(continued)

Type of cancer	Causes	Signs and symptoms
Pancreatic cancer	Old age, pancreatitis, diabetes, family history, smoking, overweight, genetic syndromes like mutation in BRCA2 gene, Lynch syndrome, familial atypical mole-malignant melanoma (FAMMM) syndrome	Abdominal pain that radiates to the back, anorexia, weight loss, diabetes, depression, fatigue, blood clots, yellowing of skin and white part of the eye
Penile cancer	Old age; phimosis; poor personal hygiene; many sexual partners; usage of tobacco; HPV infection; AIDs; UV for treatment of psoriasis	Redness, irritation, or sore on the penis; lump on the penis
Pharyngeal cancer	Using tobacco in any form (smoking, chewing tobacco, taking snuff); too much alcohol consumption; gastroesophageal reflux disease (GERD); Epstein-Barr virus (EBV); HPV; poor nutrition; salt-rich diet; diet low in vitamins A and E	Sore in mouth or lip that does not heal; red or white patch on the gums, tongue, tonsil, or lining of the mouth; growth on the lip, mouth, neck, or throat or a feeling of thickening in the cheek; sore throat that does not heal; hoarseness; numbness of mouth; bleeding or pain in the mouth; chewing and swallowing difficulty; painful movement of jaws or tongue; pain in jaw and ear; bad breath; speech change; toothache, anorexia; weight loss; fatigue
Pituitary cancer	The hereditary trait multiple endocrine neoplasia type I and type IV are risk factors	Headache; loss of vision; pituitary tumor may cause nausea and vomiting; weakness; chill; infrequent or no menstrual periods; sexual dysfunction, weight gain or loss; increased urination; thinning of the arms and legs; high blood pressure and sugar; bone weakening; and joint pain; mood swing; coarsened facial features; excessive sweating; for men there may be erectile dysfunction and lowered sperm count
Prostate cancer	Common after the age of 50 years; diet rich in red meat or high-fat dairy product; obesity; some inherited genes like RNASEL, BRCA1, BECA2, etc. are responsible factors	Frequent urges to urinate, problematic urination; hematuria; ejaculation difficulty; pain in the bone, spine, femur, pelvis, or ribs; leg weakness; fecal and urinary incontinence
Retinoblastoma	It may be inherent or sporadic. Mutation in retinoblastoma gene is a major cause. Most congenital or hereditary retinoblastomas appear within 1 year of age, while non-inherited retinoblastomas tend to be diagnosed in 1- and 2-year-olds. Retinoblastomas are rare in older children and in adults	Redness and swelling of the eye; vision problem; leukocoria, a white color at the center when light falls on it; eyes seem to be looking in different directions

(continued)

Type of cancer	Causes	Signs and symptoms
Rhabdomyosarcoma	Hardly any link with family history has been found; but having first degree, risk of embryonal type of RMS is more. Sometimes, a link with the genetic disorder neurofibromatosis has been obtained. RMS may have some correlation with inherited syndromes such as Li-Fraumeni syndrome, Beckwith-Wiedemann syndrome, or Costello syndrome	Tumor in the neck, chest, back, limbs, groin; may be painful and red; bulging of eyes due to tumors; vision affected; tumor in the ear causes ache; headache; sinus congestion; tumor in the bladder or prostate causes blood inn urine; in the vagina causes vaginal bleeding; in the abdomen causes vomiting, abdominal pain, and constipation; lump under skin may be caused; bone pain, cough, weight loss, and fatigue
Sarcoma	Family history; Paget's disease, which is a bone disorder; genetic disorders like Gardner syndrome, neurofibromatosis, retinoblastoma, etc.; exposure to radiation	Pain, when the tumor affects local tissues, nerves, and muscles; difficulty in movement of limbs
Squamous cell carcinoma of the skin	Appearance of new moles; increase in the size of mole; change in color of the mole; a spot which is increased in size due to a lump within it; rough, scaly mole; itchy moles; bleeding from moles; waxy bump; appearance of a scar-like lesion; red nodule	For squamous cell carcinoma: flat, firm, pale, or yellow areas, similar to a scar; reddish itchy patch; shiny, pearly bumps with blue, brown, or black areas; pink growth with blood vessels; sores that don't heal or reappear
Stomach cancer	Nausea and vomiting, heartburn, loss of appetite, weight loss, bloating, feeling full, blood in stool, jaundice, fatigue, stomach pain	This is common among men. Old age, smoking; *H. pylori* infection; stomach polyp; tumor in other parts of the digestive system; family history; dietary habit, e.g., those who consume too much meat and salty food and don't cook or store food properly; sedentary lifestyle; alcohol abuse; having lymphoma
Testicular cancer	Family history; cryptorchidism (undescended testicle); inguinal hernia; chances of having testicular cancer are high if there is an incidence of testicular cancer on the other; congenital abnormalities of the penis, kidneys, and testicles; Klinefelter's syndrome, where testis is not properly developed; mumps orchitis, where testicles are inflamed	Enlarged testicle; heaviness and pain in the scrotum and sudden collection of fluid; ache in the abdomen or groin; tenderness of breasts; back pain, tiredness
Thyroid cancer	Gender, common in women than men; hereditary condition; people with familial adenomatous polyposis (FAP) syndrome and its subtype Gardner syndrome may develop thyroid cancer; people with Cowden disease and Carney complex, type I, are at a higher risk; mutation in a gene named RET; a first-degree relative having thyroid cancer increases the risk	A lump in the neck growing rapidly; pain in front of the neck, moving toward ears; hoarseness; difficulty in breathing and swallowing; a nagging cough

(continued)

Type of cancer	Causes	Signs and symptoms
Ureter	Smoking most likely increases risks, as does exposure to a variety of chemicals and dyes found in factories that make leather goods, textiles, plastics, and rubbers. A previous history of bladder cancer also increases risk	Weight loss; fatigue; dark-colored urine; frequent urination with burning sensation; pain on the rib and spine; hematuria with occasional blood clots; back pain
Vaginal	Old age; presence of atypical cells in the vagina caused by HPV; usage of DES to prevent miscarriage; multiple sexual partners; early age at first intercourse; HIV infection; smoking; having cervical cancer increases the chances of vaginal cancer; radiation therapy	Vaginal bleeding after intercourse or after menopause; watery discharge; lump in the vagina; painful and frequent urination; pelvic pain; constipation; swelling and pain in the legs; back pain; abnormal bowel function
Vulvar	**Age**: over half of all cases are in mainly old age; HPV infection; women with vulvar intraepithelial neoplasia are susceptible to vulvar cancer; history of melanoma is a risk factor; smoking; immunodeficiency after kidney transplant; **Lichen sclerosus et atrophicus (LSA)** results in thickening of the skin and increases the risk of vulvar cancer; women with higher level of antibodies to herpes simplex virus to treat sexually transmitted disease are at risk; woman with HIV or AIDS	Painful sexual intercourse, painful urination, itching, wartlike growth, bleeding, pain, sensitive and burning sensation, ulceration
Intraocular melanoma	Old age; people having light eye color and white are susceptible; inherited skin disorders like dysplastic nervous syndrome, etc.; people having abnormal skin pigmentation; exposure to UV	Dark spot on the iris; light flashes, noticeable change in the shape of pupil; loss of peripheral vision, sensation of floaters
Kaposi sarcoma	HIV and AIDS are linked with Kaposi sarcoma. Infection with the Kaposi sarcoma-associated herpesvirus (KSHV), also known as human herpesvirus 8 (HHV8), is a risk factor	Normally asymptomatic; vomiting; difficulty in swallowing; pain in the belly; cough; shortness of breath; swelling of arms, legs, the face, or the scrotum; blemishes of different colors like brown, purple, pink, or red; edema may grow outward or inward in soft tissue or the bone

Appendix II

Details of Phytochemicals and Their Targets

Phytochemical	Sources	Cancer	Molecular target
Carotenoids	Carrots, plums, apricots, mangoes Cantaloupes, kale, sweet potatoes, spinach, cilantro, thyme, turnip, squash	Breast, ovary, pancreatic, prostate, leukemia, colon, gastric, adenocarcinoma, neuroblastoma, hepatic, cervical, skin, liver	Gap junctional intercellular communication, growth factor signaling, cell cycle progression, differentiation-related proteins, retinoid-like receptors, ARE, nuclear receptors, AP-1 transcriptional complex, the Wnt/β-catenin pathway, and inflammatory cytokines
Isothiocyanates	Bok choy, broccoli, Brussels sprouts, cabbage, cauliflower, horseradish, kale, kohlrabi, mustard, radish, rutabaga, turnip, watercress	Prostate, lungs, liver, forestomach, mammary gland, esophagus, small intestine, colon, bladder	Increased carcinogen metabolism and detoxification, induction of phase 1, e.g., cytochrome P450 (CYP) family, and phase 2, e.g., glutathione *S*-transferases (GST), UDP-glucuronosyl transferases (UGT), etc. Upregulation of Keap1 with concomitant repression of Nrf2. Inhibits NF-κB, iNOS, COX-2, interleukins, TNF-α. Regulates MAPK
Quercetin	Grapes, leafy vegetables, broccoli, red onion, peppers, apples, grapes, red wine, raspberry, nectarine	Lung, pancreatic, leukemia, lung, breast, neuroblastoma, colon, ovary, endometrial cancer	Quenches ROS, antioxidant defenses, including repair and detoxifying enzymes, scavenges GSH, c-FOS, c-JUN, c-MYC, PKB pathway, epigenetic modulation, p53, bax, E2F1, FOXO3, Dif1, WRN, Ku70, RAD51, β-catenin, survivin, NFκB
Anthoxanthins	Blackberries, blueberries, cherries Cranberries, eggplant, grape juice, plums, prunes, raisins, red apples, red beans, red beets, red cabbage, red/purple grapes, red onions, red pears, red wines, strawberries	Colon	ROS, caspases, attenuate Aβ oligomer-induced neuronal responses

(continued)

Phytochemical	Sources	Cancer	Molecular target
EGCG	Green tea. Strawberries, raspberries, blackberries, plums, peaches, avocado, and kiwi also contain little amount of EGCG	Breast, prostate, ovary, esophageal, leukemia, hepatocellular, oral, laryngeal, lung, cervical, melanoma, skin, esophageal, bladder	Cyclins, CDKs, CYP, PI3K, AKT, ERK, HIF1, NF-κB, VEGF, MMPs, survivin, TERT, IAP, caspases, Bcl-2
Caffeine	Cocoa beans, coffee beans, kola nuts, tea leaves	Non-melanomatous skin cancer, melanoma, gastric	PTEN, PI3K/Akt, p53, mTOR, induce apoptosis
Lignan	Flaxseeds, seeds of pumpkin, sunflower, poppy, sesame, whole grains like rye, oats, barley, wheat, oat, rye, beans, berries	Colorectal, breast, lung, prostate, cervix	Ki-67, MMPs, VEGF, caspases
Genistein	Fava beans, soybeans, lupin, kudzu	Breast, leukemia, lymphoma, ovarian, cervical, leiomyoma, melanoma, neuroblastoma, gastric, pancreatic, prostate	Targeting caspases, Bcl-2, Bax, MAPK pathway, NF-κB, Wnt/β-catenin, PI3K/Akt, apoptosis inducer
Resveratrol	Grapes, wines both red and white, peanuts, pistachios, blueberries, cocoa, cranberries	Breast, colorectal, pancreatic, hepatic, prostate, skin, cervix, ovary, stomach, thyroid	MAPK pathway, p38, JNK1, ERK2, NF-κB, cyclins, IAPs, Bcl-2, TRAF2, MMPs, COX-2, VEGF, caspases
Lutein	Kale, spinach, turnip greens, and collards, parsley, lettuce, pumpkin, corn	Breast, bladder, skin, cervical, kidney, endometrial	GADD45A, Bax, CASP3/4/8, TNFRSF10A and TNFRSF21, CD70
Curcumin	Turmeric	Breast, leukemia, bowel, stomach, skin, cervix, HNSCC, bladder, prostate, cervical, colon, esophageal	NF-κB, STAT3, EGFR, Akt, ERK, cMET cyclin D1, AP-1, Erg-1, p53, β-catenin, Notch-1, Hif-1, PPAR-α, VEGF, E-cadherin, Sp-1, FAK, and CD24, HDAC, calmodulin, MDR, PKC, mTOR, Wnt/beta-catenin signaling, COX-2
Ellagic acid	Guava, strawberries Grapes, blackberries Raspberries, pecans Cranberries, walnuts, pomegranate	Prostate, breast	G0/G1 cell cycle arrest, Bcl-2, Bax, Bid, caspases, p38, JAK-STAT

(continued)

Phytochemical	Sources	Cancer	Molecular target
Indole-3-carbinol	Brussels sprouts, cabbage, bok choy, broccoli, turnips	Breast, colon, prostate, endometrial, liver, lung, cervix, gastrointestinal tract	Akt, NFκB, caspases, cyclin-dependent kinases, cyclins, estrogen , BRCA, survivin, Bcl2, p21, p27, MAP kinases, STAT3, IL-8, MMPs
Luteolin	Oregano, fresh peppermint, sage, thyme	Breast, leukemia, pancreas, hepatoma, brain	p27, p53, caspases, fas, Bax, JNK, IAPs, NF-κB, MMPs, ILs, TNFα,PI3K/AKT,PKC, MMPs, CYP, cyclins, CDKs
Vitamin C	Lemon, lychees, papaya, strawberries, oranges, kiwis	Breast	
Selenium and vitamin supplements		Esophageal	
Lycopene	Guavas, tomatoes, watermelon, grapefruit, papaya, sweet red peppers, persimmon, asparagus, red cabbage, and mangoes	Prostate, lung, gastric, breast, hepatocellular, pancreas, colorectal, bladder, esophageal, head and neck	cytochrome P450, hepatic quinine reductase, p53, Rb, platelet-derived growth factor-BB, MMPs, scavenger of free radicals, cyclins, cyclin-dependent kinases, insulin-like growth factor (IGF), Cx43, interleukin (IL-6), LOX
Combretastatin	*Combretum caffrum*	Colon, lung, leukemia, breast	E-cadherin, beta-catenin, AKT, VEGF, MMPs
Silvestrol	*Aglaia silvestris*	Lung, breast, leukemia	NF-κB, p38MAPK, ERK
Gingerol	Ginger	Breast, colon, leukemia, lung, ovary, pancreas	ROS, p53, p27, p21, CDKs, cyclins, EGFR, cMet, PI3k, COX-2, NF-κB, AP-1
Allicin	Garlic	Forestomach, esophagus, stomach, colon, breast, lung, prostate, pancreatic	CYP, p53, Bcl-2, cell cycle regulators CDKs, cyclins, JNK, ERK1/2, and p38; p53, NF-κB, bcl-2/bax
Theaflavins	Black tea	Colon, breast, lung, oral skin, ovary, gastric, forestomach, pancreas, liver	STAT3, Bcl-2, survivin, MMPs, caspases, antioxidant enzymes, EMT markers, EGF, PDGF, JAK-STAT, AP-1, cJun NH2-terminal kinase, Erk, PKC, AP-1, Wnt/beta-catenin, COX-2, c-MYC, cyclin D1, GSK3beta, ERK, JNK, p53, p21, Bax, VEGF
Sulforaphane	Broccoli, cauliflower, kale, Brussels sprouts, cabbage, bok choy, collards	Breast, prostate, cervical, bladder, renal, non-small cell lung cancer (NSCLC), colon	G2/M phase arrest, $p21^{WAF1/CIP1}$, HIF-1α, VEGF, Nrf2, Keap1, antioxidant enzymes like NQO1, HO-1, catalase, GST, SOD. MAPK, PI3K, PKC, epigenetic modifications

(continued)

Phytochemical	Sources	Cancer	Molecular target
Homoharringtonine	*Cephalotaxus fortunei* Hook plants	Leukemia, myelodysplastic syndrome	Bcr-Abl, protein tyrosine kinase inhibitor, JAK2V617F
Kaempferol	Grapes, tomatoes, broccoli, tea, and ginkgo biloba leaves	Breast, ovary, pancreatic, prostate	ERK, Akt, ROS, VEGF, PI3K, AP1, NF-κB, p38, COX-2
Rutin	Passion flower, buckwheat, tea, apple	Leukemia, colon, neuroblastoma, hepatic, pulmonary metastasis	Bcl2, TNF-α, NF-κB, cell cycle arrest, interleukins
Capsaicin	Capsicum, bell pepper, paprika pepper, chili pepper	Pancreatic, colonic, prostatic, liver, esophageal, bladder, skin, leukemia, lung, endothelial cells	CD95, p53, β-catenin, PKC, cell cycle arrest, CDKs, cyclins, VEGF, MAPK, FAK, PI3K,
Limonene	Lemons, oranges, limes, grapefruit, and mandarins	Colon, lung, breast, skin, prostate	p-Akt, PI3K, caspases, PARP, Ras-ERK, p53, caspases, p21, Bad,
Fisetin	Strawberries, apples, onions, persimmons, cucumbers	Breast, prostate, pancreatic, colorectal, melanoma, oral	PI3K, Akt, mTOR, MMPs, Met/Src, reverse EMT
Phytochemical	Sources	Cancer	Mechanism/molecular target
Carotenoids	Carrots, plums, apricots, mangoes Cantaloupes, kale, sweet potatoes spinach, cilantro, thyme, turnip, squash	Breast, ovary, pancreatic, prostate, leukemia, colon, gastric, adenocarcinoma, neuroblastoma, hepatic, cervical, skin, liver	Gap junctional intercellular communication, growth factor signaling, cell cycle progression, differentiation-related proteins, retinoid-like receptors, ARE, nuclear receptors, AP-1 transcriptional complex, the Wnt/β-catenin pathway, and inflammatory cytokines
Isothiocyanates	Bok choy, broccoli, Brussels sprouts, cabbage, cauliflower, horseradish, kale, kohlrabi, mustard, radish, rutabaga, turnip, watercress	Prostate, lungs, liver, forestomach, mammary gland, esophagus, small intestine, colon, bladder	Increased carcinogen metabolism and detoxification, induction of phase 1, e.g., cytochrome P450 (CYP) family, and phase 2, e.g., glutathione *S*-transferases (GST), UDP-glucuronosyl transferases (UGT), etc. Upregulation of Keap 1 with concomitant repression of Nrf2. Inhibits NF-κB, iNOS, COX-2, interleukins, TNF-α. Regulates MAPK

(continued)

Phytochemical	Sources	Cancer	Molecular target
Quercetin	Grapes, leafy vegetables, broccoli, red onion, peppers, apples, grapes, red wine, raspberry, nectarine	Lung, pancreatic, leukemia, lung, breast, neuroblastoma, colon, ovary, endometrial cancer	Quenches ROS, antioxidant defenses, including repair and detoxifying enzymes, scavenges GSH, c-FOS, c-JUN, c-MYC, PKB pathway, epigenetic modulation, p53, bax, E2F1, FOXO3, Dif1, WRN, Ku70, RAD51, β-catenin, survivin, NFκB
Anthoxanthins	Blackberries, blueberries, cherries Cranberries, eggplant, grape juice, plums, prunes, raisins, red apples, red beans, red beets, red cabbage, red/purple grapes, red onions, red pears, red wines, strawberries	Colon	ROS, caspases, attenuate Aβ oligomer-induced neuronal responses
EGCG	Green tea. Strawberries, raspberries, blackberries, plums, peaches, avocado, and kiwi also contain little amount of EGCG	Breast, prostate, ovary, esophageal, leukemia, hepatocellular, oral, laryngeal, lung, cervical, melanoma, skin, esophageal, bladder	Cyclins, CDKs, CYP, PI3K, AKT, ERK, HIF1, NF-κB, VEGF, MMPs, survivin, TERT, IAP, caspases, Bcl-2
Caffeine	Cocoa beans, coffee beans, kola nuts, tea leaves	Non-melanomatous skin cancer, melanoma, gastric	PTEN, PI3K/Akt, p53, mTOR, induce apoptosis
Lignan	Flaxseeds, seeds of pumpkin, sunflower, poppy, sesame, whole grains like rye, oats, barley, wheat, oat, rye, beans, berries	Colorectal, breast, lung, prostate, cervix	Ki-67, MMPs, VEGF, caspases
Genistein	Fava beans, soybeans, lupin, kudzu	Breast, leukemia, lymphoma, ovarian, cervical, leiomyoma, melanoma, neuroblastoma, gastric, pancreatic, prostate	Targeting caspases, Bcl-2, Bax, MAPK pathway, NF-κB, Wnt/β-catenin, PI3K/Akt, apoptosis inducer
Resveratrol	Grapes, wines both red and white, peanuts, pistachios, blueberries, cocoa, cranberries	Breast, colorectal, pancreatic, hepatic, prostate, skin, cervix, ovary, stomach, thyroid	MAPK pathway, p38, JNK1, ERK2, NF-κB, cyclins, IAPs, Bcl-2, TRAF2, MMPs, COX-2, VEGF, caspases

(continued)

Phytochemical	Sources	Cancer	Molecular target
Lutein	Kale, spinach, turnip greens, and collards, parsley, lettuce, pumpkin, corn	Breast, bladder, skin, cervical, kidney, endometrial,	GADD45A, Bax, CASP3/4/8, TNFRSF10A and TNFRSF21, CD70
Curcumin	Turmeric	Breast, leukemia, bowel, stomach, skin, cervix, HNSCC, bladder, prostate, cervical, colon, esophageal	NF-κB, STAT3, EGFR, Akt, ERK, cMET cyclin D1, AP-1, Erg-1, p53, β-catenin, Notch-1, Hif-1, PPAR-α, VEGF, E-cadherin, Sp-1, FAK, and CD24, HDAC, calmodulin, MDR, PKC, mTOR, Wnt/beta-catenin signaling, COX-2
Ellagic acid	Guava, strawberries, grapes, blackberries, raspberries, pecans, cranberries, walnuts, pomegranate	Prostate, breast	G0/G1 cell cycle arrest, Bcl-2, Bax, Bid, caspases, p38, JAK-STAT
Indole-3-carbinol	Brussels sprouts, cabbage, bok choy, broccoli, turnips	Breast, colon, prostate, endometrial, liver, lung, cervix, gastrointestinal tract	Akt, NFκB, caspases, cyclin-dependent kinases, cyclins, estrogen , BRCA, survivin, Bcl2, p21, p27, MAP kinases, STAT3, IL-8, MMPs
Luteolin	Oregano, fresh peppermint, sage, thyme	Breast, leukemia, pancreas, hepatoma, brain	p27, p53, caspases, fas, Bax, JNK, IAPs, NF-κB, MMPs, ILs, TNFα, PI3K/AKT,PKC, MMPs, CYP, cyclins, CDKs.
Vitamin C	Lemon, lychees, papaya, strawberries, oranges, Kiwis	Breast	
Selenium and vitamin supplements		esophageal	
Lycopene	Guavas, tomatoes, watermelon, grapefruit, papaya, sweet red peppers, persimmon, asparagus, red cabbage, and mangoes	Prostate, lung, gastric, breast, hepatocellular, pancreas, colorectal, bladder, esophageal, head and neck	Cytochrome P450, hepatic quinine reductase, p53, Rb, platelet-derived growth factor-BB, MMPs, scavenger of free radicals, cyclins, cyclin-dependent kinases, insulin-like growth factor (IGF), Cx43, interleukin (IL-6), LOX
Combretastatin	*Combretum caffrum*	Colon, lung, leukemia, breast	E-cadherin, beta-catenin, AKT, VEGF, MMPs,
Silvestrol	*Aglaia silvestris*	Lung, breast, leukemia	NF-κB, p38MAPK, ERK
Gingerol	Ginger	Breast, colon, leukemia, lung, ovary, pancreas	ROS, p53, p27, p21, CDKs, cyclins, EGFR, cMet, PI3k, COX-2, NF-κB, AP-1

(continued)

Phytochemical	Sources	Cancer	Molecular target
Allicin	Garlic	Forestomach, esophagus, stomach, colon, breast, lung, prostate, pancreatic	CYP, p53, Bcl-2, cell cycle regulators CDKs, cyclins, JNK, ERK1/2, and p38; p53, NF-κB, bcl-2/bax
Theaflavins	Black tea	Colon, breast, lung, oral skin, ovary, gastric, forestomach, pancreas, liver	STAT3, Bcl-2, survivin, MMPs, caspases, antioxidant enzymes, EMT markers, EGF, PDGF, JAK-STAT, AP-1, cJun NH2-terminal kinase, Erk, PKC, AP-1, Wnt/beta-catenin, COX-2, c-MYC, cyclin D1, GSK3beta, ERK, JNK, p53, p21, Bax, VEGF
Sulforaphane	Broccoli, cauliflower, kale, Brussels sprouts, cabbage, bok choy, collards	Breast, prostate, cervical, bladder, renal, non-small-cell lung cancer (NSCLC), colon	G2/M phase arrest, p21WAF1/CIP1, HIF-1α, VEGF, Nrf2, Keap1, antioxidant enzymes like NQO1, HO-1, catalase, GST, SOD. MAPK, PI3K, PKC, epigenetic modifications
Homoharringtonine	*Cephalotaxus fortunei* Hook plants	Leukemia, myelodysplastic syndrome	Bcr-Abl, protein tyrosine kinase inhibitor, JAK2V617F
Kaempferol	Grapes, tomatoes, broccoli, tea, and ginkgo biloba leaves	Breast, ovary, pancreatic, prostate	ERK, Akt, ROS, VEGF, PI3K, AP1, NF-κB, p38, COX-2
Rutin	Passion flower, buckwheat, tea, apple	Leukemia, colon, neuroblastoma, hepatic, pulmonary metastasis	Bcl2, TNF-α, NF-κB, cell cycle arrest, interleukins
Capsaicin	Capsicum, bell pepper, paprika pepper, chili pepper	Pancreatic, colonic, prostatic, liver, esophageal, bladder, skin, leukemia, lung, endothelial cells	CD95, p53, β-catenin, PKC, cell cycle arrest, CDKs, cyclins, VEGF, MAPK, FAK, PI3K
Limonene	Lemons, oranges, limes, grapefruit, and mandarins	Colon, lung, breast, skin, prostate	p-Akt, PI3K, caspases, PARP, Ras-ERK, p53, caspases, p21, Bad
Fisetin	Strawberries, apples, onions, persimmons, cucumbers	Breast, prostate, pancreatic, colorectal, melanoma, oral	PI3K, Akt, mTOR, MMPs, Met/Src, reverse EMT

Index

M. Roy, A. Datta, *Cancer Genetics and Therapeutics*,
https://doi.org/10.1007/978-981-13-9471-3

H

I

J

K

L

M

N

The manufacturer's authorised representative in the EU is Springer Nature Customer Service Centre GmbH, Europaplatz 3, 69115 Heidelberg, Germany. If you have any concerns regarding our products, please contact ProductSafety@springernature.com

Printed and bound by CPI Group (UK) Ltd, Croydon, CR0 4YY
30/07/2026
02182061-0001